Praise for *What's in Your Genome?*

"A serious warning to scientists and science writers who, to seduce their public, privilege the too-fast announcement of novelties and revolutions instead of doing a cautious examination of the results. Laurence A. Moran has written a deeply honest and extremely well-documented book on one of the hottest recent controversies in science. A master class on what science is and what it must continue to be."

Michel Morange, Professor Emeritus of Biology, Institute of the History and Philosophy of Sciences and Techniques, Université Paris 1 Panthéon-Sorbonne

"Laurence A. Moran lucidly explains the science behind an ugly yet thrilling truth – the DNA sequences that specify our body's proteins and RNAs are indeed finely honed by natural selection (like a Swiss watch), but they are only tiny islands of information embedded in a swamp of evolved gibberish."

Rosemary J. Redfield, Professor Emerita of Zoology, University of British Columbia

"Rarely does a science writer respect their audience enough to bring them into the weeds of confusion the way that Dr. Moran does, but the rewards are there to be reaped by a reader with even a high school–level knowledge of biology and biochemistry. A thorough reader will never look at the genome of any living organism or the concept of junk DNA the same way after reading this and will be left wondering how so many baseless claims have made headlines in recent years. In the capable hands of Dr. Moran, we can all learn to untangle facts from unsupported assertions and come away understanding why our genome is structured in precisely the way that it is."

Ethan Siegel, theoretical astrophysicist and founder of *Starts With A Bang!*

"Junk DNA is a concept that's not well understood, even by most biologists, yet arguments about its existence are commonplace. In *What's in Your Genome?*, Moran brings together evolution, genomics, and decades of scientific history to make the case for it. Even if you don't agree with all of his conclusions, you'd benefit from the clarity he brings to the topic."

John R. Timmer, Science Editor, *Ars Technica*

"This book is a clear and fascinating guide to the exotic menagerie of elements that exists in our DNA and a no-holds-barred defense of an important but often ignored fact about the human genome: most of it has no function. But Moran shows that, even without function, junk DNA is incredibly interesting."

Michael White, Associate Professor of Genetics, Washington University in St. Louis School of Medicine

WHAT'S IN YOUR GENOME?

90% OF YOUR GENOME IS JUNK

LAURENCE A. MORAN

ÆVO UTP

Aevo UTP
An imprint of University of Toronto Press
Toronto Buffalo London
utorontopress.com

ISBN 978-1-4875-0859-3 (cloth)
ISBN 978-1-4875-3857-6 (EPUB)
ISBN 978-1-4875-3856-9 (PDF)

Figures and cover illustration courtesy of Gordon L. Moran

Library and Archives Canada Cataloguing in Publication
Title: What's in your genome? : 90% of your genome is junk / Laurence A. Moran.
Other titles: What is in your genome?
Names: Moran, Laurence A., 1956– author.
Description: Includes bibliographical references and index.
Identifiers: Canadiana (print) 20220464103 | Canadiana (ebook) 20220464308 | ISBN 9781487508593 (cloth) | ISBN 9781487538576 (EPUB) | ISBN 9781487538569 (PDF)
Subjects: LCSH: Human genome. | LCSH: DNA. | LCSH: Genes. | LCSH: Genomics.
Classification: LCC QP624 .M67 2023 | DDC 572.8/6—dc23

Printed in Canada

We wish to acknowledge the land on which the University of Toronto Press operates. This land is the traditional territory of the Wendat, the Anishnaabeg, the Haudenosaunee, the Métis, and the Mississaugas of the Credit First Nation.

University of Toronto Press acknowledges the financial support of the Government of Canada, the Canada Council for the Arts, and the Ontario Arts Council, an agency of the Government of Ontario, for its publishing activities.

Canada Council for the Arts
Conseil des Arts du Canada

Funded by the Government of Canada
Financé par le gouvernement du Canada
Canada

Contents

Preface

Lots of people are writing science books these days, and usually their goal is to explain science to the typical reader who has no scientific education. These authors assume that science is too complicated for their potential audience, so they go out of their way to dumb down the science and avoid scientific terms.

This is a problem, not a feature. We live in a scientifically challenged society, and our goal should not be to cater to those who resist, for one reason or another, learning about science. Our goal should be to teach science to those who want to learn, and you can't do that if you oversimplify and avoid difficult concepts. While I appreciate the writing skills of many authors, I insist that the top three criteria for good science writing are accuracy, accuracy, and accuracy, and that's often inconsistent with oversimplification.

Part of the problem is that most science authors are addressing the wrong audience. The audience they think they're writing for is the scientifically uneducated masses – the ones who are reading the novels on the *New York Times* best-seller lists. Well, I got news for those authors: that's not the group who is going to read your book! The real audience – the ones who are going to read this book and all the other science books – is much more knowledgeable about science than you imagine. They are eager and willing to learn more about science. They read the science magazines and what's left of the science section in newspapers. They watch science shows on television. They read about science on blogs and on Facebook.

The real audience is the people who regularly buy science books – the ones who go right to the science section of their local bookstore. The real audience doesn't want dumbed-down science; they want to be informed and educated. They want to be lifted up, not put down. They know the difference between a gene and an allele, or if they don't know, they are capable of learning this important distinction. They need to learn it in order to understand modern biology. If we do a good job of reaching this audience of science fans, then hopefully they will influence their friends and relatives and advance the cause of science literacy.

This is the audience I'm writing for. I'm an admirer of Stephen Jay Gould, and I agree with him when he writes about the noble calling of communicating science.

> [T]his worthy activity has been badly confused with the worst aspects of journalism, and "popularization" has become synonymous with bad, simplistic, trivial, cheapened, and adulterated. I follow one cardinal rule in writing these essays – no compromises. I will make the language accessible by defining or eliminating jargon: I will not simplify concepts.
>
> I can state all sorts of highfalutin, moral justifications for this approach (and I do believe in them), but the basic reason is simple and personal. I write these essays primarily to aid my own quest to learn and understand as much as possible about nature in the short time allotted. If I play the textbook or TV game of distilling the already known, or shearing away subtlety for bare bones accessible in the vulgar sense (no work required from consumers), then what's in it for me?[1]

This may seem arrogant, but when I recently reread it, I recognized that this is exactly how I feel about blogging – I do it mostly for my own pleasure in learning. I'm also a textbook author, so I recognize myself in Gould's admonitions against compromises and against simplifying difficult concepts.

Much of this book is about the failure of science journalism and popular science books to explain modern science, and this failure applies to many fundamental concepts in biology, with evolution being the most obvious. There is no simple way to explain evolution correctly, but there are many simple ways to explain it badly.

Saying that you're not going to dumb down the science any more than necessary is easy, but it's much harder to deliver on that promise than you might expect. When you are writing for a relatively informed audience, you need to be aware of the fact that they have a little bit of knowledge, and that can be a dangerous thing for a science writer. I'm reminded of something that Richard Dawkins wrote in *The Extended Phenotype*: "I myself admit to being irritated by a book that provokes me into muttering 'yes but ...' on every page, when the author could easily have forestalled my worry by a little considerate explanation early on."[2]

The Extended Phenotype is one of the books that Dawkins is most proud of, and it was not written for the same audience as *The Selfish Gene* or *The Blind Watchmaker*. It was written for the audience that I'm aiming at. Yes-buttery is a serious problem for a science writer because in order to dot all the *i*'s and cross all the *t*'s, you have to produce a text that doesn't read as smoothly as a text that ignores all "yes buts." I'm sensitive to this issue because there are a few other books on genomes that have provoked me into muttering "yes but" on every page.

Blaming science writers for the sad state of science literacy is a bit unfair because much of the problem can be traced back to the scientists themselves. In many cases, the popular press is accurately representing the views of prominent scientists who clearly don't have a firm grasp of the material they are promoting. You'll see lots of examples in this book. What this means is that there's something seriously wrong with science. I don't know if it's all of science, but I do know that biology seems to be in trouble.

I should add at this point that there are some excellent science writers who do their homework and understand the material. You'll also encounter many of them in this book.

Misconceptions

There's a lot of talk these days about the proper way to teach science. I'm on the editorial board of a science education journal, so I've been following the pedagogical literature quite closely. One of the most important issues in teaching is how to deal with the misconceptions of students. One of the best descriptions of the problem comes from the blog site of the National Center for Science Education in the United States in an article written by John Cook at George Mason University in Virginia. He was writing about climate change, but his description applies to everything else.

> To quote that famous educator Yoda, "You must unlearn what you have learned." Misconceptions interfere with new learning. They can give students false confidence that they understand a phenomenon when their understanding is actually faulty. And in the worst case, misinformation can stop people from acknowledging facts. For example, studies have shown that when people are presented with facts and corresponding misinformation that distorts those facts, if they don't have a way of resolving the conflict between the two, there's a danger that they'll disengage and not even accept the facts. Misinformation can cancel out factual information.[3]

His point is one that has been widely discussed in the pedagogical literature. You can't change students' minds by just presenting the facts; you also have to directly address their misconceptions and refute them because, otherwise, their misconceptions cause them to ignore the facts. Like Luke Skywalker, they must unlearn what they have learned.

There are a lot of misconceptions about genomes and evolution, and I've tried to refute them all. My job has gotten much more difficult over the past few decades because there are so many books and articles that promote these misconceptions.

Evolution

This book is about the evolution of genomes so, naturally, the topic of evolution will come up frequently. I'm reminded of something Stephen Jay Gould wrote more than 20 years ago when he was concerned about the public understanding of evolution in America. He recounts the story of a Victorian lady who, on hearing about Darwin's book, said, "Let us hope that what Mr. Darwin says is not true; but if it is true, then let us hope that it will not become generally known."

Gould was referring to the influence of creationism in suppressing knowledge of evolution, but we now have so many other examples of ignorance winning over enlightenment that it's no longer remarkable that knowledge of evolution is not generally known.

I think it's safe to assume that you, dear reader, are not totally ignorant of evolution and, presumably, that you know that Darwin was right. However, there may be aspects of evolution that you are not familiar with but that are essential for understanding why our genome is full of junk. This part of evolutionary theory was unknown to Darwin and not generally known today – even to most scientists. I'm referring to the neutral theory, the nearly neutral theory, and the fact that random genetic drift is the main mechanism of evolution.

Facts, opinion, and controversy

Facts and data are the currency of good science writing. As I mentioned above, accuracy is to science writing as location is to real estate, but facts aren't as straightforward as you might think. For example, I can tell you that there are between 19,000 and 20,000 protein-coding genes in our genome, and that statement may come across as a fact that's beyond dispute, but I can't be absolutely certain that the number won't drop below 19,000 at some time in the future as we learn more about genomes.

But there's even more to this "fact" than just the exact number. When I say that there are a certain number of protein-coding genes, it depends very much on how you define *gene*, so in order to avoid yes-buttery, I need to describe my preferred definition of a gene. My preferred definition is not a "fact" – it's a, hopefully informed, opinion. There are other definitions of a gene, but I consider them to be misconceptions, so it means that I must refute those other definitions in order to support my opinion.

The correct definition of "gene" is controversial, especially when it comes to noncoding genes, so it's also important to explain why some scientists have a different view and how it affects the amount of junk in our genome. Thus, a supposedly simple "fact," such as the number of genes, is complicated by opinion and controversy. Just giving you the "facts" as I see it would be so much easier and would avoid the complications. Lots of science writers do that, but I don't think that's the best way to explain science.

The very word *fact* is controversial because all scientists are obliged to say at some point that science is provisional and that "facts" can change in the future. This is the only time I will give in to that obligation. From now on I adhere to Gould's definition of a "fact" as something that's "confirmed to such a degree that it would be perverse to withhold provisional assent." According to Gould, it's a fact that humans and chimpanzees descend from a common ancestor, and I agree that this meets the definition of fact that I will use. It would be ridiculous to quibble about whether this meets the semantic objections.

What Gould failed to mention is who provides "provisional assent"? Clearly, it's not the general public because in some countries you will find a substantial percentage of the population that would withhold provisional assent to the idea that chimps and humans evolved. This brings up the question of authority – who gets to say whether something is a fact? We all know the answer: it's the experts in the field who are in the best position to make the judgment. This is where controversy comes back onto the stage because now we have to decide which experts are right if they can't agree. Much of this book is about describing the views of warring experts and trying to help you decide which ones to believe.

The historical approach

There are many ways to approach a complicated topic such as what's in your genome. There are dozens of different issues, definitions, misconceptions, opinions, and controversies, and sorting them all out can be very confusing. After several false starts, I discovered that the best way to present the topic is to adopt a historical approach in which I explain how the concept of junk DNA was developed about fifty years ago and then describe how the controversy plays out after the publication of the human genome sequence in 2001.

As you will see, one of the reasons for adopting this approach is that the history has been seriously distorted by opponents of junk DNA, and this revisionist history has become engrained in the scientific literature and in popular science writing. In my view, it's best to correct that problem before delving into the modern debate.

Style versus substance

This brings me to my final comments about science writing. You can be taught how to give a good TED Talk or an entertaining lecture. The methods include improvisational training, learning about body language, and understanding how to use the proper speaking voice. All this training can be very useful if you are able to master the tricks, but we still have the problem of style versus substance. No matter how good your style, if the substance of what you are communicating is flawed, then you are not a good science communicator. I've watched many TED Talks in which the speaker got lots of applause for a good performance in spite of the fact that most of what they were saying was wrong. If you are a scientist, then I expect you've had the same experience.

This problem also applies to books written by scientists for the general public. Many of them are quite enjoyable to read because the author is skilled at writing, but some of those popular books misrepresent science, so, in my opinion, they are not good examples of science communication. Some of these flawed books have won major awards from various organizations of science writers in spite of their inaccuracy.

I promise to concentrate on accuracy by giving you the best account of genome science while not shirking my responsibility to present all sides of a controversy. Sometimes this means that you will have to pay close attention, but it will be worth it in the long run.

Prologue

It is humbling for me and awe-inspiring to realize that we have caught the first glimpse of our own instruction book, previously known only to God.

Francis Collins (2000)[1]

Those were the words of Francis Collins, director of the International Human Genome Project (IHGP), as he announced the draft sequence of the human genome on June 26, 2000. The other two men on the stage were Bill Clinton, president of the United States, and Craig Venter, president of Celera Genomics. One of those other men, President Clinton, was probably quite comfortable with the god language. The other one probably thought those words were referring to him.

Craig Venter and his group at Celera were relative newcomers to sequencing the human genome, but they had made remarkable progress in just a few years. Their sequence – mostly of Craig Venter's own genome – was almost as good as that of the IHGP. As president of The Institute for Genomic Research (TIGR), Venter had led a team with Nobel laureate Hamilton Smith. They published the first genome sequence of a bacterial species using a "shotgun" method where short random overlapping sequences of DNA were generated and then assembled to produce the complete genome. The team then sequenced the much larger genome of the fruit fly, *Drosophila melanogaster*, demonstrating that the technique could work for more complex species.[2]

Celera Genomics was founded in 1998 to exploit this shotgun technology and produce a human genome sequence in two years. Its main financial backer was Perkin Elmer, whose subsidiary, Applied Biosystems (ABI), made the expensive DNA sequencing machines that were needed for the project. Perkin Elmer decided to produce its own version of the human genome sequence and hired Venter and most of his team from TIGR to do it.

Celera was a company based in the United States whereas the IHGP was funded by governments in several different countries and had been underway for almost a decade. I should mention that the prime minister of the United Kingdom, Tony Blair, was also at the press conference via a video link from London. President Clinton rightly thanked the ambassadors of other countries for their contributions to the final sequence.

The IHGP had taken a "slow and steady" approach to sequencing the human genome, and their efforts were just reaching the tipping point in 1998. Large blocks of cloned DNA fragments had been accurately mapped to most of the chromosomes, and everything was in place to rapidly complete sequencing of the remaining fragments using the new DNA sequencing machines from ABI. The competition from a for-profit company turned out to be just the stimulation needed to get the project finished.

The announcement in June 2000 was somewhat premature since the actual draft sequences weren't published until eight months later when the IHGP researchers published their results in the British journal *Nature* on February 15, 2001, and the Celera researchers published theirs in the American journal *Science* a day later. The IHGP sequences were deposited in public databases as soon as they were processed, and they were accessible to everyone even before the draft sequence was published. Contrast this openness with the Celera sequence, which was private – you had to pay to see it.[3]

The reason for the premature announcement was to bring about a truce between Craig Venter and Francis Collins and the two groups rushing to sequence the human genome. The idea was to make it look like a tie in order to prevent a loss of face by either group. Venter was worried because he thought that Collins was about to announce a complete a draft sequence, and this would not be good news for Celera stockholders. Collins was also worried. He was afraid that Venter was about to announce a draft sequence, turning an expensive 10-year public project into an also-ran.

As things turned out, Venter eventually lost the race because the publicly funded sequence was freely available to all researchers. Furthermore, the Celera researchers had chosen to use a good deal of IHGP data to complete their own assembly of small fragments into chromosome-sized DNA, and this upset many of the IHGP leaders, especially Eric Lander, who launched a series of attacks on the quality of the Celera genome sequence. Ironically, the actual DNA sequences produced by Celera were probably at least as good as those of the public project – possibly even a little bit better. It didn't matter because few people would pay to see the Celera genome data when free data were readily available.

Nobody ever talks about the Celera genome sequence anymore. The business plan depended on beating the publicly funded effort by several years so that scientists and large pharmaceutical companies would pay for the private sequence, and a tie didn't count.* Thus, the competition was over shortly after it began. An excellent account of the rivalry can be found in *The Genome War* by James Shreeve, in which the fierce and, at times, acrimonious rivalry does not paint a pretty picture of how scientists behave.

The IHGP was created in the United States in the late 1980s, and the project got underway in 1990 with the workload split between labs in the United States, Great Britain, France, Japan, Germany, and

* "Celera" was named after *celer*, the Latin word for speed. It wasn't fast enough.

China. Scientists realized from the very beginning that it would be necessary to sequence the much simpler genomes of bacteria, yeast, and nematodes in order to develop the technology and, eventually, to interpret the human genome sequence. In addition, parallel work on the mouse genome was essential in order to identify common features in mice and humans.

It was important to create accurate genetic maps of the human and mouse genomes before sequencing began in earnest, and most of these preliminary projects were completed in the 1990s. As a result, by March 1999, the IHGP was producing sequences of large stretches of human genomic DNA, with an overall accuracy of 99.99 percent, and those sequences could be reliably placed in the correct position on each chromosome. The sequence of chromosome 22, one of the smallest human chromosomes, was published in December 1999.[4]

The draft sequence of the complete genome covered about 82 percent of the human genome, and it had 250,000 gaps and millions of sequence errors. Later on we'll learn why it was so difficult to get to 100 percent coverage, and we'll also see why the last bit wasn't very important at the time. Preliminary results indicated that our genome contains between 30,000 and 40,000 genes, with the most likely number being closer to 30,000. The draft sequence also revealed that about 50 percent of our genome consisted of various short stretches of DNA that were present in many copies ("repeat sequences"). The evidence gathered over the previous 30 years strongly suggested that these sequences had no biological function, which means they are junk DNA.

Further work by the IHGP led to the publication of a "finished" sequence in October 2004 covering about 92 percent of the genome. It had only 300 gaps, and the error frequency was down to an acceptable level of one error every 100,000 base pairs (10^{-5}).[5]

Aside from God, who may or may not have known about this instruction book, there were many scientists who had had more than a glimpse of the human genome despite what Francis Collins said in June 2000. In fact, knowledgeable evolutionary biologists, biochemists, geneticists, and molecular biologists had a pretty good understanding of the human genome long before the complete draft sequences became available. They were pleased, for the most part, that their predictions and expectations were confirmed by the draft sequence and by the finished sequence that came out a few years later.

That's a history most people don't know about. The hype and publicity surrounding the human genome project have confused many scientists, science journalists, and the general public. The focus has been on supposedly new discoveries that have overthrown old ways of thinking about genomes and not on the slow and steady advance of knowledge that characterizes most of science. The purpose of this book is to correct the false history and give you the straight facts about what's in your genome. Along the way, we'll discover that knowledgeable scientists were predicting a genome full of junk DNA with only a few thousand genes, and as you will read in this book, that's exactly what has been found.

The junk DNA wars

This book is about the composition of your genome and whether most of it is junk or most of it has a biological function. This is a very controversial topic with two opposing sides, the junkers and the functionalists, who have each been promoting their views for more than fifty years, and there's little hope that the issue is going to be settled any time soon.

I side with the scientists who think that most of our genome is junk. The evidence was strong in 1970 and 1980, and it's even stronger today, as I will explain in the following chapters. I assume, provisionally, that 90 percent of our genome is junk, where "junk" is defined as dispensable DNA – sequences that can be deleted without significant effect on the fitness of the organism.

I cannot prove that most of our genome is junk because it's almost impossible to prove that something does *not* have a function – that's the well-known problem of "proving a negative." What I can do is to show you that the concept of junk DNA is compatible with all the evidence, consistent with our understanding of evolution and population genetics, and possesses extraordinary explanatory power. It helps us make sense of biology. I will also show you that all the arguments against junk DNA are incompatible with our present understanding of molecular biology, incompatible with evolution, and lack explanatory power. They do not make sense.

There are many ways to describe this controversy, but I have chosen a historical approach in order to explain the thinking of the scientists who first came up with the idea of a genome full of junk in the late 1960s. This concept was widely accepted by most of the experts who studied molecular evolution, but it was not widely accepted outside of that group.

The junkers were happy when the human genome sequence was published because it confirmed their predictions, but the functionalists interpreted the data very differently and soon came up with several hypotheses suggesting that most of our genome was functional despite what the sequence data seemed to indicate. These hypotheses were widely accepted by many scientists and promoted in the popular science literature so that during the past 20 years, the dominant view is that the idea of junk DNA has been refuted. I describe these hypotheses in the second half of the book and show why they are probably wrong and why the views of the experts in molecular evolution are still valid.

CHAPTER 1

Introducing Genomes

> *Nobody would claim that the first paper Watson and Crick published, "A Structure for Deoxyribose Nucleic Acid," 128 lines in Nature on 25 April 1953, has the vast power to reorder men's thinking that marked "On the Origin of Species" in 1859 or the announcement, in 1905, of the special theory of relativity. Yet the discovery of the structure – after a pause for its implications to be absorbed – was brilliantly stimulating.*
>
> Horace Freeland Judson (1996)[1]

The publication of the human genome sequence told us what's in your genome, and the trivial answer is DNA, lots and lots of DNA. There's more to the story, but we need to start with DNA and how it's packaged.

Francis Crick is often quoted as saying, "We have discovered the secret of life," over drinks at the Eagle pub in Cambridge on Saturday, February 28, 1953. He denies that he said those exact words, but there's little doubt that they reflect how he felt on that day. Watson and Crick had just deduced the structure of DNA, and a few days later they had constructed an accurate atomic model. They were confident that their model was correct. It was, in fact, correct, and this is the now-famous double helix that all of us have heard about.[2]

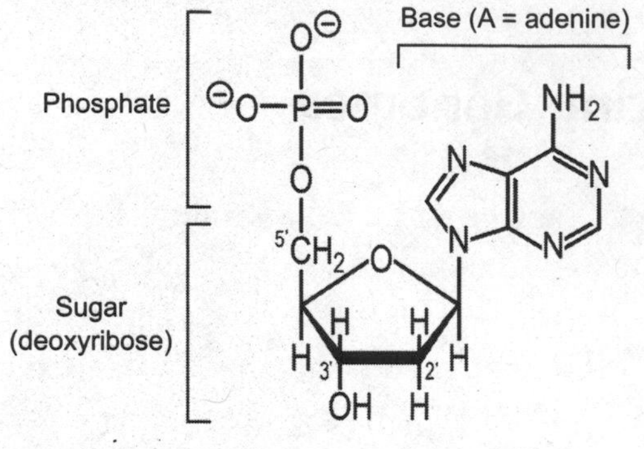

FIGURE 1.1 A nucleotide: 2′-deoxyadenosine 5′-monophosphate. This is one of four nucleotides that make up DNA. There are three parts to every nucleotide: (1) a base (A, G, T, or C), (2) a sugar (deoxyribose), and (3) a phosphate group.

Watson and Crick didn't discover the chemical structure of DNA because that was already known in 1953. What they did was to come up with an accurate prediction of the three-dimensional conformation of DNA that fit all the available data from Rosalind Franklin's X-ray diffraction studies.

The chemical structure of DNA consists of two separate strands of a *polynucleotide* made up of single *nucleotides* attached to each other, end to end. Each nucleotide consists of three parts: a sugar called deoxyribose, a phosphate group ($-PO_4^{2-}$), and a cyclic molecule called a base. The base is a complex molecule containing nitrogen, carbon, hydrogen, and sometimes oxygen atoms. The one shown in Figure 1.1 is adenine (A), but other nucleotides have guanine (G), cytosine (C), or thymine (T) at the same position.

Remember that we don't show all the carbon atoms in organic chemistry drawings, so every time two of those lines meet, there's a "C" (for "carbon") at that position. The lines represent bonds between adjacent atoms, and the double lines are double bonds. The other atoms are oxygen (O), hydrogen (H), nitrogen (N), and phosphorus (P).

The molecule shown here is 2′-deoxyadenosine 5′-monophosphate. That's the name my students have to know for the exam, but we can call it by its more familiar nickname, adenosine, or we'll just call it "A" when we write out the sequence of a DNA strand.

The nucleotides are attached to each other via their phosphate groups. Each phosphate at the top of a nucleotide forms a bond with the –OH group at the bottom of the next molecule. Thus, the backbone of a strand of DNA consists of alternating sugar–phosphate–sugar–phosphate and so on, creating a giant chemical polymer. The negative charges on the phosphate groups make the molecule acidic, and that's why the formal name of DNA is deoxyribonucleic acid.

The information component of DNA is defined by the order of the four nucleotides (A, T, G, C) along the DNA strand. This information determines whether the DNA will direct the formation of a mouse, a mycobacterium, a mushroom, or a maple tree.

Figure 1.2 shows part of a polynucleotide that has all four bases. This DNA strand has distinctly different ends called the "5-prime" (5′) and "3-prime" (3′) ends. This is important because there's a rule about writing the sequence of DNA. The rule is that you always write it from the 5′ end to the 3′ end (5′ → 3′). Thus, the DNA shown in the figure has the sequence AGTC and *not* CTGA even if you turn it upside down.[3]

The double helix

The DNA strand in Figure 1.2 looks like a complicated chemical that's not much different from those you saw in chemistry class in high school, and that's important because that's what DNA really is – it's just a complex chemical. Note that Watson and Crick could have drawn the structure shown in Figure 1.2 before they even began building their model because everyone knew that this is what the chemical structure looked like. So where is the famous double helix?

In order to understand the double helix, you have to realize that the complete DNA molecule actually consists of two different strands that are attached to each other via weak bonds between the

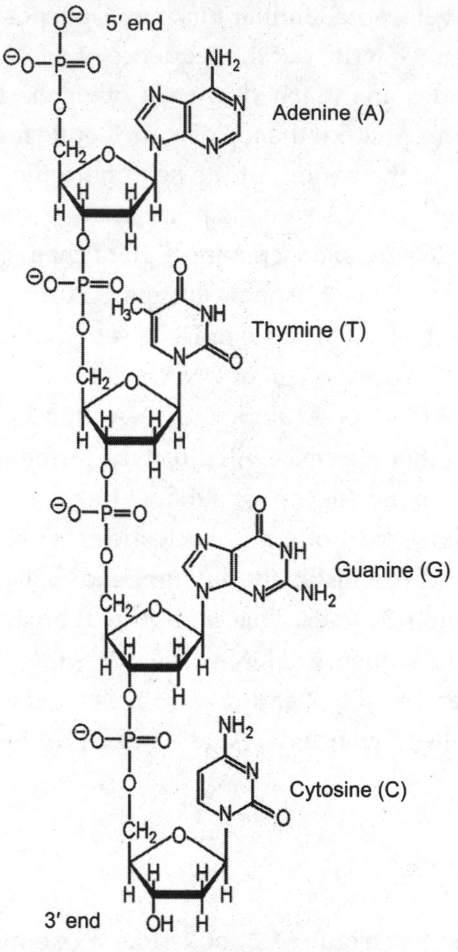

FIGURE 1.2 Part of a single strand of DNA. Individual nucleotides, A, T, G, and C, are joined by covalent bonds between the 5′ phosphate group and the 3′ –OH group to form a single-stranded polynucleotide. The "top" end is the 5′ (five prime) end and the bottom end is the 3′ (three prime) end. By convention, the sequence of the polynucleotide is read from 5′ to 3′. The sequence of this strand is ATGC.

bases. The structure of the bases is such that adenine (A) can pair up with thymine (T) and guanine (G) can pair up with cytosine (C) to form A–T base pairs and G–C base pairs. Because of these interactions, the two strands of DNA are *complementary*, meaning that the

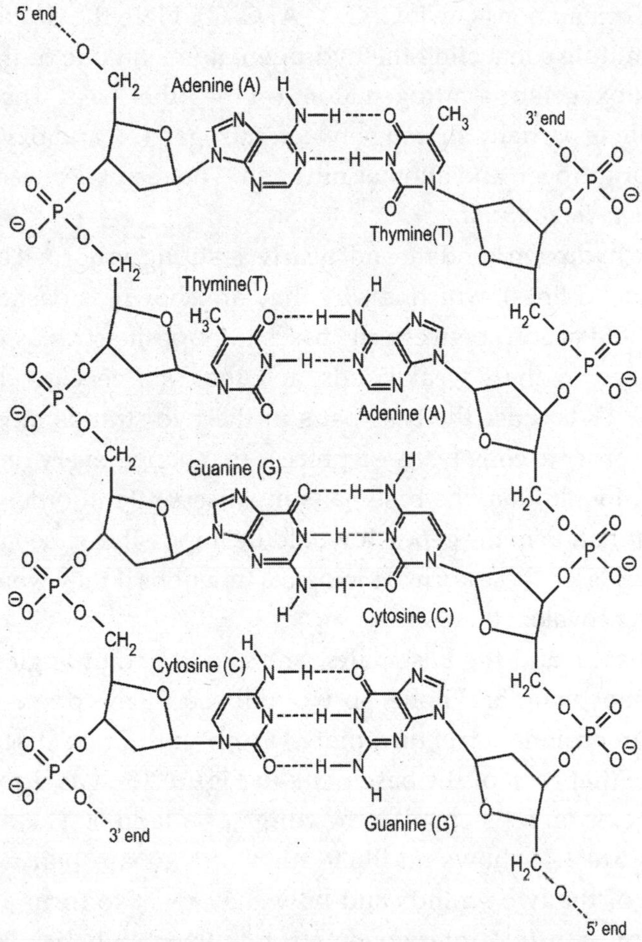

FIGURE 1.3 Double-stranded DNA. Two single-stranded polynucleotides come together to form double-stranded DNA. The bases on the interior are connected by weak hydrogen bonds between complementary base pairs. Base A pairs with base T, and G pairs with C. The two strands run in opposite directions.

bases along one strand complement, or pair with, the bases on the other strand.

In order for the two strands to fit together properly they have to run in opposite directions – this was one of the key insights of Watson and Crick that led to them building a correct model. Figure 1.3 shows the two strands with the two standard base pairs in both orientations (A–T, G–C, T–A, C–G). Note the dashed lines in the middle connecting the hydrogen atoms on one of the bases with an oxygen or a nitrogen atom on the other base. The hydrogen atom is partially shared between nitrogen (N) and oxygen (O) atoms or nitrogen and another nitrogen. The result is a weak bond called a *hydrogen bond*.

These hydrogen bonds are not nearly as strong as normal covalent bonds (solid lines), which is why they are shown as dashed lines. All the interactions between the bases in opposite strands of DNA are formed by these weak bonds, and there's a very good reason for this: it's because the base pairs in the two strands must come apart in order to copy DNA – a process that occurs every time a cell divides. In addition, the base pairs must separate in order to read the information in the genes. It would be impossible to separate the two strands for these normal biological functions if they were joined by strong covalent bonds.

The bases, and the base pairs, are tilted at right angles to the sugar-phosphate backbone, so we will see them edge on in the three-dimensional structure that Watson and Crick discovered. Imagine that each of the base pairs in Figure 1.3 was tipped into the page, forming steps like the rungs of a ladder. The next figure (Figure 1.4) shows an illustration of the base pairing interactions of the two strands and how they twist to form a helical structure with two interwoven strands (double helix). You can see that there are gaps between the base pairs in the ladder-like structure shown in the middle image, but these gaps are smaller in the more complex helix structure (the right image). The formation of the compact helical structure is driven by the interactions between the tops and bottoms of the flat base pairs as they attempt to stack tightly on top of one another, and it's these *stacking interactions* that make the double-stranded helix so stable. The stacking interactions are also weak compared to normal covalent bonds because they, too, have to be disrupted when DNA is replicated or copied.

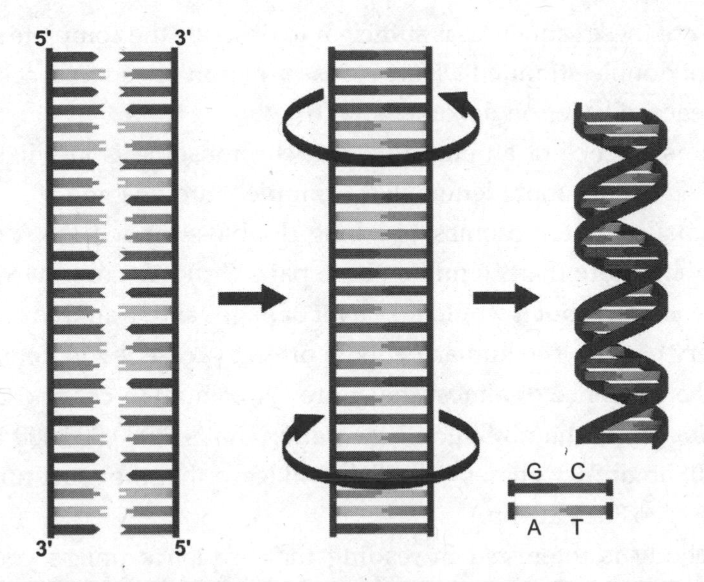

FIGURE 1.4 Forming the double helix. Imagine two complementary single strands coming together to form base pairs. The ladder-like structure has gaps between the base pairs. The top and bottom surfaces of the base pairs attract one another via stacking interactions, and this causes the entire structure to twist into the familiar double helix.

The goal of the human genome project was to sequence of all the base pairs

The sequence of the double-stranded molecule in Figure 1.4 is

GAGCTAGTCAAGGCTCATCTGAGCTAG

according to the convention of writing the sequence from the 5′ end to the 3′ end. But that's only if you start at the top-left corner of the molecule. You could also start at the bottom right and write the sequence of the other strand from the bottom 5′ end to the top 3′ end. That sequence is

CTAGCTCAGATGAGCCTTGACTAGCTC

Either of these sequences is sufficient to describe the complete structure of double-stranded DNA because you don't need to specify the sequence of bases on the other strand.

The sequence of an entire human chromosome is just like this only with a lot more letters. For example, chromosome 20, one of the smallest chromosomes, is a long double-stranded DNA molecule with more than 60 million base pairs (bp). We could write the sequence here, but it would take a lot of pages and, frankly, it would be very boring. The human genome project succeeded in determining the sequence of almost the entire set of human chromosomes, totaling more than 3 billion base pairs. That's 3,000,000,000 bp or 3×10^9 bp or 3 gigabases (Gb). (We will learn that the exact number is closer to 3.2×10^9 bp.)

Nobody is interested in reading that sequence unless you can interpret it. Where are the genes and the other bits of information required to make a human? That's really what we want to know when we ask, "What's in your genome?" By the time you finish reading this book, you will have a pretty good idea about the role of all 3 billion base pairs.

Prokaryotes and eukaryotes

This is an appropriate time to review the main features of the two different kinds of organisms: prokaryotes and eukaryotes. Prokaryotes are quite small and their DNA floats freely in the *cytosol* (*cytoplasm*). They all have a cell membrane called a *plasma membrane*, and most of them have a cell wall. Many of them have *flagella*, or tiny whip-like propellers, that propel them through the medium they inhabit.

Prokaryotes are commonly called bacteria,[4] and there are millions of species of bacteria. They inhabit all possible environments on our planet from high in the atmosphere to deep underground and everywhere in between. The total mass of bacteria is 35 times greater than all the animals on the planet, including all elephants, all blue whales, and all insects.[5]

Modern bacteria and modern humans are both highly evolved species – both of them are the end products of 3.5 billion years of evolution. It's important to understand that life on Earth began with the Age of Bacteria, and we are still in the Age of Bacteria; that's because dinosaurs and mammals are minor additions to the diversity of life if you count the total number of organisms or the total number of species.

Do not think of bacteria as "primitive" just because they look simple. The structures inside bacterial cells are the products of a long history of evolution just like the structures inside your own cells. In fact, don't ever use the word *primitive* to refer to *any* modern species. I also avoid using the adjectives *higher* and *lower* to refer to modern species following the lead of Charles Darwin, who often reminded himself to avoid these terms.[6]

One of the reasons for this little tangent is that I'm about to spend an enormous amount of time explaining the structure and organization of just one genome among the millions of species that I could have chosen. That one genome is ours, of course. I know full well that readers will not be as interested in the genome of *Drosophila melanogaster* (fruit fly), *Mus musculus* (mouse), or any plant species. A book with the title *What's in the Fruit Fly Genome?* would probably not have attracted your attention; more important, it probably wouldn't have attracted the attention of my publisher. But now that I have your attention, please forgive me for proselytizing a bit from time to time. Biology is not just about humans.

Eukaryotic cells are much larger than bacterial cells – a typical animal cell is about 1000 times larger than a typical bacterial cell. The distinguishing feature of eukaryotes is that the genomic DNA is confined to a membrane-bound nucleus.* Another difference is that eukaryotic cells have lots more internal membranes than prokaryotic cells. In addition to the nuclear membrane, there are various vesicles, a *Golgi apparatus* for sorting proteins, and a complex system

* The word *eukaryote* was constructed from the Greek *eu*, meaning "well," and *karyon*, meaning "nut." The "nut" is the nucleus – the defining characteristic of eukaryotes.

called the *endoplasmic reticulum*. Furthermore, almost all eukaryotic cells have small bacteria-sized organelles called *mitochondria* that produce most of the energy the cell needs. (Some species have lost the classic version of the mitochondrion but retain a few remnants.) Mitochondria are the evolutionary descendants of bacteria that invaded another primitive cell about 2 billion years ago.

Eukaryotes that are capable of photosynthesis, such as plants and algae, have *chloroplasts* for converting sunlight into chemical energy. Chloroplasts are descendants of a certain class of photosynthetic bacteria called cyanobacteria.

How big is your genome?

The "genome" of any species is roughly defined as the sum total of all the pieces of DNA that are required for the survival of the species. This definition works well for bacteria because most of them have only a single piece of DNA that carries all the genes.[7]

In eukaryotes, the strict definition of a genome is usually restricted to the chromosomes within the nucleus. In our case, we have 46 chromosomes in our nuclei, and nothing else is going to count as part of the genome. However, there aren't 46 *different* strands of DNA in the human genome sequence. Instead, there are only 23 strands because most eukaryotes are *diploid*, meaning they have two sets of chromosomes. We have 23 pairs of chromosomes (23 × 2 = 46), and the human genome is one complete set of chromosomes – this is the *haploid* set, and it's the number found in the germ cells (eggs and sperm).

There's an additional complication in some eukaryotes, including animals like us. We have two different *sex chromosomes*, a big X chromosome and a much smaller male Y chromosome. The complete human genome has to include one of each of the sex chromosomes plus one copy of each of the 22 *autosomes* (nonsex chromosomes). Thus, the human genome sequence has 24 separate pieces of DNA corresponding to chromosomes 1 to 22, plus the X chromosome and the Y chromosome.

There's more DNA in your cells than just the DNA found in your chromosomes. Mitochondria are the descendants of ancient bacteria, and because of that ancestry they still retain a small bacteria-like chromosome with a dozen or so important genes. There are hundreds of mitochondria in most of your cells, so this is a significant amount of DNA, but it doesn't count as part of your genome according to the standard definition. Similarly, chloroplast DNA doesn't count as part of the plant or algae genomes. There are separate online databases for mitochondrial and chloroplast genomes.

We've known the dimensions and the molecular weight of DNA ever since Watson and Crick published the structure of DNA. The distance between each stacked base pair along the double helix is 0.33 nanometers (nm, or one-billionth of a meter). The weight of a *base pair* averages 650 daltons, or 1086×10^{-24} grams (g).

We've known since the 1950s that there's a lot of DNA in the nucleus. The earliest studies used dyes to color the DNA, and some of these stains are quantitative – the amount of dye bound to DNA depends directly on the amount of DNA. We know from staining standard amounts of DNA that the nuclei of human cells have a *haploid* DNA content of 3.5 picograms (pg, or 0.0000000035 g), so we can calculate the number of base pairs by simple arithmetic:

$$3.5 \times 10^{-12} \text{ g divided by } 1086 \times 10^{-24} \text{ g/bp}$$
$$= 3.5 \text{ pg}/1086 \times 10^{-12} \text{ pg/bp} = 3.2 \times 10^{9} \text{ bp}$$

Thus, the human genome must be about 3.2 billion base pairs – a value that was known for decades before the human genome project was started. By 1991 – 10 years before publication of the draft human genome sequence – the approximate amount of DNA in each chromosome was known, and the total amount of DNA was estimated to be 3.227 Gb in females and 3.174 Gb in males (Gb = gigabase pairs).[8]

A bunch of DNA molecules totaling 3.2×10^9 bp of DNA will be quite long if all the molecules are stretched out end to end. Recall that the distance between each stacked base pair is 0.33 nanometers, or 0.33×10^{-9} meters, and the arithmetic here is pretty simple. It's a little more than 1 meter of DNA (3.2×10^9 bp $\times 0.33 \times 10^{-9}$ m per bp = 1.06 m). Since you are diploid, you have twice as much DNA in your nuclei, so that's about 2 meters' worth of DNA if all 46 pieces are stretched out end to end (2 meters is about 6.5 feet in the old British Imperial system). If you stood a single cell's worth of my DNA on one end, it would be taller than I am and probably taller than you.

You have about 10 trillion cells in your body, so imagine how tall the stack of DNA would be if every single piece of DNA was stacked end to end one on top of the other. It would be 20 trillion meters tall, or 20 billion kilometers. That's about 10 times farther than the distance from the Earth to Pluto when they are at their closest approach, but you wouldn't be able to see this mass of DNA because it's far too thin to be seen with the naked eye. I hope that gives you an idea of what we mean when we say that DNA is a long, skinny molecule.

Michael Lynch has calculated the approximate length of all the DNA molecules in all humans alive today. It comes out to far more than the distance to the nearest galaxies. When you add in all the other species it turns out that the length of all DNA molecules on Earth is about 10 times the diameter of the known universe!

The size of genomes in other species is still measured by staining cells and the techniques for measuring the amount of stain have gotten quite sophisticated over the past few decades. My friend Ryan Gregory at the University of Guelph (Ontario, Canada) is very interested in genomes, and he has built an Animal Genome Size Database that you can look at to see the genome size of your favorite animal. We're going to hear a lot more about Ryan Gregory because he invented the Onion Test, and he's a major player in the junk DNA wars.[9]

Packaging DNA: nucleosomes and chromatin

At this point, you may be wondering how all that DNA can be squeezed into a tiny nucleus, and back in the late 1960s there were many scientists who wondered the same thing. They realized that typical chromosomes contain lots of protein in addition to DNA, so it was natural to think that the proteins were responsible for packaging the DNA into some sort of useful three-dimensional structure.

All species contain packaging proteins, but let's focus on eukaryotes since this book is about your genome and not that of bacteria. Most of the proteins bound to DNA in eukaryotes belong to a special class of proteins called *histone proteins*. Four of these proteins (histones 2A, 2B, 3, and 4) combine to form a disc-shaped *core particle*. The core particle helps package DNA because the outside is covered with positively charged amino acids. Since opposites attract, the negatively charged DNA molecules will bind to the positively charged core particle, forming a disc-shaped structure called a *nucleosome* (Figure 1.5). This first level of packaging is enough to reduce the overall length of the DNA molecule to about one-tenth the length of stretched-out naked DNA.

The second level of packaging involves the association of individual nucleosome particles to form more complex structures. This requires the fifth histone protein, histone H1, and it results in further condensation. The nucleosome structures shown in Figure 1.5 were deduced about 45 years ago, and they have had a profound impact on our understanding of how the genome functions in gene expression.[10]

The combination of DNA plus structural proteins is called *chromatin*, and we need to understand some more basic principles of DNA packaging in order to evaluate arguments for and against junk DNA.

The length of a bit of typical chromatin is more than 40 times shorter than the length of the DNA it contains, but there are even

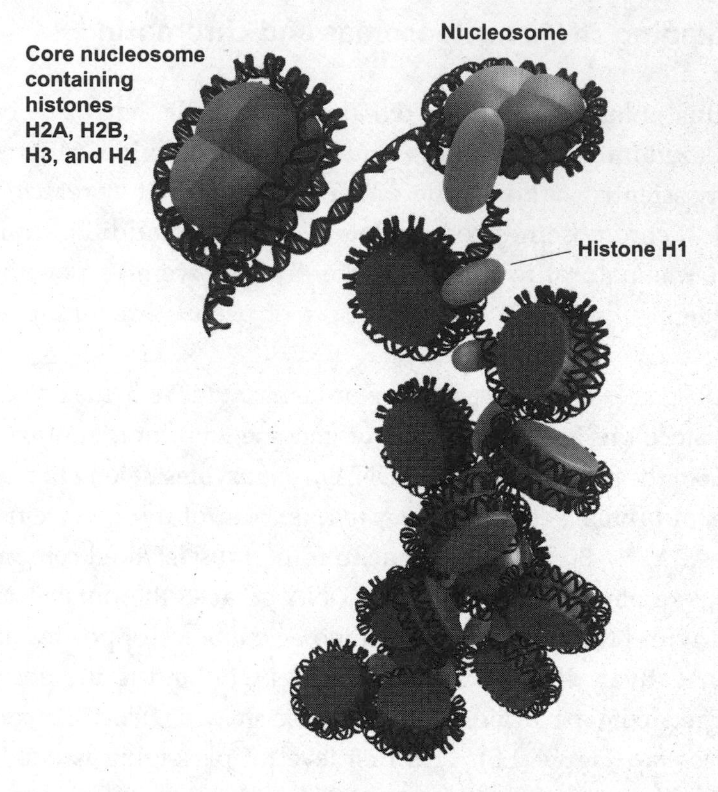

FIGURE 1.5 Chromatin. The core particle is made up of four histone proteins; H2A, H2B, H3, and H4 present in two copies each. DNA wraps around the core particle to form a nucleosome. Individual nucleosomes associate with each other through histone H1 to form the more densely packaged structure shown at the bottom of the figure.

higher levels of packaging. For example, the DNA in a chromosome is organized in large loops that are attached to a complex RNA–protein scaffold at the base of each loop. There are about 100,000 loops in human chromatin, and each loop can be twisted and compacted to give an even more complex structure.[11] The ultimate form of DNA packaging is visible in dividing cells, where the compact chromosomes can be easily seen in the light microscope. These are the metaphase chromosomes that you see so often in figures and drawings.

The compact structure is called *heterochromatin*, and the DNA in heterochromatin is inert because information in that sequence cannot be easily read and the DNA can't be replicated. The DNA is more accessible when it is in a somewhat more open structure called *euchromatin* – think of it as loops of DNA in the form of nucleosome complexes. The information in euchromatin can be read much more easily because the DNA is more accessible. Most of the DNA in our genome can flit back and forth between euchromatin and dense heterochromatin or "open" and "closed" chromatin structures depending on whether the information needs to be read or not. We'll see later that this is an important part of regulating gene expression and that certain regulatory sequences in our genome play a crucial role in this transition.

Some of the DNA in our genome is always in a compact heterochromatic state. These DNA sequences are usually quite boring stretches like ACACACACACAC or other highly repetitive sequences that go on and on for thousands of base pairs. Some of this DNA has important functions, but the problem for the original sequencers was that they couldn't assemble such repetitive sequences into a complete genomic sequence. This is why the International Human Genome Project concentrated on sequencing the "euchromatic" part of the genome and not the entire genome that includes the heterochromatic part, hence the title of the 2004 paper: "Finishing the Euchromatic Sequence of the Human Genome."

Transcription

I'm going to be talking a lot about genes, so it's a good idea to describe how genes work and how we define them. Here's a good working definition of a gene: **A gene is a DNA sequence that's transcribed to produce a functional product**. I discuss some controversies about the definition in chapter 6 but, for now, let's just accept this definition. It's what most biochemists and molecular biologists mean when they talk about genes.

We need to understand some basic biochemistry before we move on so that I can refer to it in the next chapters. *Transcription* is the process whereby part of the double-stranded DNA is opened up and one of the strands is copied into ribonucleic acid (RNA). RNA is another polynucleotide like DNA except that it's not "deoxy" because RNA molecules have ribose, the normal sugar, in their backbone and not the deoxyribose version, which is missing an oxygen at the 2′ position. The other difference between DNA and RNA is that RNA has the base uracil (U) instead of the thymine (T) found in DNA. Uracil behaves very much like thymine (T) – it forms a base pair with A to form A–U pairs that are very similar to A–T base pairs in DNA.[12]

Transcription requires an enzyme called *RNA polymerase*. The first step in transcription is to form an *initiation complex*, where RNA polymerase binds to a short stretch of DNA at the very start of a gene (Figure 1.6). The binding site is called a *promoter* because it promotes transcription.

RNA polymerase first binds to any random stretch of DNA and then scans the DNA by sliding along the chromosome until it bumps into a promoter sequence. When it recognizes a promoter, it clamps down on that particular stretch of DNA and unwinds a bit of the DNA double helix, resulting in the formation of a small *transcription bubble*. The next step is to make a little piece of RNA by stringing together nucleotides that are complementary to one of the DNA strands. (Remember that "complementary" means they form base pairs.)

At this point the transcription complex rearranges, so it can leave the promoter sequence and move down the gene copying the DNA strand as it goes. Eventually it reaches the end of the gene, where a termination event releases the RNA and the transcription complex dissociates from the DNA.

There are three steps in transcription: initiation, elongation, and termination. The important step is the initiation step because it requires specific sequences of DNA that are not part of the gene.

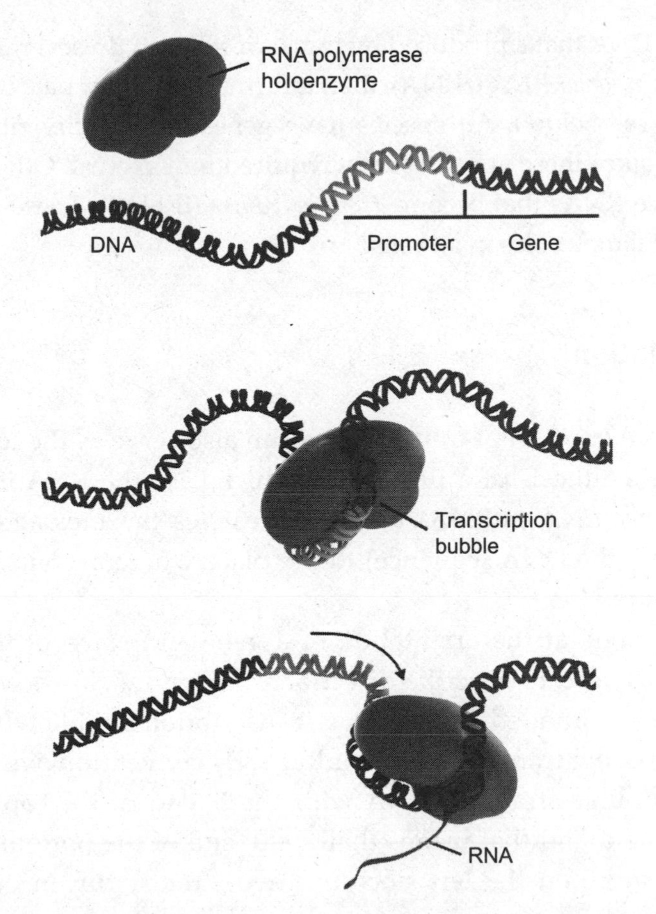

FIGURE 1.6 Transcription. Transcription begins when RNA polymerase binds to DNA and searches along the strands until it encounters a specific sequence called a *promoter*. When it finds a promoter, it opens up the DNA double helix to form a transcription bubble. Synthesis of RNA is initiated, and the RNA polymerase complex moves down the gene, making RNA that's complementary to one of the strands of DNA.

I referred to one of them as a *promoter* but, as we shall see, there are many other sites to which proteins bind in order to control proper transcription initiation. Together, these sites define the *regulatory sequences*.

The RNA that's produced by transcription can do several things. If it's *ribosomal RNA* (rRNA), then it forms the central part of a ribosome (see below). All species have genes that specify ribosomal RNAs since these are absolutely required for survival. Other genes produce RNAs that become *transfer RNAs* (tRNAs). These are also essential molecules in all cells.

Translation

The RNA product of transcription can also serve as the template for the synthesis of a protein, in which case the RNA is called *messenger RNA* (mRNA) because it carries the message in the genome (the DNA sequence) to the place where protein synthesis occurs.

Let's look at the orientation of a gene and some of the conventions used to describe directions. Figure 1.7 shows a stretch of double-stranded DNA with a transcription bubble where the gene is being transcribed into mRNA. By convention, we always draw double-stranded DNA with the 5′ end of the top strand on the left, and that means that the 3′ end of the bottom strand will also be on the left since the two strands run in opposite directions.

We define the orientation of the gene by reference to the top strand. Thus, by convention, the 5′ end of the *gene* is on the left and the 3′ end of the *gene* is on the right. Transcription always begins, by definition, at the 5′ end of the gene and proceeds left to right until it reaches the termination site at the 3′ end of the gene. We'll see later that the regulatory sequences that determine when and where transcription begins are almost always just to the left of the 5′ end of the gene.

The bottom strand is copied by RNA polymerase to produce an mRNA that's complementary to this bottom strand, or *template*

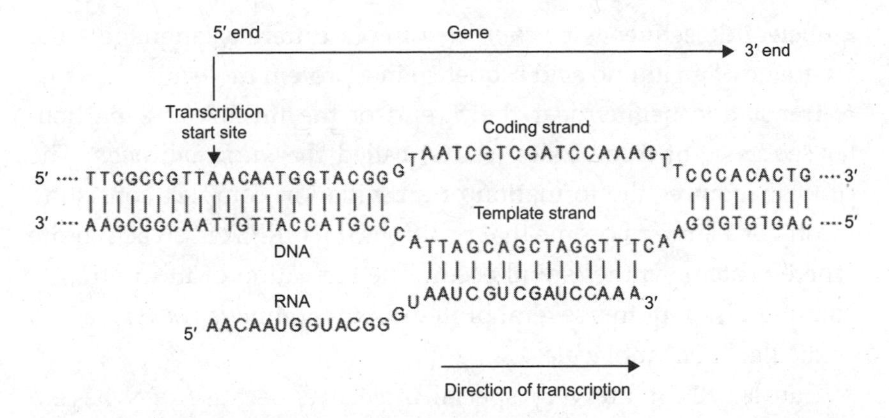

FIGURE 1.7 The orientation of a gene. Double-stranded DNA is drawn with the 5′ end of the top strand on the upper left. The bottom strand is the *template strand* that is copied to produce complementary RNA. The upper strand corresponds to the sequence of the mRNA, so it is called the *coding strand*. Transcription begins at the transcription start site located relative to the top strand. Thus, the gene is defined as beginning at the 5′ end and ending at the 3′ end.

strand. As you can see from the diagram, this mRNA will be identical in sequence to the top strand of the DNA molecule except that T is replaced by U. The mRNA contains the information that codes for the protein, and since the top strand of the gene has the same sequence, it is called the *coding strand*.

You don't need to memorize any of this – it won't be on the test! However, you do need to understand that genes are made of double-stranded DNA and that they have a specific orientation relative to the two strands of DNA.[13] I'm going to use the scientific terms for the front ends and the back ends of genes and RNA molecules, so when I talk about the 5′ end of a gene or the 3′ end, you'll know what I mean.

The process where the message in RNA is interpreted to produce protein is called *translation* because it really does require that a nucleotide sequence be *translated* into a different language – the language of an amino acid sequence in a protein molecule.

Translation begins near the 5′ end of the mRNA at a particular sequence of three bases (AUG) called the *initiation codon*. The process requires the formation of an initiation complex consisting mostly of a large ribosome that positions the mRNA in an active site where protein synthesis will occur. The formation of this initiation complex also requires several protein factors (*initiation factors*) and a particular *tRNA* molecule.

Transfer RNAs are very special molecules because they have a short sequence of RNA at one end called the *anticodon*, and at the other end they are attached to an amino acid. The first tRNA to bind to the initiation complex has an anticodon that's complementary to the initiation codon. This initiation codon is almost always the methionine codon AUG, and that's why almost all newly synthesized proteins begin with methionine.

The goal of translation is to line up all the different tRNAs so that their anticodons match the three nucleotide codons on the mRNA molecule (Figure 1.8). Then the amino acids that are attached to each tRNA are removed from the tRNA molecules and joined to each other to make a protein. The bonds that are formed between the amino acids are called *peptide bonds*, and the technical name for the product of this reaction is a *polypeptide* because it contains lots ("poly") of peptides. I'll usually refer to it as a protein.

As we saw in transcription, there are three distinct steps in translation: initiation, elongation, and termination. The termination step is triggered when the ribosome encounters one of three specific codons called *termination codons* (UAA, UAG, or UGA). There's no tRNA molecule with an anticodon that matches a termination codon; instead, termination occurs when a protein factor enters the ribosome complex, recognizes the termination codon, and triggers the disassembly of the complex.

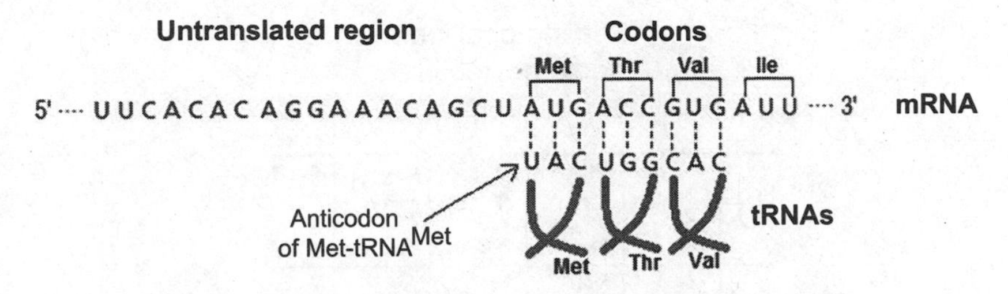

FIGURE 1.8 Translation. Translation begins when the ribosome (not shown) binds to the initiation codon AUG. A transfer RNA (tRNA) carrying methionine (Met) binds to the initiation codon (AUG) through complementary base pairing between the codon and the anticodon of the tRNA. The next aminoacyl tRNA carrying threonine (Thr) binds to the next codon (ACC), and the ribosome catalyzes the formation of a peptide bond between Met and Thr to form the first bit of the protein. Translation then proceeds down the mRNA reading one codon at a time and adding one amino acid at a time to the growing chain.

The genetic code

Aminoacylated tRNA molecules are the key intermediates in translating a nucleotide sequence into an amino acid sequence. The crucial step is making sure that the correct amino acid is attached to the correct tRNA. The first tRNA in Figure 1.8 is a specific transfer RNA called $tRNA^{Met}$ because it carries the anticodon that will pair with the codon for methionine (Met), and it is absolutely essential that the correct amino acid (Met) be attached to this particular RNA. Making that connection is the job of an enzyme called *aminoacyl-tRNA synthetase* – in this case it's *methionyl-tRNA synthetase* – and this step is so important that it is often called the second genetic code.

Since there are 20 different amino acids that can be incorporated into proteins, there must be at least 20 different tRNA molecules and at least 20 different aminoacyl-tRNA synthetases.

The translation process is referred to as decoding the mRNA, and the one-to-one correspondence between a three-nucleotide codon

Second position (middle nucleotide)

First position (5' end of the codon)	U	C	A	G	Third position (3' end of the codon)
U	Phe	Ser	Tyr Y	Cys	U
	Phe F	Ser S	Tyr	Cys C	C
	Leu L	Ser	STOP	STOP	A
	Leu	Ser	STOP	Trp W	G
C	Leu	Pro	His H	Arg	U
	Leu	Pro	His	Arg	C
	Leu L	Pro P	Gln Q	Arg R	A
	Leu	Pro	Gln	Arg	G
A	Ile	Thr	Asn N	Ser S	U
	Ile I	Thr T	Asn	Ser	C
	Ile	Thr	Lys K	Arg R	A
	Met M	Thr	Lys	Arg	G
G	Val	Ala	Asp D	Gly G	U
	Val	Ala	Asp	Gly	C
	Val V	Ala A	Glu E	Gly	A
	Val	Ala	Glu	Gly	G

FIGURE 1.9 The standard genetic code. The codons are read from 5′ to 3′ beginning with the nucleotide in the left-hand column. The next nucleotide is one of the rows along the top of the box, and the final nucleotide is from the right-hand column. The 20 standard amino acids corresponding to each codon are shown in their three-letter short form and their one-letter code: phenylalanine (Phe, F), leucine (Leu, L), isoleucine (Ile, I), methionine (Met, M), serine (Ser, S), proline (Pro, P), threonine (Thr, T), alanine (Ala, A), tyrosine (Tyr, Y), histidine (His, H), glutamine (Gln, Q), asparagine (Asp, N), lysine (Lys, K), aspartate (Asp, D), glutamate (Glu, Q), cysteine (Cys, C), tryptophan (Trp, W), arginine (Arg, R), and glycine (Gly, G). There are three termination (STOP) codons (UAA, UAG, and UGA).

and an amino acid is the genetic code. The *standard genetic code* is shown in Figure 1.9. You read the codons from 5′ to 3′ beginning with the nucleotide in the left-hand column (U, C, A, or G) followed by the one along the top row and ending with the one in the right-hand column. For example, UCG is the codon for serine (Ser), and AAG is the codon for lysine (Lys).

There are 64 different codons since each of the three positions can be occupied by one of four different nucleotides ($4 \times 4 \times 4 = 64$), but there are only 20 standard amino acids, so this means that some amino acids have multiple codons. There are two codons for phenylalanine (Phe), four for proline (Pro), and six each for serine (Ser), leucine (Leu), and arginine (Arg).

Introns and exons

Let's summarize what we have learned so far. A gene is a DNA sequence that's transcribed to produce a functional product. There are two kinds of genes. The first kind includes those genes that produce a functional RNA molecule like rRNA, tRNA, and a host of other kinds of RNA that we'll discuss later. We'll refer to these genes as *noncoding genes*. The second kind of gene is transcribed to produce mRNA that's translated to make a protein. These are the *protein-coding genes*.

The next figure (Figure 1.10) outlines the features of the two kinds of genes using the standard icons that you'll find in most textbooks. The solid gray line is double-stranded DNA, and the boxes represent the genes found in a typical genome. Transcription begins at a promoter (P) and ends at a termination signal (t). The sequences that control when and where the gene is transcribed are most often located in regulatory regions upstream (5′ end) of the gene.

Genes for noncoding RNAs are shown as open boxes, and protein-coding genes are shown as open boxes with the protein-coding region colored black. The codons in mRNA are translated from the start codon to the stop codon, and there are noncoding parts of these genes located between the beginning of the mRNA and the start codon and between the stop codon and the 3′ end of the transcript.

This was pretty much the state of knowledge when I graduated from university in 1968. James Watson's book *The Molecular Biology of the Gene* had been published three years earlier in 1965, the genetic code had been cracked, and the basic principles of gene regulation had been, or were being, worked out.

By the time I finished graduate school, six years later, everyone was confident that these principles were correct, at least for bacterial

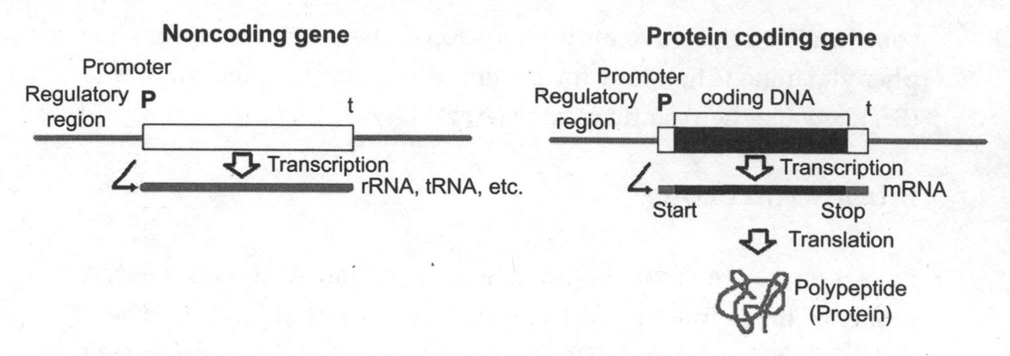

FIGURE 1.10 Gene icons. A gene is a transcribed region of DNA shown here as a rectangular box. Double-stranded DNA on either side of the gene is drawn as a thick gray line. The gene on the left specifies a functional RNA such as ribosomal RNA, transfer RNA, and so on. The gene on the right contains a coding region shown in black. The RNA product is messenger RNA that is translated to produce a polypeptide (protein).

genes. The best-studied organism was the bacterium *Escherichia coli*, and we now know that typical strains of this species contain about 4500 genes, of which 4400 are protein-coding genes (98 percent). The ratio is a bit different in eukaryotes; in humans, for example, about 20 percent of the genes specify functional RNAs and "only" 80 percent are protein-coding.

An extra layer of complication had been added to our understanding of genes and gene function by the time I finished my postdoc. A lot of experiments in the 1970s had shown that the primary transcripts of many eukaryotic genes were much bigger than the final products, so an additional step, called *RNA processing*, had to be added to the scheme shown in Figure 1.10. The extent of processing was surprising, especially the fact that, in many cases, there were bits of RNA chopped out of the middle of the initial transcript.

The next figure (Figure 1.11) shows the updated view of typical eukaryotic genes. The boxes outlining the gene (transcribed stretches of DNA) have been modified to include two classes of sequence. *Exons* specify the parts of the gene that will be retained in the final

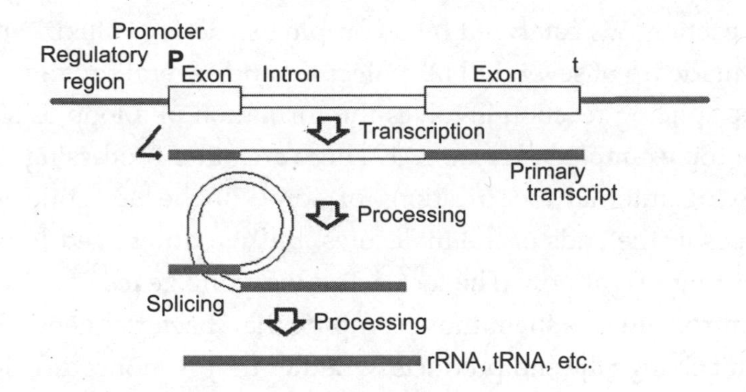

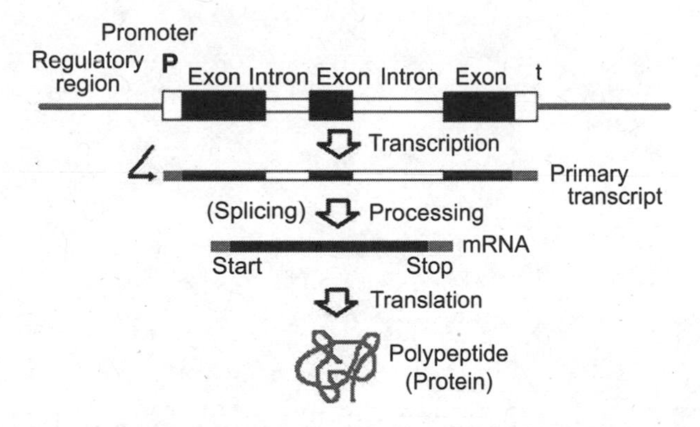

FIGURE 1.11 Typical eukaryotic genes. The top gene specifies a functional RNA. It consists of three regions of transcribed sequence: two exons and one intron. The intron sequence is removed from the primary transcript by splicing, giving rise to a final product that only contains the exon sequences. The bottom gene is a protein-coding gene with three exons and two introns. The final mRNA product is derived from the primary transcript by splicing out the two intron sequences.

functional RNA product, and *introns* specify the parts that will be removed and discarded.

The excision step of RNA processing is when the intron sequences are removed, and the two flanking exon sequences are joined together. This is called *splicing* since it requires the joining together of two separate RNA molecules – a process that's similar to the splicing together of two pieces of rope. We now know a great deal about this reaction: it's catalyzed by a complex structure called a *spliceosome* made up of several RNA molecules and several proteins.

The splicing reaction involves the formation of a loop as shown in the top example in Figure 1.11. There are other processing events involving internal modifications of some of the nucleotides and changes at the ends of the molecules, but we don't need to worry about them right now. The fact that eukaryotic genes usually contain introns makes them more complex than bacterial genes, but it doesn't change the end products, whether they be noncoding RNAs or messenger RNAs.

CHAPTER 2

The Evolution of Sloppy Genomes

> *This story begins more than 20 years ago with the observation ... that different species contain different amounts of DNA in their nuclei. This harmless information caused some discomfort when it was learned that primitive amphibians and fish contained more than 20 times as much DNA per nucleus as did man. It was argued that mammals display a greater developmental complexity than primitive fish, therefore, they must have more genes, yet why should the lower forms have more DNA, if DNA is the chemical basis of the gene?*
>
> Charles A. Thomas Jr. (1971)[1]

If you are ever in a fancy Japanese restaurant you might be tempted to order fugu sashimi. The dish consists of thin slices of fish, usually from the species *Fugu rubripes* (also known as *Takifugu rubripes*). This is one of a class of species known as pufferfish, and before eating it, you better make sure that the cook is qualified to prepare it safely because *Fugu* contains tetrodotoxins that can kill you.

We are not interested in *Fugu rubripes* because of Japanese eating habits. We are interested in its genome because the 390 Mb *Fugu* genome is one of the smallest vertebrate genomes – only one-eighth the size of our genome. Despite its small size, it has about the same number of genes as humans, so it's obvious that genome size doesn't reflect the number of genes.

You should also avoid eating lungfish but not because they are poisonous. Lungfish probably don't taste very good, and they are a threatened species. The genome of the lungfish, *Protopterus aethiopicus*, is one of the largest vertebrate genomes known at 133 Gb (133,000 Mb). That's more than 40 times larger than our genome. What's it doing with all that extra DNA?

These are just a few examples of the huge variation in genome sizes seen in related species. There was no obvious explanation for such variation when these data began to be compiled in the middle of the last century, so the problem became known as the C-Value Paradox, where "C" stands for the haploid genome size.[2]

The C-Value Paradox used to describe a real paradox. It looked like there was no correlation between the size of a genome and the complexity of an organism, and many scientists thought this was a serious problem because they expected that more complex organisms would need more genes and a larger genome.

You could argue that fish are very different than mammals and that pufferfish and lungfish aren't very closely related. However, the real paradox became obvious with the discovery that some very similar species had very different sizes of genomes. For example, the frog *Xenopus laevis* has a genome that's about the same size as ours, but there are other frogs that have genomes 30 times larger than *Xenopus*. Within the frog genus, *Rana*, we have *Rana erythraea* (leaping frog) with a genome a bit smaller than ours at about 2.7 Gb and *Rana esculenta* (green frog) with a genome size of 10 Gb. That's a threefold difference within the same genus, and it's hard to imagine that the green frog could be three times more complex than the leaping frog.

The complexity of genomes

Beginning with the pioneering work of Britten and Kohne in 1968, early genome studies relied on measuring the reassociation kinetics of DNA fragments – a type of experiment that involves separating the two strands of DNA (*denaturing*) and observing the time it takes for the stands to come back together to form double-stranded DNA (*renaturing* or *reassociation*). The experiments generated C_0t curves ("cot" curves) that correlate the initial concentration of DNA (C_0) and the time (t) it takes for renaturation.[3]

If a particular sequence of DNA was present in multiple copies, then those pieces would come together to form double-stranded DNA faster than sequences that were present in only a single copy because the concentration of repetitive DNA is higher. The experiments were difficult to perform but highly informative in the hands of experts.

The results indicated that a typical mammalian genome consists of a small fraction of highly repetitive DNA (about 10 percent), a lot of moderately repetitive DNA (about 40 percent), and the rest is unique sequence DNA (about 50 percent). Larger genomes had more repetitive DNA, and smaller genomes had less.

Attempts to identify functional regions of the genome, such as genes, involved hybridizing RNA to DNA and assessing their concentrations using similar techniques (R_0t curves). These results showed that mRNA hybridized to only a few percent of typical eukaryotic genomes confirming that most eukaryotic genomes contain 15,000 to 30,000 protein-encoding genes.

These studies established that large eukaryotic genomes contained a great deal of repetitive DNA and that there were fewer than 30,000 genes, but the technique of C_0t analysis fell out of favor, and consequently, it was not taught in university courses after the mid-1980s. Thus, there's a whole generation of scientists who are unaware of these important discoveries.

Variation in genome size

The C-Value Paradox applied to all the major classes of eukaryotes as shown by the data in Figure 2.1, where the values for genome sizes are given in picograms (pg). (One picogram is approximately

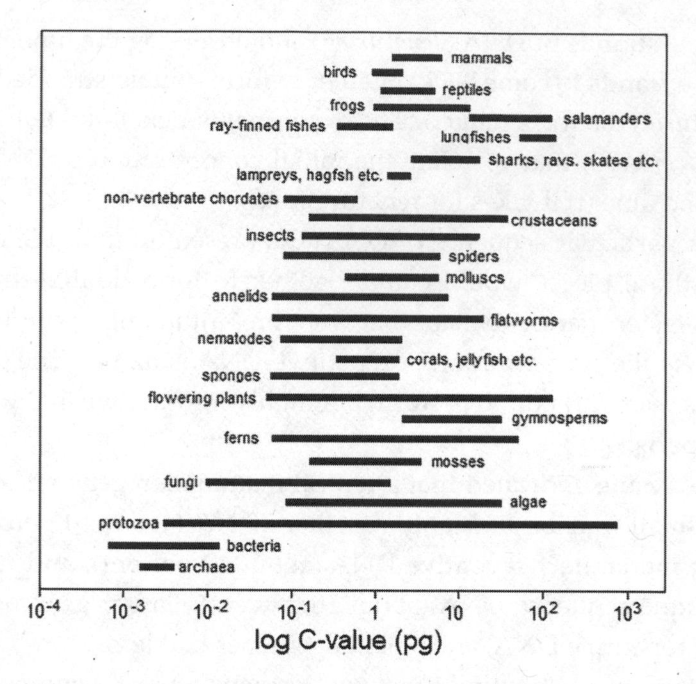

FIGURE 2.1 Genome sizes. Each bar represents the approximate range of genome sizes in picograms for each group. Note that the *x*-axis is a log scale, so the range of values in mammals is about 10-fold, and the range in flowering plants (angiosperms) is 1000-fold (0.1 pg–100 pg). (Original figure courtesy of Ryan Gregory.)

980 Mb, or 980 million base pairs.) In single-cell eukaryotes (*protists* or *protozoa*), the genome sizes range from 0.55 Mb for *Guillardia theta* to 100,000 Mb in some amoeba. That's a difference of five orders of magnitude (10^5). Similarly, there's a 1000-fold difference between the genome sizes of various flowering plants.

There's considerable variation even within mammals. The little brown bat (*Myotis mystacinus*) has a genome size of about 1.9 Gb, which is considerably smaller than our genome (3.2 Gb). The house mouse (*Mus musculus*) has a genome size similar to ours, but the Venezuelan spiny rat (*Proechimys trinitatus*) has a genome size of about 6.2 Gb, which is almost twice as large as ours. The African elephant (*Laxodonta africana*) also has a large genome (4.4 Gb).

By the late 1960s, knowledgeable scientists were used to the idea that genes occupied only a small part of the genome, and in 1974 the editor of the journal *Cell*, Benjamin Lewin, was expressing the consensus view of the experts when he wrote that the C-Value Paradox could be resolved by assuming that much of the genome is composed of nonfunctional repetitive DNA (junk DNA). The idea is that the amount of DNA in different species is due to varying amounts of junk DNA. Although this explanation resolves the paradox, we still need to explain *why* the amount of junk DNA is different in related species. Ryan Gregory suggests that we stop talking about paradoxes and use the term *C-Value Enigma* to refer to this part of the problem.[4]

The junk DNA explanation fits very nicely with the discoveries of developmental biologists that were being published at the same time.[5] The basic idea behind the evolution of new and complex organisms is that you don't need new genes – all you need is to change the way existing genes are expressed in different species. The idea is quite old, but it was enunciated clearly in Stephen Jay Gould's book *Ontogeny and Phylogeny*, published in 1977. Subsequent work with fruit flies and other insects showed that you could get many different species, and many different morphologies, using the same genes and a small number of transcription factors and regulatory sequences.

We now know that the differences between mice, elephants, bats, rats, and gorillas follow the same rules. All mammals have pretty much the same genes, and the differences in morphology and complexity are due to developmental differences affecting when the genes are turned on and off.

This is a robust explanation of the data on genome sizes. There are no other explanations that come close to explaining the data, but that hasn't stopped some scientists from denying that there's junk DNA and denying that humans are similar to other mammals.

Instantaneous genome doubling

Cell division is a highly regulated process whereby all the chromosomes are copied and one complete set is segregated into each of the two daughter cells. However, sometimes the cells do not divide following chromosome duplication, resulting in a cell with twice as much DNA and twice the number of chromosomes. The general term for this is *polyploidy*.

Polyploidy is a natural occurrence in many species; for example, human liver cells and placental cells are often polyploid. The most reasonable explanation is that polyploidy increases the number of copies of certain genes so that each cell can produce more proteins.

The first cell formed when a sperm cell invades an egg cell is called the *zygote*, and it can accidentally fail to divide following the first DNA duplication, resulting in offspring with twice as many chromosomes. This is a very rare event, but it has been documented in many lineages; for example, the big differences in the sizes of frog genomes are due to genome doubling events. Polyploidy can also occur by hybridization between two related species.

Recent examples of polyploidy are easy to recognize because the new species will have twice as many chromosomes as the parental species. The classic example is several varieties of wheat that have been artificially bred by humans. The ancestral plant was diploid (2C), but several modern varieties are tetraploid (4C) or hexaploid (6C).

The plant genus *Brassica* provides a well-documented case of polyploidy by hybridization. You are probably familiar with the species *Brassica oleracea*, which has been bred to produce a multitude of common vegetables such as cauliflower, broccoli, cabbage, kale, and brussels sprouts. The related species *Brassica rapa* produces Chinese cabbage. Canola (from *Can*ada and *oil*) is a major crop for the production of canola oil, and the species that produces it is *Brassica napa*, also known as rapeseed. The sequence of the rapeseed genome clearly shows that it is a hybrid of *Brassica oleracea* and *Brassica rapa*.[6] Rapeseed plants have twice as much DNA as either ancestor, and it's very unlikely that all this DNA is required to differentiate rapeseed from related species. Most of it must be superfluous.

Over time it becomes less obvious that a polyploidy event has occurred because chromosomes are lost and rearranged and redundant DNA segments within chromosomes are spontaneously deleted. A lot of detective work is required to detect ancient polyploidy events, but the data is persuasive: we now know, for example, that most modern vertebrates descend from a species that arose by polyploidization in ancient fish.

We can conclude two things from the data on polyploidy. The first is that species can often tolerate a doubling of their genome content even though most of the excess DNA must be redundant. The second is that over millions of years, the amount of DNA usually decreases, lending further support to the idea that much of the excess DNA was dispensable junk that was discarded over time.

The Onion Test

Ryan Gregory has created the Onion Test as a way of checking to see if you have an explanation of genome size variation that's better than junk DNA. Gregory points out that the onion *Alium cepa* has a 16 Gb genome that's 5 times larger than ours. Related species of onion have genomes that range from about 7 Gb to 30 Gb. Here's how he described the Onion Test on his blog in 2007:

> The onion test is a simple reality check for anyone who thinks they have come up with a universal function for junk DNA. Whatever your proposed function, ask yourself this question: Can I explain why an onion needs about five times more noncoding DNA than a human?[7]

The Onion Test is widely misunderstood, so let's make sure we understand what Gregory is saying. He is not proposing that onion genomes are proof of junk DNA. What he is saying is that the best explanation of bloated genomes is junk DNA, and if you want to propose an alternative explanation, then you should be able to explain onion genomes using your favorite hypothesis. It's a reality check that aims to avoid frivolous suggestions that only apply to a single organism, such as humans. It's a check on the explanatory power of your favorite hypothesis.

Throughout this book, we will encounter many attempts to dismiss junk DNA in the human genome and offer another explanation for all that extra DNA. All of them fail the Onion Test.

It may seem strange that species could tolerate a large amount of junk DNA since you might think that all that extra DNA is deleterious and should be rapidly purged by natural selection. We need to refresh our knowledge of evolution in order to see why this doesn't happen.

Modern evolutionary theory

> *Another curious aspect of the theory of evolution is that everybody thinks he understands it. I mean philosophers, social scientists, and so on. While in fact very few people understand it, actually, as it stands, even as it stood when Darwin expressed it, and even less as we now may be able to understand it in biology.*
>
> Jacques Monod (1975)

This book is about the evolution of genomes, so let's make sure we are all on the same page when we talk about evolution. Most of us learned about Darwin's theory of natural selection when we were in school, and that understanding has been reinforced by popular books and documentaries over the years. My focus in this book is on evolution at the molecular level, so we can view natural selection as a competition between two variants of the same gene. For example, one variant of a bacterial gene may encode a standard metabolic enzyme, but a mutation can change the sequence of the gene so that the enzyme can now degrade a drug used to kill the bacteria. The variants are called *alleles*, and in this case, one of the alleles confers drug resistance and the other one doesn't.

A population of bacteria consists of cells carrying a large number of different alleles of each gene. In the presence of the drug, the bacterial cells containing the drug-resistance allele will survive, and the ones containing the parent allele will die. This is the essence of natural selection, and it applies to all cases in which one allele confers a more fit phenotype than another.

Evolution is defined as a change in the frequency of alleles in a population over many generations, and in this example, we see that the drug-resistance allele will increase in frequency in the population meaning that the population has evolved.[8]

Natural selection is responsible for transforming reptilian scales into feathers, and it's the mechanism behind the evolution of eyes and many other features of complex species. These are examples of *adaptations* leading to a complete replacement of one allele by another that confers greater fitness. This process is known as the *fixation* of an allele in a population. Thus, we can think of the change in frequencies of an allele in a population as a process that occurs in two steps: (1) the creation of a new allele by mutation and (2) the gradual fixation of that allele by natural selection. This is *positive selection*.

The creation of a new beneficial allele by mutation is a rare event. It's much more common for mutations to be detrimental, in which case they will be rapidly eliminated by natural selection, a process called *negative selection* or *purifying selection*.

Random genetic drift

I am an unrepentant "beanbag geneticist."

J.B.S. Haldane (1963)[9]

Scientists in the 1960s were able to look at the amount of variation in a typical population using the new tools of biochemistry and molecular biology. The results were surprising because there was much more variation than anyone suspected; for example, a typical population of fruit flies or humans contains dozens and dozens of alleles of various genes.[10] As far as anyone could tell, the different variants produced by these alleles were all about the same with respect to fitness; in other words, they were neither beneficial nor detrimental but neutral.

It's easy to understand why beneficial alleles can become fixed in a population by natural selection or why deleterious alleles can be eliminated, but what about neutral alleles? What's the fate of a new neutral allele that arises by mutation?

The answer requires an understanding of random genetic drift – the other mechanism of evolution. The easiest way to understand drift is to use the analogy of beanbag genetics – a term coined by Ernst Mayr to disparage the great work of the early twentieth-century population geneticists. (Mayr was not a fan of mathematics or population genetics.[11])

Consider a beanbag that contains 100 black beans and 100 white beans representing the alleles of a particular gene in a population of 100 diploid individuals. Imagine that whenever you shake the bag the beans will spontaneously reproduce – a process that simulates the actual reproduction of individuals in a real population. The bag now contains 200 black beans and 200 white beans.

Most biological populations are relatively stable, meaning that the number of individuals neither increases nor decreases. In order to simulate this in our beanbag analogy, we have to grab 200 beans and discard them, leaving only 200 in the bag ready for the next shakeup. Imagine putting your hand in the bag and picking 200 beans at random. What will be the frequency of white and black beans (alleles) left in the bag?

It's possible that there will be 100 white beans and 100 black beans remaining, but it's also possible that the ratio will be 99:101, 98:102, or even 97:103 just by chance. This change in the frequency of black beans and white beans is analogous to the change in the frequency of alleles in a population by random genetic drift, where individuals die by accident or disease unrelated to their genes. If you keep shaking your beanbag and discarding half the beans you will eventually end up with a bag full of either white beans or black beans, although you may be shaking for many days. Similarly, in a real population the frequency of neutral alleles will vary just by chance until, after many generations, one of the alleles becomes fixed and the other is eliminated. This is evolution by random genetic drift.

Neutral theory

> *The main tenet of the neutral theory is that the great majority of evolutionary changes at the molecular level are caused not by Darwinian selection but by random fixation of selectively neutral (or very nearly neutral) alleles through random sampling drift ...*
>
> Motoo Kimura (1989)

The basic mathematics of population genetics were worked out in the first half of the twentieth century, and although the original emphasis back then was on natural selection, the role of random genetic drift was well understood. Back then, drift was not thought to be very important except when it interfered with, or modified, natural selection, but the discovery of widespread variation in populations called for a different explanation since natural selection should rapidly eliminate variation. The new explanation was that most of the variation must be due to the segregation of neutral alleles by random genetic drift – a process that takes a much longer time than selection.[12]

The person most often associated with the development of the neutral theory is the Japanese geneticist Motoo Kimura, but many other scientists contributed to the widespread acceptance of the neutral theory. Those scientists were well aware of the significance of this new perspective on evolution because they knew that it conflicted with the standard view that attributed almost all change to natural selection. That standard view is often referred to as neo-Darwinism or the Modern Synthesis (synthetic theory). I believe that this shift in thinking – from an adaptationist view of evolution to one that recognized the importance of chance and accident – represents a major change in how we look at molecular evolution and is intimately connected to the view that much of our genome is junk. We now recognize that the great majority of change is due to random genetic drift when you examine evolution at the molecular level.[13]

It's important to keep in mind that this shift in emphasis met with considerable resistance 50 years ago and still isn't widely accepted by many of today's scientists who prefer an adaptationist view of evolution based on natural selection. King and Jukes knew this in 1969 when they published their famous paper on "Non-Darwinian Evolution": "The idea of selectively neutral change at the molecular level has not been readily accepted by many classical evolutionists, perhaps because of the pervasiveness of Darwinian thought."[14]

Kimura was also acutely aware of the fact that his neutral theory went against the common view of evolution, and he was still making this point three decades later: "I would like to emphasize the importance of random genetic drift as a major cause of evolution. We must be liberated, so to speak, from the selected constraint imposed by the neo-Darwinian (or the synthetic) theory of evolution."[15]

Nearly neutral theory

Recent advances in molecular genetics have had a great deal of influence on evolutionary theory, and in particular, the neutral mutation-random drift hypothesis of molecular evolution has stimulated much interest. The concept of neutral mutation substitution in the population by random genetic drift can be extended to include random fixation of very slightly deleterious mutations which have more of a chance of being selected against than being selected for.

Tomoko Ohta (1973)

The idea that large amounts of our genome could be strictly neutral and therefore be invisible to the process of natural selection wasn't completely satisfactory because common sense suggests that big, bloated genomes are probably bad. It seems likely that populations should select against excess DNA over time – a result that's compatible with the gradual loss of DNA following polyploidization.

These issues were resolved with the development of the nearly neutral theory, an advance attributed mainly to Tomoko Ohta, who worked initially with Motoo Kimura. Kimura and Ohta were population geneticists at the National Institute of Genetics in Mishima (Japan), but, after publishing with Kimura on the neutral theory, Ohta became convinced that strictly neutral mutations were uncommon. Her concept of nearly neutral alleles included alleles that could be slightly deleterious under some conditions but not so deleterious that they could be eliminated by natural selection.

Slightly deleterious alleles can be fixed in a population, but in order to see how this happens, we need to understand a bit of the mathematics behind population genetics. It may seem strange to include population genetics in the list of things that you need to know in order to understand junk DNA, but it turns out to be extremely important.

Population genetics

Basic principles of Mendelian genetics, supported by decades of empirical observation, tell us that long-term evolutionary change is a consequence of the origin of new variation by mutation and subsequent modifications of allele frequencies. This fundamental fact has provided a solid foundation for the development of a mechanistic framework for understanding evolutionary processes with a level of mathematical rigor that has few rivals in the life sciences. Indeed, the general principles of population genetics are now so well established that the credibility of any proposed scenario for genomic evolution must remain in doubt until it has survived this theoretical gauntlet.

Michael Lynch (2007)

Most readers will be familiar with the basic tenets of transmission genetics, where traits are passed from parents to children. You may even have drawn a Punnett square or two in order to estimate the probability that dominant or recessive alleles will be passed on from one generation to the next. The laws governing this version of genetics were worked out by Gregor Mendel more than 160 years ago, and they apply to diploid species that contain two copies of each gene. If both copies are identical, then the individual is *homozygous* for one particular allele, but if the copies differ, then the individual is *heterozygous* because they have two different alleles of the gene.

Mendel's laws work very well if you want to know how many short pea plants will be produced when you cross a single plant that's homozygous for the tall allele with a plant that's homozygous for the short allele. However, a more sophisticated analysis is required if you want to understand the evolution of a population of pea plants containing a large number of tall and short plants. This is the domain of population genetics – a field of study that grew up in the early twentieth century when the mathematical equations were developed by R.A. Fisher, J.B.S. Haldane, and S. Wright.

Before continuing, I want to emphasize that what I'm about to describe is the consensus view of the experts in the field of molecular evolution. It's not new, and it's not radical despite the fact that it is not widely known. There were, and still are, biologists who are skeptical of using mathematical equations to describe the evolution of populations – that was the position of Ernst Mayr in his debate with J.B.S. Haldane over beanbag genetics – but the tide is against them. As Michael Lynch says, paraphrasing a famous quotation from Theodosius Dobzhansky,

"Nothing in evolution makes sense except in the light of population genetics."[16]

Much of the debate about junk DNA would go away if the opponents of junk DNA had a deeper appreciation of modern evolutionary theory, as Lynch implies in the quotation at the beginning of this section.

One of the important features of a population is its size (N).[17] Population sizes vary over a huge range, from 10^{20} in some bacteria species to only a few dozen in species that are on the verge of extinction. As a general rule, bacteria and single-cell eukaryotes have huge populations and large multicellular species have much smaller populations. Most vertebrate species have small populations of many thousands.

The *selection coefficient(s)* of a given allele is a measure of the selective advantage of the allele compared to other alleles. Sophisticated mathematical equations have been worked out to describe the behavior of an allele in a population given the population size and the value of the selection coefficient. For our purposes we can look at simplified versions of these equations that give us approximate estimates of the probability of fixation.

If a new beneficial mutation arises with a selection coefficient of s, then its initial frequency in the population will be $1/2N$ in a diploid population. The probability that it will eventually become fixed depends on the size of the population, but for large populations and small values of s, the probability is approximately $2s$.

We are used to thinking of traits that confer very large selective advantages such as drug resistance in bacteria. In such cases, the value of s can be very high, and the probability of fixation approaches 100 percent. However, in most cases the selection coefficient is much less than 1.0, and a new allele will be lost by chance long before it can be fixed by natural selection. Imagine, for example, that a mutation in one of our early ancestors gave rise to a slight increase in intelligence. The selective advantage of such an allele might be quite small; for example, if $s = 0.01$, then the probability that it will be fixed is only 2 percent. In other words, there's a 98 percent chance that humans would not get any smarter.

This is a very important point because it's not what most of us were taught in school, and it's not how evolution is described in TV documentaries and popular books. The old-fashioned view of evolution is that once a beneficial allele occurs, no matter how slight the benefit, it will sweep through the population in just a few generations. The truth is that those beneficial alleles will usually be lost by chance unless the selective benefit is quite large.

Completely neutral alleles ($s = 0$) are invisible to natural selection, so they can only be fixed by random genetic drift. In this case the population size really matters; for example, in a small population it takes only a few chance events for a neutral allele to become established, but in a large population it's almost impossible. Population geneticists have shown that the probability of fixing a neutral allele is approximately $1/2N$ in a diploid population.

This is a very low probability even in small populations of only 10,000 individuals, but the most interesting calculation comes from considering alleles with very small values of the selection coefficient. If $s = 0.0001$, then the probability of fixation by natural selection is only 0.02 percent, but the probability of fixation of a neutral allele in a population of 10,000 is $1/20{,}000 = 0.00005$ or 0.005 percent. These probabilities aren't very different, so as s approaches a value of $1/2N$ a beneficial allele becomes effectively neutral.[18]

What this means is that in very large populations natural selection can lead to fixation of alleles with very small beneficial effects, but in small populations selection can be overwhelmed by drift and the beneficial allele will be lost.

Now let's consider a detrimental allele with a negative selection coefficient. The same calculations apply to detrimental alleles, namely, that in large populations, slightly deleterious alleles will be eliminated by negative selection but in smaller populations the probability of fixation by random genetic drift becomes significant, and slightly deleterious alleles can be fixed by chance. Thus, when the absolute value of s

(positive or negative) approaches $1/2N$ the alleles become effectively neutral, and natural selection loses its power to affect evolution.

This is the essential point in the nearly neutral theory and its modern form, the drift–barrier hypothesis, promoted by Michael Lynch and his collaborators. The models say that genome architecture is critically affected by the size of the evolving population. Species with large populations will have streamlined genomes with very little junk DNA because selection will remove most junk before it can become fixed in the genome. Species with smaller populations will accumulate junk DNA because it is not harmful enough to be purged by natural selection.[19]

The fundamental concept in population genetics is that evolution is a *stochastic* process meaning that it involves a certain amount of chance or randomness. What this means is that the fixation (and elimination) of beneficial, neutral, and detrimental alleles is governed by probabilities and not certainties, and that's a very different view than the traditional adaptationist view that often sees evolution as a strictly algorithmic process. That old-fashioned view is often referred to as "Darwinian" because Darwin's major emphasis was on natural selection. Thus, the views of Kimura and modern population geneticists are often seen to be in conflict with traditional Darwinism, as described by Mark Ridley in his 1997 book, *Evolution*. "Kimura's original radical claim, that most molecular evolution proceeds by drift, not selection, remains intact in the nearly neutral theory. It still contrasts strongly with the view that molecular evolution is powered by Darwinian natural selection."[20]

ARE HUMANS STILL EVOLVING?

The success of modern medicine means that people with deleterious mutations can survive and reproduce in today's societies whereas they might have died had they been born

(*continued*)

ARE HUMANS STILL EVOLVING? (continued)

100,000 years ago. The classic example is diabetes, but the argument applies to many other genetic diseases.

This fact often leads to a discussion about whether humans are still evolving, and many people conclude that the answer is "no" because they think that natural selection is no longer selecting for the fittest individuals. This answer is wrong because it's based on an incorrect view of evolution.

Think of it this way. Evolution is a change in the frequency of alleles in a population. In the past, the frequency of the allele causing diabetes was kept at a very low level by negative selection, but in modern societies that frequency may be increasing because diabetes can be treated. Thus, modern medicine leads to more evolution, not less.

Furthermore, most mutations produce neutral or nearly neutral alleles, and their frequencies are determined by random genetic drift. Neutral alleles are constantly being eliminated or fixed in the population, and this process can't be stopped – evolution can't be stopped. In addition, new beneficial alleles are still being created and are still being fixed by natural selection. Humans are still evolving, and we will continue to evolve until we become extinct.

On the evolution of sloppy genomes

Dear Francis, I am sure that you realize how frightfully angry a lot of people will be if you say that much of the DNA is junk. The geneticists will be angry because they think that DNA is sacred. The Darwinian evolutionists will be outraged because they believe every change in DNA that is accepted in evolution is necessarily an adaptive change. To suggest anything else is an insult to the sacred memory of Darwin.

letter from Thomas Jukes to Francis Crick
on December 20, 1979[21]

The genomes of evolving populations can expand or contract over long periods of time – usually millions of years. The driving forces are insertions and deletions of large segments of DNA. If the rate of insertion is greater than the rate of deletion then the genome will gradually become larger; conversely, if the rate of deletion is greater, then the genome will shrink.

In species with small population sizes, the slightly deleterious effect of a single insertion event does not meet the threshold value that allows it to be purged by natural selection. In such species, insertion mutations can be fixed by random genetic drift unless they disrupt a gene.

Deletion mutations can remove important parts of the genome, such as genes, and these mutations will have severe consequences and will be rapidly eliminated by purifying selection. Other deletions may be slightly beneficial or slightly deleterious but not enough to be recognized by natural selection in small populations. Deletion alleles can be fixed in the population by random genetic drift.

It looks like the rate of insertions was greater than the rate of deletions in most eukaryotes, and that's why most of these genomes increased in size hundreds of millions of years ago. This process led to modern species with large genomes that differ in C-value, and this result is consistent with modern evolutionary theory provided that much of the excess DNA is junk and population sizes are relatively small. The result is sloppy genomes that do not look like they are fine-tuned by natural selection.

If we imagine that the expansions and contractions of eukaryotic genomes are due mostly to a stochastic process, such as random genetic drift, then it's not surprising that we see variations from species to species, including some outliers with very large genomes and others with very small genomes. For example, the small genome of the pufferfish and the large genome of the lungfish can be viewed as just the extremes of a normal distribution.

Many insertions are true insertion events due to the incorporation of new DNA from viruses and transposons. Over time, this gives rise to large amounts of repetitive DNA in the genome. Repetitive DNA will be covered in more detail in the next chapter. Another mechanism of genome expansion is called *segmental duplication,* whereby errors in recombination cause large pieces of DNA to be duplicated. This results in an expansion of the genome by adding a big hunk of DNA, and that increase will be seen as an insertion when comparing genomes.

We know that duplications are an ongoing process in all species. We even know a bit about its frequency in humans thanks to projects such as the 1000 Genomes Project. The latest results show that the average person has about 1000 such duplications relative to the standard reference genome. Altogether, there are thousands of different segmental duplications in human populations. This tells us that such increases in genome size are relatively common.[22]

There's another mechanism for altering genome size. Chromosomes are often broken – the technical term is double-stranded breaks – and these breaks need to be repaired. The process involves both recombination, a form of DNA repair, and actual repair mechanisms involving the synthesis of new pieces of corrected DNA.

The double-strand break is repaired by matching the ends to homologous DNA on the other chromosome. (Recall that we are diploid.) The gap region is filled in by repair DNA synthesis. If the homologous chromosome has a bit of extra DNA due to an insertion, then this DNA will be copied during the repair process, and both chromosomes will now have the insert. The process is called *gene conversion* even though it need not involve genes, per se.

There's some evidence that this process is biased in favor of copying insertions and this biased gene conversion results in the gradual expansion of the genome by spreading small insertions. If this is correct, then there's an intrinsic mechanism for the expansion of the genome that has nothing to do with selection. If there's no selection for either insertions or deletions, then genomes could gradually expand due to biased gene conversion until they reach a size where the rate of deletions becomes significant. I won't mention this again because it's beyond the scope of this book, but you should be aware of this hypothesis.

Errors in DNA replication can cause small deletions (less than 10 bp). The error rate is significant, on a population timescale, and that's why most human genomes differ from each other at thousands of sites where small deletions have occurred. The fact that these deletions persist in the human population suggests that they are neutral, and that's evidence that the DNA that has been deleted is junk.

Larger deletions are due to errors in recombination. Recombination occurs when two very similar stretches of DNA on two different chromosome line up and both chromosomes are cut in two and the ends are swapped. The "left" end of one sister chromosome is attached to the "right" end of the other and vice versa. This is a way of rearranging the alleles (genes) on a single chromosome, and it's one of the characteristics of genetics and sex. But allele shuffling is not the reason for recombination – the real reason is that it can be used to repair mutations.

Occasionally the process makes a mistake by bringing together two very similar stretches of DNA on the same chromosome. Imagine that there are two similar regions about 100 kb apart, and when the recombination enzymes bring them together it forms a big loop with the recombination site at the base of the loop. Now when the enzymes break and rejoin the DNA the loop of DNA will be removed resulting in a deletion of 100 kb worth of DNA. This is the most common mechanism for generating large deletions. Its frequency will be enhanced in large, bloated genomes containing lots of repetitive DNA because there's an increased probability that the recombination enzymes will bind to similar repetitive DNA sequences on the same chromosome and delete the DNA between them.

In the absence of selection, the balance between rates of insertion and deletion will determine the size of the genome. If a genome is growing by DNA insertions, then the increase in genome size is unrelated to the fitness of the individuals. Eventually, the rate of deletions will catch up because deletions are favored when there's lots of opportunities for recombination to make mistakes. An equilibrium will be reached where the genome size remains relatively stable, and this has nothing to do with whether the excess DNA is functional and whether larger genomes are better than smaller ones.

CHROMOSOME DYNAMICS

Chromosome insertions, deletions, fusions, breakages, and rearrangements are common features in the evolution of sloppy genomes. A recent comparison of gorilla, chimpanzee, and human genomes revealed a total 614,186 deletions, insertions, and rearrangements. There were 17,789 fixed insertions and deletions that were specific to humans.

Many more insertions and deletions are polymorphic in humans, meaning that they have occurred relatively recently and haven't been fixed or eliminated. The data is consistent with the idea that most insertions, deletions, inversions, and other chromosome rearrangements have little or no effect on the fitness of the individual and their frequency in the population is due to random genetic drift.[23]

Bacteria have small genomes

This is all very interesting, but why do bacteria have such small genomes? We know that there is very little junk DNA in bacterial genomes because most of their chromosomes consist of genes and associated regulatory sequences and very little else.

The standard explanation was that bacterial genomes are kept junk-free by selection. Bacteria divide very rapidly, so the argument goes; thus, bacteria with smaller genomes will divide faster and outcompete their neighbors with larger genomes. This is the argument often seen in textbooks and review articles, but it's almost certainly wrong. Michael Lynch, among others, has pointed out the problems.

I'm embarrassed to say that for years I taught my students that bacterial genomes were under strong selection to reduce the size of their genomes because they grew and divided very rapidly. I was vaguely aware of some problems with this argument because I knew that the average doubling time of bacteria was not 20 to 30 minutes as we saw in the laboratory under ideal conditions but more like several days or weeks in the real world.

I also knew that DNA replication could not be limiting cell division because most bacteria could begin a new round of DNA replication before the first one was complete. That's why *E. coli* can divide every 30 minutes in laboratory flasks even though it takes 40 minutes to replicate its DNA. This has been in my textbooks for 30 years, but I never put two and two together.

I met Michael Lynch at a Society for Molecular Biology and Evolution meeting in Lyon, France, in 2010. He told me in his usual straightforward manner that I was dead wrong and probably not very smart. As he points out in his 2007 book, *The Origins of Genome Architecture*, there is no evidence whatsoever that cell division in bacteria is limited by the time it takes to replicate DNA.

There's a much better explanation, and it has to do with the modern view of population genetics that I have just described. There is a slight cost associated with carrying junk DNA in a genome. While it may not affect the rate of cell division, making that extra DNA does take energy, and cells with large genomes will be at a very slight disadvantage in terms of their ability to survive in challenging environments. Although the cost of junk DNA applies to all species, including us, bacterial species have very large populations, so natural selection works very effectively with small differences in fitness. This explains why bacterial genomes don't accumulate junk DNA.[24]

The take-home lesson is that modern evolutionary theory has led to a dynamic view of genome evolution where much of the DNA may not be subject to natural selection. This explanation accounts for the variation in genome sizes that have been observed – variations that are not correlated with the number of genes or the complexity of a species – and it also accounts for the differences between prokaryotes and eukaryotes.[25] But, as Thomas Jukes predicted, this makes a lot of people "frightfully angry" because it's counter to the old-fashioned, and still persistent, view of evolution as equivalent to natural selection. This ongoing resistance is why some prominent evolutionary biologists like Michael Lynch are still trying to make this point as forcibly as possible.

> Recent observations on rates of mutation, recombination, and random genetic drift highlight the dramatic ways in which fundamental evolutionary processes vary across the divide between unicellular microbes and multicellular eukaryotes. Moreover, population-genetic theory suggests that the range of variation in these parameters is sufficient to explain the evolutionary diversification of many aspects of genome size and gene structure found among phylogenetic lineages. Most notably, large eukaryotic organisms that experience elevated magnitudes of random genetic drift are susceptible to the passive accumulation of mutationally hazardous DNA that would otherwise be eliminated by efficient selection.[26]

CHAPTER 3

Repetitive DNA and Mobile Genetic Elements

A concept that is repugnant to us is that about half of the DNA of higher organisms is trivial or permanently inert (on an evolutionary timescale).

Roy Britten and David Kohne (1968)

Roughly half of our genome is composed of repetitive DNA that can be subdivided into highly repetitive DNA and moderately repetitive DNA. As I explained in the previous chapter, the insertion and deletion of repetitive DNA is one of the most important contributors to the expansion and contraction of genome sizes over time, but the important question in this chapter is whether this repetitive DNA is functional or junk. Its discoverers, Britten and Kohne, thought that the idea of junk DNA was "repugnant," and they speculated that "it will ultimately be found important to the genome." Many other scientists disagreed.

The highly repetitive DNA category consists of very short DNA segments that are repeated hundreds or thousands of times in a row. A simple example is a two-nucleotide repeat such as ATATATATAT and so on, but there are more complicated examples with repeats of three, four, or more nucleotides. The main characteristic of these repeats is that they are contiguous, producing long stretches of tandemly repeating DNA units.

Highly repetitive DNA was discovered many years ago when DNA was isolated and broken into small fragments that were then separated in an ultracentrifuge according to their density. Most of the DNA fell into a large bell-shaped cluster of average density since the ratios of A–T and G–C base pairs are similar throughout the genome. However, there was a distinct "satellite" peak with a different density because it was enriched for one of the base pairs. This is why highly repetitive DNA is often called satellite DNA.

The presence of highly repetitive DNA was confirmed by the reassociation kinetics experiments (C_0t analysis) described in the last chapter. Those experiments established that, depending on the species, somewhere between 3 and 10 percent of the genome is composed of highly repetitive DNA. We now know a great deal more about this DNA.

Centromeres

Centromeres are the part of a chromosome required for proper segregation after the DNA has been copied and the cells are dividing. Each one of the pair of newly replicated chromosomes has to be segregated into one of the daughter cells, and the centromeres are the places where the spindle fibers attach in order to pull the chromosomes to opposite ends of the dividing cell. It's easily recognized as a constricted region in the middle of a metaphase chromosome.

We've known about centromeres for fifty years, but it wasn't until the 1980s that the DNA content of these regions was revealed. Each centromere contains highly repetitive DNA sequences of several different types, but most of them fall into two clusters called human-satellite sequences (Hsats) and alpha-satellite sequences.[1]

Telomeres

Telomeres are found at the ends of all chromosomes, and they consist of multiple repeats of the sequence TTAGGG interspersed with very short stretches of unique sequence DNA. The average length of a typical telomere is about 2000 bp. Telomeres are essential functional DNA that is required for proper DNA replication because the DNA replication enzymes need help in replicating to the very ends of chromosomes.

Telomeres get shorter every time a chromosome is replicated, so they are constantly being expanded by adding more TTAGGG repeat units. In most cases, the expansion keeps pace with the shortening, but there's a general decline in the length of telomeres in the cells of older people.

DEAD CENTROMERES AND TELOMERES

Our closest relatives are chimpanzees and bonobos, but they have 24 pairs of chromosomes while we have only 23 pairs. The difference is due to an ancient fusion event that joined two smaller chromosomes in our ancestors to make one large human chromosome 2. The end-to-end joining left a stretch of telomere sequences in the middle of chromosome 2, and it also resulted in the formation of a degenerative centromere since after the fusion event only one of the two centromeres was required.

There are many degenerate (dead) centromeres in our genome, but this one is particularly important since its recent origin is quite clear. The others presumably arose similar mechanisms that occurred many millions of years ago.

Note that there's no strong correlation between the size of a genome and the number of chromosomes; for example, the Indian muntjac (a small deer) has only three pairs of autosomes plus a Y chromosome, while the various members of the dog family have 39 pairs of chromosomes. The record for the largest number of chromosomes in mammals is held by the red viscacha (a small rodent) with 54 pairs. It's

(*continued*)

DEAD CENTROMERES AND TELOMERES (continued)

possible to reconstruct the original chromosome set of the ancestor of all mammals by comparing all the breakages that have occurred in various lineages – an analysis that has only become possible in recent years with the advent of rapid genome sequencing. The results suggest that the mammalian ancestor had 22 pairs of chromosomes, but only seven of these have been conserved in humans. About 100 different breakages can be identified in the primate lineage, but most of them are rearrangements within a chromosome (inversions) or exchanges between chromosomes (recombination), and these will not change the number of chromosomes.[2]

Short tandem repeats

As the name suggests, short tandem repeats (STRs) consist of short stretches of repetitive DNA where the repeat unit can be anywhere from 2 to 6 bp. There are about 350,000 STRs in our genome, and they are widely dispersed with a median length of about 25 repeats.[3] Most of these short repeats are due to errors during DNA replication.

Huntington's disease is caused by multiple repeats of the sequence CAG that interfere with the expression of the Huntington gene. Normal individuals have fewer than 30 repeats, and the disease becomes apparent with more than 30 repeats. It becomes quite severe when

DNA FINGERPRINTS

Each of us has a unique genome that differs at millions of sites from other individuals. One of the most common differences is in the number of repeat units in stretches of short tandem repeats. At any one site you may have 15 copies of a repeat, and I may have 18. It's easy to detect this difference by cutting out the STR region and measuring its length.

Standard DNA fingerprints are based on 20 different STRs located on 15 different chromosomes. Each of these sites is known to be highly variable. Most of them contain 4 bp repeats, such as AGAT or TCTA.

The chances of any two individuals having exactly the same number of repeats at all 20 sites are minuscule. That's why a DNA fingerprint uniquely identifies an individual.

there are more than 40 repeats present in the Huntington gene. Several other genetic diseases are caused by the presence of STRs.

Mobile genetic elements

Moderately repetitive DNA consists of multiple copies of sequences that range from 50 to 5000 bp in length. There was a lot of speculation about what they were doing when they were first identified, but by the end of the 1970s, it became clear that most of this DNA is related to mobile genetic elements, such as viruses and transposons, that jump around in the genome. The two most important differences between moderately repetitive DNA and highly repetitive DNA are that the moderately repetitive repeats are much longer and that they are dispersed at multiple sites within the genome instead of forming tandem arrays.

Hidden viruses in your genome

[T]he bodies of extinct viruses still litter our genome.

Richard Dawkins and Yan Wong (2016)

We all know about viruses – they are little bits of DNA or RNA enclosed in a protein shell. The virus genome can have anywhere from 3 genes to more than 100, and during a typical viral infection the virus particle enters the cell and releases its nucleic acid. The viral genes are then expressed, making new virus particles that are subsequently released to infect other cells.

There are two types of viruses: DNA viruses and RNA viruses. As the name implies, DNA viruses contain DNA as their genetic component, but that's just about the only thing that these viruses have in common. Some DNA viruses carry dozens of genes, while others have only three or four genes. The most common human DNA viruses are papillomavirus (causes warts) and herpes virus.

Some of these viruses can be maintained in infected cells as minichromosomes that replicate whenever the cell divides so they can be passed on from cell to cell. These viruses can be reactivated from time to time to produce more virus particles. Other viruses enter a dormant state in certain cells where they can remain for many years. The classic example is a type of herpes virus called varicella zoster virus that causes chickenpox in children but can reactivate later in life, causing shingles in adults.[4]

On rare occasions these viruses can integrate, by accident, into the genome of germ cells (eggs and sperm) and the viral DNA will be passed on to the next generation. About 3 to 4 percent of our genome consists of the remnants of various DNA virus sequences that have been with us for millions of years.[5] All of them have accumulated mutations that disrupt their genes, so they are unable to produce virus particles, and many of them have been disrupted by insertions and deletions so that all that's left is a fragment of the original DNA virus.

RNA viruses contain RNA as their genetic material. This is a large group of different viruses, and some of the ones affecting humans can be devastating, such as a coronavirus (COVID-19), influenza virus (flu), polio virus, measles virus, and lentivirus (HIV).

A typical RNA virus, like coronavirus, contains a single strand of RNA that's injected into the cell on infection, where it functions as an mRNA that's translated to produce viral proteins. The RNA is copied by an RNA-dependent RNA polymerase to make new

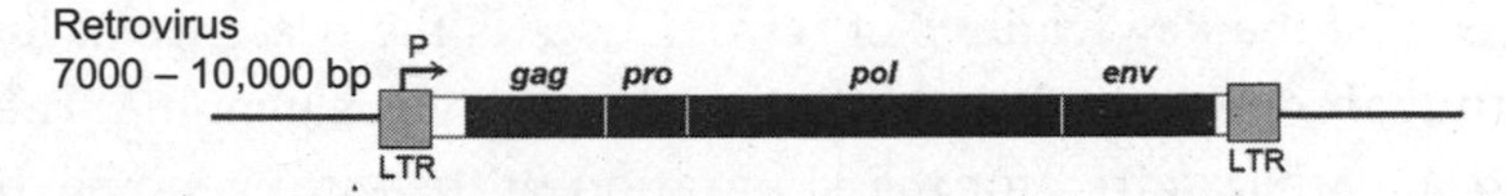

FIGURE 3.1 Retrovirus genome. A typical retrovirus contains genes for reverse transcriptase (*pol*), envelope protein (*env*), and another virus structural protein called group-specific antigen (*gag*). The initial translation product is a very large polyprotein that is subsequently cleaved into smaller proteins by a protease enzyme (*pro*). The ends of the virus genome consist of regions called long terminal repeats (LTRs) that contain regulatory elements, including a promoter. (See Johnson [2019] for a recent review of retroviruses.)

strands of RNA that are then packaged into new viral particles. This type of RNA virus rarely becomes part of our genome because it never has a DNA intermediate.

Part of the normal life cycle of other RNA viruses requires that the RNA genome is copied into DNA using the enzyme *reverse transcriptase* and the DNA produced by reverse transcriptase is then copied by DNA polymerase to produce a double-stranded DNA version of the viral genome. The viruses were called "retro"viruses because this is the reverse of the normal DNA-to-RNA information flow seen in transcription. The gene for reverse transcriptase occupies a significant fraction of the retrovirus genome (Figure 3.1).

The double-stranded DNA intermediate recombines with nuclear DNA, resulting in the insertion of retroviral DNA. The viral genes are then transcribed from viral promoters, and the resulting mRNA is translated to produce viral proteins. If this recombination event occurs in a germ cell, then the insertion can be passed on to future generations as part of the human genome as long as the virus becomes dormant and stops producing new virus particles. The insertion allele may subsequently become fixed in the population by drift in which case we see it today as part of the moderately repetitive DNA component of the genome. Such sequences are known as endogenous retroviruses (ERVs). (The human endogenous retroviruses are called HERVs.)

One of the key features of retroviruses is the presence of long terminal repeats (LTRs) at either end of the virus genome. These repeats are required for the integration of the viral genome, but they also contain a strong promoter for transcribing the viral DNA.

Once a retrovirus inserts into the genome, it begins to acquire mutations every time a cell replicates its DNA. Because the virus is dormant, there is no selective pressure to maintain its sequence so mutations, including insertions and deletions, accumulate and become fixed at the neutral rate. About 9 percent of our genome is composed of bits and pieces of retrovirus-related DNA fragments – there are only a handful of intact retroviruses.[6]

There are 31 different retrovirus sequences in our genome, and most of them are present in multiple copies at different sites. The total number of sites is difficult to determine because some of the remaining fragments of the virus are quite small, but it looks like there are about 200,000 separate insertions. Some of these consist of a single LTR.

Because the LTR contains a promoter sequence, the ERV sequences will often be transcribed even if the virus is defective. This can lead to transcription of adjacent human genome sequences, resulting in a low level of spurious transcription from many of the 200,000 sites in the genome.

There are examples of currently infectious viruses that are closely related to the ERV sequences in the genomes of several species, but this is not true of humans: no examples of exogenous viruses corresponding to HERV sequences have been identified. This could be due to the fact that we haven't yet discovered the parental virus, but it could also indicate that the parental virus is now extinct. Keep in mind that most HERVs are quite ancient – usually millions of years have passed since they were first integrated into the human genome. We know this because mutations in their DNA have become fixed at the neutral rate and the date when these mutations began accumulating can be calculated based on the molecular clock.[7]

If the ERVs really are ancient, then you would expect them to be present in many different species and this is exactly what we observe. For example, some human ERVs are also present in the chimpanzee and gorilla genomes. Some ERVs are about 20 to 40 million years old, and they are found in the genomes of all Old World primates but not New World primates. This is consistent with the split between Old World and New World primates that occurred around this time.

A handful of ERVs retain the sequences needed to make the envelope protein of the virus (*env*). They are present in the genomes of several primates, indicating that they must be under selection because otherwise the *env* coding sequence would have accumulated mutations. One famous example is the ERV–WE1 sequence that makes an *env* protein in the placenta, where it plays a role in fusing maternal and embryonic cells to form large cells containing multiple nuclei. These cells, called *syncytia,* facilitate the transfer of nutrients from mother to fetus. The gene is now called *syncytin-1*. There are a few other examples of ancient viral *env* proteins that have been co-opted to aid in cell fusions.

What do we need to know about transposons?

The other class of mobile genetic elements is transposable elements, or transposons, and they are a major source of excess DNA in most genomes. Transposons were originally discovered in corn by Barbara McClintock in the 1940s, and she subsequently received the 1983 Nobel Prize for this work. The details of transposon structure and function were worked out in the 1970s by studying bacterial transposons, and it turns out that they are basically little bits of selfish DNA that can copy themselves and insert into the genome.

We need to know about transposons because they account for most of the moderately repetitive portion of the genome. There are many different types of transposons, but we don't need to learn all the details. Let's just concentrate on generic examples of the two main types: *DNA transposons* and *retrotransposons*.

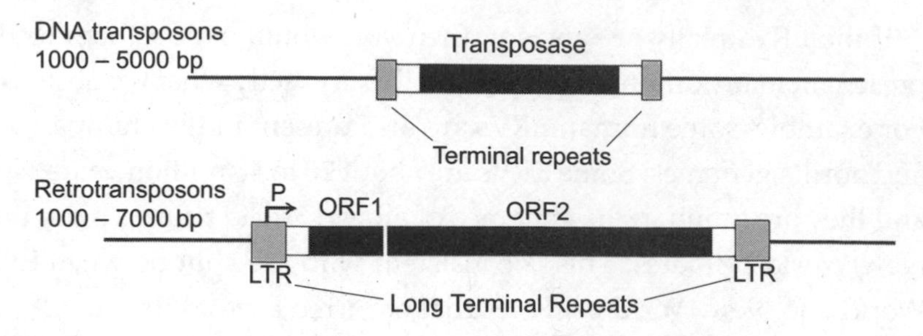

FIGURE 3.2 Transposons. There are two main types of transposons. DNA transposons (top) consist of a gene for transposase flanked by terminal repeats. The DNA is excised from the genome when the repeats come together to form a loop and the transposase enzyme cuts out the transposon, which can then reinsert at another location by reversing the excision process – a cut-and-paste mechanism. Retrotransposons (bottom) contain two open reading frames (ORF1 and ORF2). The first one encodes an RNA-binding protein, and the second one encodes a protein with endonuclease activity, integrase activity, and reverse transcriptase activity. The two genes are flanked by long terminal repeats (LTRs) containing a strong promoter that transcribes the transposon sequences. Reverse transcriptase copies the RNA into DNA, which can then reinsert into the genome at another location via a copy-and-paste mechanism.

DNA transposons (Figure 3.2) jump from location to location by excising and reinserting at the site of short repeated stretches of DNA using a "cut-and-paste" mechanism catalyzed by a transposase enzyme that's usually the only gene encoded by the transposon. DNA transposons that have integrated into germ cell DNA will become part of the heritable genome, and as is the case with viral DNA, the integrated transposons will gradually become inactive over time as they acquire mutations. About 3 percent of our genome consists of various fragments of once-active DNA transposons derived from integration events that occurred millions of years ago. There are no active DNA transposons in humans, and they are very rare in other mammals.[8]

The bacterial transposons of this type are famous for carrying multiple drug-resistance markers that can pass from species to species, but active DNA transposons are common in many eukaryotes, and they go by several amusing names such as *hobo*, *gypsy*, *jockey*, *H.M.S. Beagle* (my favorite), and *mariner*. The Ac/Ds transposons in corn that were discovered and characterized by Barbara McClintock in the 1940s and 1950s are DNA transposons. DNA transposons are so common that the transposase gene is the most abundant gene in the world.[9]

The other class of transposons is retrotransposons, and they are related to retroviruses. They jump from one site to another by transcribing their DNA and converting the RNA into DNA using reverse transcriptase. The DNA copy then reinserts into the genome by a copy-and-paste mechanism.

A typical retrotransposon has two genes or open reading frames (ORFs). The first one encodes an RNA-binding protein that assists in replicating the RNA (Figure 3.2). It is analogous to the *gag* protein of retroviruses. The second ORF encodes a multifunctional protein with reverse transcriptase activity and an endonuclease activity that's required to make complete copies of the genome. It also contains an activity called integrase that catalyzes the integration of the viral DNA into the genome of the host. The second ORF is analogous to the *pol* genes of retroviruses.

LINEs and SINEs

Most of the transposons in the human genome are variants of retrotransposons known as short interspersed elements (SINEs) and long interspersed elements (LINEs; Figure 3.3). LINEs are clearly related to retroviruses and retrotransposons because they still contain a gene for reverse transcriptase. Most of them also have another gene that can be one of several different types of RNA-binding proteins. The region upstream of the first gene contains a strong promoter, so LINEs are often transcribed in a variety of human cells. The LINE insertion sites are flanked by short target site duplications

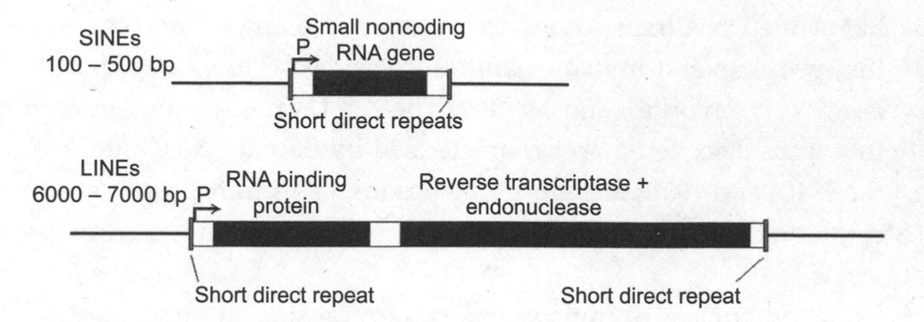

FIGURE 3.3 SINEs and LINEs. Short interspersed elements (SINEs) are usually less than 500 bp and they contain a copy of a small gene for a noncoding RNA. The gene is flanked by short direct repeats. Long interspersed elements (LINEs) are more than 10 times larger than SINEs, and they contain two genes: one that encodes an RNA-binding protein and one that encodes a protein with reverse transcriptase activity and endonuclease activity.

produced when the transposon inserts into the genome. Unlike true retrotransposons, LINEs do not have LTR sequences.

LINE sequences occupy about 21 percent of the human genome, but most of those sequences are defective versions that are no longer capable of transposing – many of them are just fragments of the original transposon. They can be subdivided into families based on sequence similarity. Most of the families do not contain any active members, but the LINE-1 family is an exception. There are about 500,000 copies of LINE-1 elements, and a few of them are still quite active, with about 500 recent insertions. The LINE-1 family alone takes up about 17 percent of the genome, but the vast majority of the family members are just fragments of the original transposon. It's important to note that active LINE transposons produce a lot of reverse transcriptase so that most of the RNAs expressed in a cell will be accidentally copied into DNA at some time or another.

SINEs do not contain a gene for reverse transcriptase, but they do have a promoter, and they are transcribed. They require reverse transcriptase – supplied by LINE elements – in order to copy the transcript into DNA that can be reintegrated into the genome.

SINEs are called *nonautonomous transposons* because they can't copy themselves.

SINEs originate when a small RNA is copied into DNA that inserts near repeat elements that allow it to become mobile. The genes for these small RNAs contain *internal* binding sites for proteins that can activate the upstream promoter so that the DNA copy can promote its own transcription. (Most genes have *upstream* binding sites that are not transcribed, so the DNA copy of the transcript lacks the ability to promote its own transcription.)[10]

Many species have abundant copies of SINEs derived from tRNA, but in primates the most abundant SINE is derived from the small 7SL RNA that forms the core component of signal recognition particle. (This particle helps transport cytoplasmic protein into the endoplasmic reticulum.) These abundant SINEs were recognized in the 1970s because they contain a site for the DNA-cutting restriction enzyme called *Alu*I, and they are now called *Alu* elements.

There are more than 1 million *Alu*s in the human genome, but almost all of them are defective because they contain inactivating mutations and/or deletions. They are subdivided into several distinctive families that appear to have spread in primates in several bursts dating back millions of years. Together they make up about 11 percent of the entire genome.

Science writer Carl Zimmer had his genome sequenced and analyzed by experts who found 1,024 *Alu* sequences at sites that aren't in the standard human reference genome. What this means is that a few active *Alu* SINES have been constantly jumping around in our genome for the last several million years so that many of these sites are polymorphic in the human population. In other words, they have not yet been eliminated or fixed by evolution, so Carl's genome is different from yours and mine and from the standard reference genome. Carl is not unusual. The 1000 Genomes Project has discovered hundreds of additional *Alu* insertions that are polymorphic in the human population.[11] This is what sloppy genomes look like.

How much of our genome is composed of transposon-related sequences?

Believe it or not, nobody looks by eye at all 3 billion base pairs of the human genome to see which segments look like transposons! That task is turned over to computers that are programmed to recognize sequences resembling known transposons. The program scans the genome, testing each segment to see if it aligns with the sequences in its transposon database, and the algorithm is designed to accept only those matches that are obviously similar in order to cut down on the number of false positives. That's because genome annotators don't want to label some sequence as transposon-derived when it might be something else.

But this approach comes with a cost, namely, false negatives, meaning that some transposon-related sequences are not identified by the computer program. The results of such annotations indicate that about half of our genome – and the genome of most mammals – is derived from transposons and repetitive DNA.

What about the rest of the genome? It's very likely that some of this DNA is also related to transposons. It could contain remnants of very ancient transposons insertions that have become inactivated and whose sequence has diverged so much from the original active transposon that it is barely recognizable in the alignment, so the annotation program doesn't count it. Small fragments of transposons might also be rejected by the program because the match to known transposon sequences is below the threshold for acceptance.

Some scientists have developed more sensitive computer programs to look for repetitive DNA sequences in the human genome by eliminating the false negatives missed by the standard programs.[12] The standard programs assign about 50 percent of the genome to repetitive DNA, but the new programs estimate that up to 67 percent of the genome is repetitive DNA.

It's important to note that these studies lump together all repetitive elements, including transposons, retroviruses, DNA viruses, and simple repeats, such as those in centromeres and telomeres. It also included tRNA genes, ribosomal RNA genes, and the genes for several other functional RNAs since these genes are present in multiple copies. The important point is that about half of our genome consists of just virus- and transposon-related sequences, and this explains the total amount of moderately repetitive DNA discovered by C_0t analysis in the 1960s.

Transposons insert pretty much at random, so they are perfectly capable of inserting into the active part of genes and disrupting their function. The effects of such deleterious insertions are severe enough to be subjected to negative selection. This is why transposon sequences are scattered all over the genome except in the active regions of genes and other functional elements.

The flip side of this observation is that in regions where transposon insertions are common, we can assume that the insertion is either neutral or only slightly deleterious – too slight to be significantly affected by natural selection. For example, transposon insertions are common in introns suggesting that the length and sequence of introns have no effect on fitness. An analysis of recent *Alu* insertions shows no evidence of positive or negative selection, which supports the view that they are "neutral residents of the human genome."[13]

These observations contribute to a consistent model, indicating that most transposon sequences are junk DNA and that most of the regions containing these transposons are also junk because they tolerate insertions. Thus, by this criterion, at least half of our genome appears to be junk.

Some genomes have many fewer transposon sequences than others; for example, the pufferfish genome I described in the previous chapter has only 6 percent transposon sequences. By contrast, 90 percent of the lungfish genome is repetitive DNA (mostly transposons). Similar differences are seen in plants where the corn (maize) genome is 85 percent transposons, but the thale cress

(*Arabidopsis thaliana*) genome has only 10 percent. Rapid DNA sequencing technology has helped elucidate our understanding of the C-Value Enigma by showing that differences in the genome sizes of closely related species are directly due to the expansion and contraction of transposon-related sequences as described in the previous chapter.[14]

In species with small genomes, there could have been more effective selection against transposons, or alternatively, they could have smaller genomes just by chance.

WHAT DOES THE HUMPED BLADDERWORT TELL US ABOUT JUNK DNA?

The humped bladderwort (*Utricularia gibba*) is a small carnivorous flowering plant found all over the world. Its genome was sequenced in 2013, and everyone was amazed to discover that it contained only 82 Mb of DNA, making it the smallest flowering plant genome. The genome sizes of most other flowering plants are between 3,000 and 4,000 Mb.

The bladderwort genome contains about 28,000 genes, and this is about the same number of genes found in other plants. The remarkable thing about the bladderwort genome is that only 3 percent of the genome consists of transposons and transposon fragments whereas in other flowering plants more than half the genome is transposon-related.[15] This is further evidence that most of the DNA in large genomes is dispensable and that much of the excess DNA is repetitive sequences that appear to be junk DNA.

We don't know why the bladderwort genome is so small; it could just be a lucky accident of evolution.

Selfish genes and selfish DNA

Highly and moderately repetitive sequences may be considered to be "selfish," in that they spread by forming additional copies of themselves within the genome even if they are not useful to the organism.

Tomoko Ohta (1983)

Transposons are examples of "selfish DNA," a term that was popularized in 1980 when it was used to describe DNA that exists for the sole purpose of propagating itself without regard for the fitness of the organism it inhabits.[16]

This led to a debate over whether we should refer to transposon sequences as selfish DNA or junk DNA, but the debate is easily resolved because we now know that almost all transposon-related sequences are not active – they are defective. This is important because back in 1980 the idea of a genome full of selfish DNA was often interpreted by opponents of junk DNA to mean that there was very little junk, and this idea has persisted for the past 40 years.

Note that the discussion about transposons and selfish DNA has nothing to do with the main idea promoted by Richard Dawkins in his book *The Selfish Gene.* Dawkins is referring to bits of the genome that are conserved because they confer a fitness benefit on the organism. Such "genes" will increase in the population by natural selection, and Dawkins was just using a gene-centric view to describe ordinary evolution by natural selection. Selfish DNA, on the other hand, has no effect on the fitness of the individual because it is due to selection at the level of DNA and not the organism.[17]

Exaptation versus the post hoc fallacy

As I pointed out earlier, lots of scientists have difficulty coming to grips with a genome full of junk DNA; prime examples include Roy Britten and David Kohne, who were quoted at the beginning of this chapter. Today's anti-junk scientists have developed some creative adaptive hypotheses to explain the presence of transposon-related sequences, and one of these hypotheses is that transposons promote variation by increasing recombination. The idea is that similar transposon sequences in different chromosomes can align, thereby allowing for recombination events that wouldn't occur if the chromosome lacked transposons. The result is that species with lots of transposons will be able to shuffle their genomes more frequently, creating more variation for evolution to act on.

There are several problems with this hypothesis, but the most important is that there's no evidence to support it in mammals. Most lineages show only a small number of chromosome rearrangements over millions of years (see the earlier box "Dead Centromeres and Telomeres"), and there's no evidence that these rearrangements have contributed to an increase in fitness as required by the hypothesis.

Several other scientists believe that transposon sequences have a direct function; for example, Barbara McClintock believed that they helped cells cope with stress. Others point out that some transposon sequences have evolved to play a role in the expression of nearby genes. This second idea is not surprising since many transposons contain active promoters that enhance transcription, so if a transposon lands near a gene, it could have a beneficial effect in promoting the expression of that gene; for example, there are several dozen transposons in *Drosophila* that affect transcription of nearby genes.[18] There are also cases in which new genes have arisen by incorporating parts of the transposase gene from transposons to create new proteins. Many of these new genes encode novel transcription factors that make use of the DNA-binding properties of the transposase.

These are examples of *exaptation*, when a sequence that evolved for one purpose is co-opted to perform another function as in the evolution of *syncythin-1* from a viral gene. Exaptation is a real phenomenon that was first described by Stephen Jay Gould and Elisabeth Vrba in 1980, but the problem with using exaptation as a general explanation for the existence of millions of transposons is that there are only a few proven examples of transposon-related sequences that have secondarily acquired a new function.[19] Most of the scientists who promote this hypothesis want to ascribe a function to all, or almost all, transposon-related sequences in order to counter the more obvious conclusion that they are junk. In order to do this they have to assume that we will eventually discover a function for the majority of defective transposons in our genome, but this seems like wishful thinking to me. It also appears to be inconsistent with what we know about evolution since in order for this hypothesis – better described as speculation – to be true there would have to have been thousands of new functions that evolved in the last few million years of human evolution. Natural selection is not capable of such a feat according to the modern view of evolutionary theory.

The fallback position of those scientists who oppose junk DNA is that most mobile genetic elements don't currently have a function but that they are preserved in the genome because they act as a reservoir for the evolution of new functions in the future. This sort of explanation is similar to a form of faulty logic called the *post hoc fallacy*. It assumes that because something happened (A) after something else (B), then B must be the cause of A.[20]

In the context of this chapter, the fallacy manifests itself in the following way. Because some transposons have been co-opted to produce functional DNA, then transposons must be there in order to facilitate the evolution of new functional elements. It supposes that species have selected for a genome full of bits and pieces of mobile genetic elements so that they can evolve in the future.

One of the most vocal proponents of this view is James Shapiro, who argues that species are designed for "natural genetic engineering" and that transposons play an important role in this process by allowing genomic alterations to facilitate their future evolution.[21]

Stephen Jay Gould explains why this kind of reasoning is faulty:

> It is a clear, though lamentably common, error to assume that the current utility of a feature permits an inference about the reasons for its evolutionary origin. Current utility and historical origin are different subjects. Any feature, regardless of how or why it first evolved, becomes available for co-option to other roles, often strikingly different. Complex features are bursting with potentialities: their conceivable use is not confined to their original function (I confess that I have used a credit card to force a door). And these evolutionary shifts in function can be as quirky and unpredictable as the potentials of complexity are vast. It happens all the time; it virtually defines the wondrous indefiniteness of evolution.[22]

This flawed way of thinking about evolution relates to something called "evolvability," or the idea that species can prepare for future evolution by selecting for features that currently have no use but can evolve a function later on. I like the way Gould dismisses the evolution of evolvability in his book *The Structure of Evolutionary Theory*, where he explains that evolution may *utilize* such features (e.g., mobile genetic elements) but evolution cannot *fashion* these features. He concludes that the ability to utilize and the inability to fashion such features is compatible with modern evolutionary theory but that the idea that evolution could select for evolvability is not.[23]

He also points out the irony in such reasoning. Proponents of this idea attribute immense power to natural selection and its ability to see into the future. However, that "power" is ascribed to huge amounts of DNA that appear to be junk by any reasonable criteria. If natural selection is as potent as they imagine, why then, doesn't it eliminate the junk?

Given such strong opposition, you may be wondering why anyone would ever propose that our junky genome was a way of preparing for future evolution. I can't be certain of everyone's motives, but I tend to agree with Sydney Brenner, and many others, who attribute it to adaptationist bias:

> There is a strong and widely held belief that all organisms are perfect and that everything within them is there for a function. Believers ascribe to the Darwinian natural selection process a fastidious prescience that it cannot possibly have and some go so far as to think that patently useless features of existing organisms are there as investments for the future.
>
> I have especially encountered this belief in the context of the much larger quantity of DNA in the genomes of humans and other mammals than in the genomes of other species.
>
> Even today, long after the discovery of repetitive sequences and introns, pointing out that 25% or our genome consists of millions of copies of one boring sequence [*Alu*s], fails to move audiences. They are all convinced by the argument that if this DNA were totally useless, natural selection would already have removed it. Consequently it must have a function that still remains to be discovered. Some think that it could even be there for evolution in the future – that is, to allow the creation of new genes.[24]

Words like this remind us that there is a genuine scientific controversy about junk DNA and that the two sides tend to have different views of evolution. Those who oppose the idea of junk DNA see natural selection as an all-powerful, almost exclusive force in evolution, and the idea that most of our genome could be junk is incompatible with such a worldview. The assumption that most of our genome is functional is consistent with a strictly Darwinian interpretation of evolutionary theory – a view that some of us refer to as an adaptationist perspective.

Proponents of junk DNA, on the other hand, tend to see evolution as a messy and inefficient means of evolving organisms that are "good enough" to survive but far from perfect. That view is compatible with modern population genetics and the neutral theory and with a major role for random genetic drift. If you count yourself in the first group, then I hope to convince you to join the second one.

Mitochondria are invading your genome!

There's another kind of repetitive DNA that I haven't yet mentioned: it comes from mitochondria. A typical human genome contains numerous fragments of DNA derived from the mitochondrial genome. They are called *nu*clear *mit*ochondrial *s*equences or *NumtS* (pronounced "new mites"), and their sizes range from about 30 bp to almost full-length mitochondrial sequences of 16,500 bp. This may seem like a lot, but it only accounts for a tiny percentage of the entire genome (<0.01 percent). All of it is junk DNA because mitochondrial DNA is not supposed to be in the nuclear genome, and once inserted, it rapidly acquires mutations.

The human genome contains about 600 fragments of mitochondrial DNA, and the chimpanzee genome has about 750. About 550 of the chimpanzee *NumtS* are the same as those in the human genome, indicating that they colonized the nuclear genome before the split into human and chimpanzee lineages. There are fifty-three human-specific *NumtS* representing pieces of mitochondrial DNA that inserted into nuclear genomes along the lineage leading to modern humans. Some of these are polymorphic, meaning that some groups of modern humans have them and some don't. This polymorphism shows that such colonization is an ongoing process, and it's consistent with the idea that these mitochondrial DNA inserts are not under selection and have no biological function.[25]

Billions of people are alive today, so that means millions of newborn babies have a new piece of mitochondrial DNA in one of their chromosomes. This conjures up an image of mitochondria breaking up and fragments of mitochondrial DNA raining down on the cell nucleus. Most of them never get incorporated into the genome, and most of those that do get incorporated will be lost in a few generations.

This is all very interesting, but it doesn't make much of a difference in our genome. It's just a little bit more junk DNA that comes and goes. Remember that we're talking about small, random pieces of mitochondrial DNA – most of them are only a few hundred base pairs in length, and they don't include any full-length mitochondrial genes. However, if we look far back into the earliest ancestors of all eukaryotes, we see a very different story. Early on in the evolution of eukaryotes, two primitive bacterial cells fused, giving rise to the first true eukaryotes in a process called *endosymbiosis*. One of the ancestors was an *alphaproteobacterium*, and the other was an *archaebacterium*. The alphaproteobacterium eventually became mitochondria, retaining a small fragment of its original genome. Most of the genes in the original proteobacterial genome colonized the genome of the archaebacterial partner, which became the nuclear genome. These genes were subsequently lost from the mitochondrial genome, and that's why most mitochondrial proteins are made by genes that now reside in the nucleus. More than half of the genes in our nuclear genome are more closely related to our mitochondrial ancestor than to our archaebacterial ancestor because those genes started out as *NumtS*.

On the origin of junk DNA

Recall that the false view of evolution sees natural selection as the powerful and dominant mechanism of evolution, leading to the conclusion that junk DNA should be purged from the genome because it is useless or even harmful. In order to counter this conclusion, junk DNA proponents have proposed reasonable explanations for both the retention of junk DNA and its origin.

Back in 1980, Leslie Orgel and Francis Crick wrote, "In summary, then, there is a large amount of evidence which suggests, but does not prove, that much DNA in a higher organisms is little better than junk. We shall assume, for the rest of this article, that this hypothesis is true. We therefore need to explain how such DNA arose in the first place and why it is not speedily eliminated, since, by definition, it contributes little or nothing to the fitness of the organism."[26]

What was true in 1980 is just as true today. One could argue that the *existence* of enormous amounts of junk in our genome is a separate issue from how it got there. One could argue that the evidence for junk DNA stands and falls on its own merit and that we don't necessarily need to have a good theory about why it's there. The problem is that we can't ignore the case for the creation of junk DNA because it seems to fly in the face of widely held, but false, notions of evolution.

As I said in chapter 2, we now know that most of the junk DNA in our genome comes from selfish DNA insertions in the form of viruses and transposons.[27] In the short term (hundreds of generations), the insertion of transposons is roughly balanced by deletions, giving rise to an "accordion" model of genome evolution.[28] This dynamic expansion and contraction is one of the reasons why individuals within a population have slightly different genome sizes.

The slow increase in genome size that we see in mammals and other lineages is thought to be due to small shifts in favor of transposon insertion over deletion over millions of years. The model I'm promoting attributes this slow increase to chance, keeping in mind that genome expansion may be slightly deleterious but not enough to prevent it in small populations.

The argument against junk DNA assumes that genome expansion is due to positive selection for new transposon insertions that have a function, but this explanation cannot account for the fact that transposons accumulate mutations at the neutral rate. There is very little evidence to support such a model; however, both models agree that a great deal of excess DNA is due to the expansion of transposons.

If it walks like a duck...

When I see a bird that walks like a duck and swims like a duck and quacks like a duck, I call that bird a duck.

James Whitcomb Riley (1849–1916)

The "Duck Test" became popular during the McCarthy years in the United States when it was applied inappropriately to persecuting suspected communists, but despite that tainted history, it is still a valid figurative expression that describes a form of abductive reasoning. Abductive reasoning is a form of reasoning that seeks to find the most likely explanation based on a set of observations. It's not a rigorous example of logic because there's always another possible explanation even if it's unlikely. Nevertheless, we use this form of reasoning all the time because it's very effective.

Here's how the Duck Test works when we're thinking about mobile repetitive elements. Almost all the sequences look like defective viruses or defective transposons. Most of them are just bits and pieces of once-active viruses, transposons, or mitochondrial genomes. They look like junk, they behave like junk, and they evolve like junk, so let's just call them junk.

CHAPTER 4

Why Don't Mutations Kill Us?

[A]lthough the mammalian genome consists of roughly 3×10^9 base pairs of DNA, the actual number of still functioning gene loci could be less than 10^5. Thus, more than 90% of the genomic DNA of mammals is ignored by natural selection, and may, therefore, accumulate all manner of randomly sustained mutational base substitutions, deletions, and insertions.

Susumo Ohno (1985)

Gregor Mendel (1822–1884) is the father of genetics, but he knew nothing about mutations when he was crossing his pea plants in the garden of St. Thomas Abbey in Brno (now in the Czech Republic). All he knew was that there were several varieties of peas that differed in characteristics, such as the height of the plants, the color of the flowers, and the shape of the peas. He discovered the fundamental principles of genetics by showing that these characteristics behaved as discrete entities that could be passed on to the next generation in a predictable manner.

We now know that each of Mendel's characteristics is due to a mutation that changes the sequence of base pairs in a gene resulting in a new, distinct, phenotype. For example, one of the traits he studied was wrinkled peas; the normal pea is smooth but certain plants produce peas with a wrinkled surface. The smooth pea is the normal wild-type phenotype, and the wrinkled phenotype is due to a mutation in a gene that produces starch, which is stored in the seeds (peas). The starch-producing enzyme is disrupted by a mutation caused by the insertion of a transposon.[1]

Mendel's traits were all spontaneous traits that he discovered or collected from others, but by the beginning of the 1900s scientists knew that they could generate more varieties by treating their favorite organism with various chemicals, ultraviolet light, or X-rays. This was the beginning of our understanding of mutations and mutagenesis.

External mutagens can greatly increase the rate of mutation – that's why smoking, X-rays, and atomic bombs are bad – but it may surprise you to learn that mutagens are not the major source of mutation. Most mutations are spontaneous mutations caused by errors occurring in DNA replication every time a cell divides. These mutations are rare, but the more genes you have, the greater the risk that at least one of them will acquire a mutation.

Why aren't we extinct? A 100-year-old problem

J.B.S. Haldane was worried about mutations back in the middle of the last century. He knew that mutations were essential for evolution, but a simple back-of-the-envelope calculation suggested that having too many mutations could lead to extinction. Haldane didn't know about DNA, but he did know a thing or two about population genetics,[2] and he calculated that a species could not tolerate a mutation rate of one or two deleterious mutations per generation. The resulting mutation load could lead to extinction.

Haldane wasn't the only one who was worried. A number of other geneticists were also grappling with the concept of mutation load in the 1940s. The most prominent of these geneticists was Nobel Laureate (1946) Hermann Muller, who published a classic paper on "Our Load of Mutations" in 1950.[3]

He estimated that the rate of severe deleterious mutation in humans was about one in 50,000 genes per generation, based on the spontaneous appearance of genetic diseases. The implication is that humans can't have more than 50,000 genes because then every child would likely be born with a genetic disease mutation. This rate of mutation would drive our species to extinction.

By 1966 Muller had collected more data and came to the conclusion that we must have only 30,000 genes. He knew that our genome was about 3×10^9 bp, so he concluded that our genes are very large. However, other scientists reached a slightly different conclusion: they recognized that a typical gene might be only 2000 base pairs long, and they also realized that not all of our genome had to be functional. This conclusion led to a series of papers by Motoo Kimura, Susumu Ohno, Tomoko Ohta, James Crow, Thomas Jukes, and others who promoted the idea that we had only 30,000 genes and much of the rest of our genome was nonfunctional junk that must be unaffected by mutation.[4]

We've learned lots of things since these early estimates of mutation load, so let's update the mutation load argument beginning with the latest data on mutation rate.

Biochemical mutation rate

We can calculate the human mutation rate based on the size of the genome and the error rate of DNA replication. The first estimate of this biochemical mutation rate was published by Motoo Kimura in 1968, but since then the estimate has been substantially improved.

We now know that the error rate of DNA polymerase is one mistake in every 10^8 bp. Most of these errors (99 percent) are corrected by repair enzymes to give an overall error rate of 10^{-10} or one in 10 billion base pairs. Our genome contains 3.2 billion base pairs, but because we are diploid, each cell has 6.4×10^9 bp of DNA. What this means is that every time our cells divide, each daughter cell will acquire an average of $6.4 \times 10^9/10^{10} = 0.64$ mutations.

In order to estimate the mutation rate per generation, we need to know how many cell divisions have occurred from one generation to the next. Since eggs and sperm are the relevant cells, we need to know how many divisions occur between the formation of the *zygote* (the first cell) to the production of the egg and sperm that will give rise to the zygote of the next generation.

Egg production in females begins early in the embryo so that by the time a baby girl is born, she already has all the eggs she will ever produce. There are about 30 cell divisions between zygote and egg, so each haploid egg cell will have an average of $30 \times 0.32 = 9.6$ mutations.

The calculation in males is more difficult since men produce new sperm cells throughout their adult life. On average, it looks like there are about 400 cell divisions from zygote to mature haploid sperm cells, so this means $400 \times 0.32 = 128$ mutations. Adding the two numbers together gives 138 new mutations per generation in each zygote.

Note that men are responsible for most of the mutation load in humans. Furthermore, older men contribute far more mutations than younger men because there are more cell divisions required for sperm production.

Phylogenetic mutation rate

Population genetics and the nearly neutral theory give us another way of estimating the mutation rate. The probability of a mutation occurring anywhere in the genome is given by μ (mu, mutation rate). Diploid populations of size N will have $2N$ genomes, so they will accumulate $2N\mu$ new mutations per generation. Recall that the probability of fixation of a single neutral mutation by random genetic drift is $1/2N$, which means that the overall probability of fixation of all $2N\mu$ mutations is $2N\mu \times 1/(2N) = \mu$, provided that all those mutations are effectively neutral.

The important conclusion is that the rate of fixation of neutral alleles is equal to the mutation rate and therefore all we need to do is count the number of neutral mutations in the DNA sequences of two different species in order to get the mutation rate.

The earliest data were based on examining pseudogenes (defective genes) in humans and chimpanzees. This gave a neutral mutation rate (μ) of 2.5×10^{-8} mutations per base pair per generation. Extrapolating that to the number of mutations gave a value of 160 – not much different than the value of 138 from the biochemical method.[5] Lots of other analyses give similar results.

Let's see how we can estimate the phylogenetic mutation rate now that we have a relatively complete genome sequence of humans and chimpanzees. If you align the human and chimpanzee genomes, you'll discover that they are very similar; they differ by only 1.4 percent when you count up all the substitutions and small deletions.[6]

This difference of 1.4 percent corresponds to 44.8 million fixed mutations, or 22.4 million in each lineage since the time of the last common ancestor. If the two species diverged 5 million years ago, then we need to calculate the number of generations in order to know the mutation rate. The average generation time for humans is 30 years, and for chimpanzees in the wild, it is about 25 years. Taking the average generation time (27.5 years), this corresponds to 182,000 generations and a mutation rate of 134 mutations per generation.[7]

This value is remarkably similar to the mutation rate calculated by using the biochemical method. It will depend on the date of the last common ancestor of humans and chimps, but even if that date is 7 million years, the mutation rate doesn't change that much (96 mutations per generation).

Let's pause for a minute and think carefully about what this calculation means in the context of our discussion about junk DNA. The fact that the phylogenetic mutation rate agrees with the biochemical calculation means that almost all the differences between the human and chimp genomes are due to the fixation of neutral mutations by random genetic drift. Since these 44.8 million mutations are scattered throughout the genomes of the two species, then that must mean that most of their genomes are evolving at the neutral rate. That's exactly what you would expect if most of their genomes

TICK, TOCK, THE MOLECULAR CLOCK

[B]ecause the majority of mutations fixed in the populations are more or less neutral, the evolution of proteins may be regarded as a neutral process. This is the essence of the neutral theory of evolution.

Masatoshi Nei (2013)

Modern evolutionary theory explains a puzzling result that was first observed when protein amino acid sequences were compared in the early 1960s. It was possible to construct robust phylogenetic trees from this data, and it was observed that the rates of amino acid substitutions were relatively constant on each branch. This gave rise to the idea of a molecular clock that could be used to measure the time of divergence of various lineages – a result that is widely used today. However, at the time, this result was unexpected since there was no reason to suspect that adaptive changes in amino acid sequences should occur at a constant rate. The issue was resolved with the realization that most of the substitutions were neutral and were fixed by random genetic drift. Since this fixation occurs at a rate equal to the mutation rate and since that rate is very similar in different species, it gives rise to the molecular clock.

It's important to note that the molecular clock is a stochastic clock. It does not tick at a constant rate because there are many variables that influence the timing of fixation. The proper way of viewing the molecular clock was described by Masatoshi Nei: "It is quite 'sloppy' but useful for obtaining a rough idea of evolutionary time when fossil records are absent or unreliable."[8]

were junk, and it's clearly not what you would expect if most of their genomes were conserved by negative selection.

The direct method of calculating the mutation rate

The third method of calculating the mutation rate sounds extremely simple: just sequence the genomes of a mother, a father, and their child and count the number of new mutations in the child's genome. This is the direct method.

Unfortunately, the direct method isn't as simple as it seems. One of the problems is that you don't have the original egg, sperm, and zygote genomes for obvious reasons. You have to make do with the DNA from somatic tissues like cheek cells in saliva or tissue culture cells that have been growing for hundreds of generations. All these cells have acquired somatic cell mutations over the years, and it's difficult to distinguish those mutations from mutations that arose spontaneously in the germ cells.

There are additional complications because everyone is diploid and many sites in the genome sequence contain two different base pairs because the individuals are heterozygous. This makes it difficult to sort out which copy the child inherited. But the main problem with the direct method is much more serious than either of these two issues. It's that the error rate in sequencing DNA is much higher than the number of new mutations, so a huge number of differences in the genomes of parent and child are due to mistakes in the sequences and not to new mutations.

Despite these difficulties, quite a few studies have dealt with the false negatives and false positives and arrived at a mutation rate. The values range from about 60 mutations per generation to about 130 mutations per generation, but most of them seem to cluster around 70 to 80 mutations per generation. I suspect that the direct method underestimates the real mutation rate, so I will use 100 mutations per generation as a rough average of the three methods. It's good enough for our purposes.[9]

You are not Craig Venter

James Watson was the first person to have his genome sequenced – the data was deposited in the public database in June 2007. In September of that year, Craig Venter became the second person to have a complete genome sequence, but his genome sequence was much better than Watson's because separate sequences were generated for each of Venter's 46 chromosomes.[10] This gave a much more complete profile of his genome and provided us with some insights into the amount of heterogeneity within an individual.

There are three kinds of mutations that are due to DNA replication errors. The most common is mis-incorporation of a single nucleotide, giving rise to a *single nucleotide polymorphism* or SNP (pronounced "snip"). The second kind of mutation is a small deletion of one or two base pairs due to slippage in the replication machinery. The third kind is a small insertion of a few base pairs due to stuttering. Many of these small insertions and deletions are, in short, tandem repeat regions – the ones that are used in DNA fingerprinting (see chapter 3). Venter's chromosome pairs differed from each other at 4.1 million sites, and 78 percent of these were SNPs. The rest were small insertions and deletions (indels).

Mutations can also be caused by recombination errors that give rise to larger duplications and deletions, and Venter's genome contained 62 of these copy number variations.

Many of the variants in Craig Venter's genome were already known in 2007, but there were more than 1 million new variants that had not previously been seen. Each new genome that is sequenced today contains only a few thousand new variants that have never been seen before, and this number will decline as more and more genomes are sequenced.

We now know that the human population harbors tens of millions of variant alleles (SNPs, deletions, insertions), which is further support for sloppy genomes and the idea that most of our genome is junk. The genomes of two unrelated individuals are likely to differ by about 0.14 percent – this may seem like a lot of variation within the species, but keep in mind that the average difference between us and our closest relative (chimpanzees) is 10 times greater.

As a general rule, people of European descent have less variation, and people of African descent have more variation.[11] This is consistent with our understanding of human evolution since most models suggest that Europeans and Asians are a subset of a much larger African population because the lineage leading to Europeans and Asians split from Africans relatively recently. What we know for certain is that no two genomes are alike with the exception of identical twins.[12] Your genome is not the same as mine and neither of us is Craig Venter.

Revisiting the genetic load argument

The mutation load argument – also known as the genetic load argument – was originally based on the frequency of deleterious mutations estimated from the frequency of genetic diseases in humans. Now that the actual mutation rate is known, the argument can be extended to take this into account.

The key parameter is the fraction of all mutations that are harmful since this is what contributes to the "load" that the population must tolerate. It's actually very difficult to obtain accurate values for this fraction, but most population geneticists think that a value of one-tenth (10 percent) is likely to be a good estimate. This value is based largely on the number of deleterious mutations in protein-coding regions, but it's reasonable to assume that it applies to all functional regions of the genome, including regulatory sequences, centromeres, and the like.

This number means that only 10 percent of all mutations in functional DNA regions (e.g., genes) are deleterious and that the other 90 percent are effectively neutral, although there are a small number of beneficial mutations occurring from time to time. If the entire genome is functional and there are 100 mutations in every newborn baby, then 10 of these mutations will be harmful, and nobody thinks that our species could tolerate such a high genetic load.

In theory it's possible to calculate the amount of functional DNA if you know the minimum number of deleterious mutations that can be tolerated, and until very recently, geneticists agreed with Haldane and Muller that this number is close to one.[13] This means that only one of the ten potentially deleterious mutations is actually bad, and the rest must be harmless. Thus, the studies from 50 years ago concluded that only 10 percent of our genome is functional and about 90 percent of our genome has to be junk, and that's the same conclusion we would reach today if all those assumptions are correct.

Accurate calculations of genetic load are somewhat more complicated than this estimate because we are dealing with probabilities (not every child has exactly 100 new mutations) and because fertility is important. A population can survive a heavy mutation load by having many offspring when only a few survive to adulthood. Historical data suggest that each human couple produces somewhere between 2.1 and 3.5 children who can contribute to the next generation, and this must be incorporated into the calculation.

Using these values, Dan Graur estimated that at least 75 percent of our genome has to be junk, and it's likely that the actual amount of junk DNA is closer to 90 percent. However, a more recent analysis shows that calculating the fraction of junk DNA is a lot more difficult than Graur thought and certainly a lot more complicated than the simplistic calculations that I presented earlier.[14]

The debate over genetic load involves complex mathematical formulae that are far beyond the scope of this book. Part of the problem is related to the fact that the human population has increased dramatically in the past 10,000 years, and it's difficult to model the evolution of an expanding population. Most of the deleterious mutations that we carry arose in the past few thousand years, and there hasn't been sufficient time for them to be purged by natural selection, which means that the modern human population is very heterogeneous.[15] But the most important contentious issue is the overall cost of these deleterious mutations. Since the average person already carries hundreds of recessive deleterious mutations, the cost of adding one or two isn't as crucial as previous calculations suggested. In other words, our species can theoretically tolerate a much higher mutation load so that 2, 3, or even 4 of the 100 new mutations could, in theory, reduce fitness without necessarily leading to the extinction of our species, although I should point out that one population geneticist, Michael Lynch, thinks that our species is doomed. He estimates that we are each acquiring between one and four deleterious mutations per generation, and this mutation load is not sustainable. He argues that because of the success of modern medicine, we are removing the traditional selective constraints against slightly deleterious mutations and increasing our mutational load.[16]

The important point here is not that we are doomed to extinction but that the genetic load argument is still as valid today as it was 50 years ago. It still suggests that most of our genome has to be junk even though it can't be used to establish exactly how much is junk because there are too many variables whose exact values can't be determined. We must look to other data to establish exactly how much of our genome is functional and how much is nonfunctional.

HUMAN GENE KNOCKOUTS

Humans have thousands of protein-coding genes, and most of them are likely to be essential because they produce important proteins and enzymes that are required for metabolism and information transfer. Almost all these genes will have several different alleles segregating in the population and some of these will be loss-of-function (LoF) alleles that interfere with the synthesis of functional protein. The mutations can affect any of the steps in gene expression: transcription, RNA processing, or protein synthesis. The latest data suggests that the average person carries about 400 LoF mutations.[17]

Some of these alleles may cause genetic diseases when two copies are present but may have no effect when only one gene is inactivated – these are the classic recessive alleles first identified by Mendel. Typical LoF alleles are present at low frequencies in the population where they don't seem to have much effect on individuals that are heterozygous for one LoF allele and one normal allele.

We would like to know how many human genes are truly essential, but the way to test this in other species is to disrupt both copies of the gene, creating a "knockout" to see if it's lethal. These experiments have been carried out in yeast and mice, and the results show that many gene knockouts have no observable effect. The knocked-out gene isn't essential.

It would not be ethical to deliberately delete genes in humans, but nature makes its own knockouts when couples who are heterozygous for LoF mutations produce offspring. The probability of producing a child with homozygous LoF alleles is greatly enhanced when first or second cousins marry because they are more likely to carry the same recessive alleles. By studying the offspring of many such marriages, scientists can generate a database of all human genes that have been knocked out without lethal or severe consequences. There have been several large studies, and so far, more than 4000 gene knockouts have been detected. It looks like a substantial number of our genes are dispensable.[18]

Before leaving this topic, let me say a word about somatic cell mutations. They cause cancer and other problems, and they are part of the mutation-load burden. By the time you reach the age of 15, most of the cells in your body have acquired hundreds or thousands of new mutations depending on whether they were replicating frequently or not. By the time you reach 60 years of age, your proliferating cells (e.g., blood cells, cells in your intestine) will have somewhere between 4000 and 40,000 mutations, according to a calculation by Michael Lynch.[19] These mutations can cause cancer, and that's why cancer is more common in proliferating cells and more common in older people.

It's true that external mutagens such as radiation can increase the mutation rate and cause cancer, but recent studies suggest that more than half of all cancers are just bad luck due to spontaneous mutations that you could not have prevented no matter how much you watched your diet and worked out in the gym.[20]

How much of our genome is conserved?

Most of our genome is free to accumulate random mutations, which is why it appears to be evolving at the neutral rate. This is evidence that most of our genome is junk, but the flip side of that argument is that some of our genome must be functional and that that part cannot accumulate mutations. We identify such regions by comparing different species and looking for stretches of DNA that do not change; in other words, they are conserved by negative selection or purifying selection. Sequence conservation is the very best way of recognizing function.

Sequence conservation is determined by aligning the DNA of two different species and counting the number of differences in the sequences. When the sequence similarity is high (greater than 30 percent or so), the region is said to be "conserved," but this can be misleading. For example, short stretches of DNA that are very similar – or even identical – in humans and chimpanzees may, in fact, be junk DNA where, by chance, no neutral mutations have been fixed in either lineage. These regions may not be under negative selection as the word *conserved* implies; instead, they just may not have diverged in the relatively short time since humans and chimpanzees last shared a common ancestor.

Let's agree that similar sequences are not necessarily conserved and let's not use the conclusion *conserved* as a synonym for *similar* unless we have good evidence that negative selection is playing a role in maintaining a specific sequence.

It's quite easy to align protein-coding regions and estimate the degree of sequence conservation when looking at more distantly related species, such as humans and mice. The mouse and human genomes have at least 15,000 conserved protein-coding genes in common, and these represent three-quarters of the known genes.[21] However, aligning the remaining relevant parts of the genomes is often difficult because they do not contain known genes, and in such cases, there is a significant probability that short stretches of DNA will be artificially aligned because they just happen to have similar sequences. The degree of sequence similarity is very sensitive to the window being used (e.g., 20 bp, 100 bp, 1000 bp) and whether gaps (indels) are allowed in the alignment.

Keeping these limitations in mind, the identification of sequences that are really and truly evolutionarily conserved should be a good approximation of the amount of functional DNA in the genome. In the case of the human genome, the amount of truly "conserved" DNA is between 8 and 10 percent of the entire genome.[22]

This value of 8 to 10 percent is consistent with the value expected from genetic load arguments. As we shall see in the next chapter, only a fraction of this conserved DNA has a known function associated with genes, regulatory sequences, centromeres, telomeres, and so on; the rest is assumed to be functional simply because it is conserved.

Defining function

We need to be clear about how we define function and junk. I've already stated that sequence conservation is the best way we have of identifying functional sequences in the genome, but that statement requires a bit of tweaking. Philosophers talk about two different kinds of function: causal role (CR) function and selected effect (SE) function. A CR function can be crudely described as a thing that does something; for example, if a defective transposon is still transcribed, then the DNA has a CR function (producing RNA). As the name implies, an SE function identifies a feature that has arisen by natural selection. CR functions are not necessarily biologically relevant functions, but SE functions are because they contribute to fitness.

What I'm promoting is a version of SE function that's used by molecular biologists. It's a version that relies on evolution as the most reliable indication of function, but there are several variations of that definition. The one I prefer is the one that Dan Graur uses in his textbook *Molecular and Genome Evolution*:

> Functional DNA refers to any segment in the genome whose selected-effect function is that for which it was selected and/or by which it is maintained. Most functional sequences in the genome are maintained by purifying selection.[23]

But there's a problem with adopting the terminology of the philosophers because they use "selected effect" in a different way. We don't need to get into nitpicky arguments about the exact meaning of "selected effect" and whether the philosophers and molecular biologists are using it in the same way (they aren't). What we need is a good working definition of what we mean by "function" and "junk" and a practical way of experimentally distinguishing between the two – even if it's only a thought experiment.

Hence, in my opinion, it's best to avoid the term *selected effect* because it just irritates the philosophers and leads to more papers being published in the philosophical literature – papers that, quite frankly, are practically unintelligible to anyone not well versed in the fine details of philosophical discourse.[24]

The most important point in Graur's definition is that functional sequences are maintained by purifying selection, which means that they are currently under selection. That's important because it's possible to have sequences that appear to be conserved but that are no longer functional. For example, you can have newly duplicated genes where one of them has recently acquired a mutation, rendering it nonfunctional. Both copies will be very similar to sequences in other species, but only one of them will be maintained by purifying selection in the future. Conversely, a new gene can evolve from junk DNA, in which case it is functional and under purifying selection but not conserved.

Let's use the term *maintenance function* to describe the molecular function that emphasizes purifying selection as the definitive criterion.[25] This avoids the ambiguity associated with SE function and whether a functional stretch of DNA must have a history of selection.

In addition to new genes, there is one other class of functional DNA sequences that are not conserved: spacer DNA. Spacer DNA is DNA that's required to keep some functional regions apart. The most obvious example is the DNA in introns that's necessary to form a loop between the splice sites (see Figure 1.11). An intron must be big enough to form the loop, but the DNA sequence of the spacer will not be conserved even if it's essential. Similarly, there are DNA loops in the promoter regions of most genes that allow transcription factors to interact with RNA polymerase at the promoter even though the transcription factor binding site might be some distance away. The minimum size of such loops is only 30 bp, so even if we add up all the introns and all the promoter regions, it's not likely to amount to more than 300 bp per gene, and that's not going to make a big difference in the amount of functional DNA in our genome. (Most loops will be much larger than 30 bp, but the extra DNA isn't necessary.)

What this means is that, because of spacers, the total amount of functional DNA is definitely greater than the total amount of conserved DNA identified by sequence similarity, but this difference isn't going to change the big picture.

Spacer DNA does not conflict with the definition of function because the amount of the spacer is under purifying selection even if the exact sequence is not. It does, however, conflict with any definition of function that relies exclusively on the conservation of a specific DNA sequence as the only way of identifying function.

This leads to an operational definition of molecular function that goes like this: **Functional DNA is any stretch of DNA whose deletion from the genome would reduce the fitness of the individual.** This definition covers stretches of DNA that exhibit sequence conservation as well as spacer DNA. If you want to get very specific, then the conservation definition of function relates only to sequences that are currently under purifying selection. This is what we mean by *maintenance function*.

Experimentally testing for function by deleting sequences is very difficult (but see the box "Deleting DNA to Prove That It Is Junk"), which is why sequence conservation is still the easiest and fastest way to assess function. However, in theory it should be possible to look at purifying selection as a reliable indicator of function, but this requires very large databases of accurate human genome sequences.

The United Kingdom Biobank project is committed to sequencing the genomes of hundreds of thousands of UK citizens, and some of the data for 150,000 genomes has been published. By looking at sequence variants (mutations), the project researchers can identify regions that have fewer than the expected number of mutations, indicating that selection has removed some of them. As expected, many of these functional regions are within or close to known genes, and a large part of the genome appears to be evolving at the neutral rate.[26]

LEVELS OF SELECTION

The operational definition of function that I'm proposing focuses on selection at the level of the individual. This is the level that we are most familiar with, and it results in allele changes within a population. However, there are other levels of selection that are part of hierarchical theory, an expansion of modern evolutionary theory promoted by Stephen Jay Gould and others.[27]

Selfish DNA is an example of selection at a lower level than the individual. In this case, the propagation of things like transposons depends on selection for their survival within an individual as opposed to between individuals. Group selection or species selection (also called species sorting) is an example of selection at a higher level than individuals where there's selection or competition between different groups or species within the same environment.

Some of these other levels impinge on our discussion of junk DNA. For example, transposon sequences are conserved if they remain active, but to a first approximation (see below), they can be deleted without affecting the survival of the individual. This is why bacteria have very few transposons in their genomes and why the humped bladderwort and pufferfish have fewer transposon sequences than most of their relatives. Thus, while selfish gene transposons are free to jump around in the human genome, we still must explain why they haven't been suppressed by selection at the level of the individual. That explanation relates to the nearly neutral theory and the idea that individual negative selection isn't powerful enough to suppress selfish genes.

Selfish DNA explains the origin of junk DNA while the nearly neutral theory and the drift–barrier hypothesis explain why transposon-related sequences are retained in

(*continued*)

LEVELS OF SELECTION (continued)

our genome long after the active transposon has mutated and is no longer selfish.

Some junk DNA skeptics promote evolution at the species level. They suggest that the presence of large amounts of excess DNA gives a species an advantage over other species because the extra DNA provides lots of opportunities to evolve new genes and new functions. Thus, the species with a large genome are more likely to evolve and outcompete those with smaller genomes. Often this proposal is connected to the existence of transposons, but it applies to all other types of excess DNA. The phenomenon is connected to the idea of evolvability that I discussed in chapter 3 (in the "Exaptation Versus the Post Hoc Fallacy" section).[28]

Species-level selection conflicts with the definition of function that I'm proposing because it claims that specific stretches of DNA might easily be deleted without affecting the fitness of the individual but large amounts of excess DNA could not be deleted without affecting the survivability of the species. This raises the question of whether it's appropriate to call that excess DNA "junk."

As I explained in the previous chapter, I tend to side with Gould (and Dawkins[29]) on the issue of evolvability: there probably are some examples of higher-level selection when it comes to some new features that give a species and all its descendants an advantage over other species, but the concept is often abused. We don't have any evidence that carrying around a huge amount of excess DNA in the genome gives a species a significant advantage over other species with smaller genomes and we don't have any evidence that the invention of new genes from junk DNA is a driving force in the survival of species in the short term (tens of millions of years). (This point is discussed further in chapters 8 and 10.)

Selection at higher or lower levels might play a role in explaining some aspects of evolution, but it's very unlikely that you can account for a genome that's 90 percent junk by invoking other levels of selection. Thus, defining function at the level of individual selection remains a good operational definition that will help us clarify much of the debate over junk DNA.

Why is the evidence of sequence conservation so hard to accept?

For many of us, the fact that only 10 percent of our genome is conserved is powerful evidence that 90 percent of our genome is junk – especially considering all the other evidence for junk. However, there are still many other scientists who reject that evidence and argue in favor of a mostly functional genome. Their rationales for rejecting the sequence conservation argument fall into several categories.

The most important counterargument is that there is non-conserved DNA that is functional, so it's wrong to think that all non-conserved DNA is junk. This is certainly true for spacer DNA, so let's cover that part first.

In addition to the spacers required for loops, there's the extra DNA located between genes. Many scientists have argued that eukaryotic genes need to be separated by a considerable amount of intergenic DNA in order to function properly. The sequences of these separator DNAs will not be conserved, but they may be functional.

The most famous proponent of this model is Emile Zuckerkandl, one of the most important founders of the field of molecular evolution. He pointed out that genes are found in large chromatin loops that must be physically separated from each other. This model of genome organization postulates that much of the DNA between genes is required to position the genes on the chromatin loops. There are several variants of this model, but all of them are similar in that they require a considerable amount of extra functional DNA that's required for the three-dimensional organization of genes.[30]

There are two problems with this model. The first is that there's very little direct evidence that a large amount of intergenic DNA is necessary, and there's no evidence that the amount of DNA between any two genes is conserved in different species. The second objection is that the model fails the Onion Test: it doesn't account for the huge differences in the amount of extra DNA in different species.

Another common objection has nothing to do with spacer DNA or separator DNA. It's a much more fundamental attack on the basic concept of using sequence conservation as the determining signature of function. The argument goes like this. There are huge amounts of functional DNA in our genome that aren't detected by conservation because they have evolved very recently. This might include thousands of new genes and new regulatory sequences that have evolved from junk DNA within the past several million years. They don't appear to be conserved because their sequences aren't similar in other species, but according to this argument, they represent a substantial amount of new functional DNA in our genome.

I call this the "high-speed evolution" argument because it requires a rate of evolution of new functions that's so far out of line with modern evolutionary theory that it would require scrapping the entire field of population genetics. Furthermore, the argument lacks any basis of experimental support other than wishful thinking – we'll see in subsequent chapters that there's no solid evidence that massive numbers of new genes and/or new regulatory sequences have arisen in the human lineage.

I don't mean to argue that no new genes and no new functions have evolved recently. Of course they have, but the tiny number of examples falls into the "exceptions that prove the rule" category and not the "slaying of a beautiful theory by an ugly fact" category. Despite these objections, high-speed evolution is frequently invoked as an argument against junk DNA, and we'll encounter it several more times in this book. In essence, it's an example of *ad hoc rescue*, a version of logical fallacy that attempts to avoid an unpleasant conclusion by making up excuses.

There's one last argument against sequence conservation that I have to mention for completeness. It's the argument used by John Mattick, one of the most prominent opponents of junk DNA, but a few others have made the same argument. They believe that the sequence conservation data is misleading because the wrong controls are being used. Specifically, they argue that scientists are using the evolution of defective transposon sequences as a measure of the neutral rate of evolution and judging all other sequences by that standard. But, they claim, lots of transposons have a function, so they are not evolving at the neutral rate, just evolving slowly. Thus, according to their logic, much of the genome appears to be evolving at the neutral rate when, in fact, it's actually somewhat conserved, meaning that a large proportion of the genome exhibits weak sequence conservation that is being overlooked because of a false premise.[31]

Several scientists have addressed this criticism and concluded that it has no merit. With few exceptions, degenerate transposon sequences are, in fact, evolving neutrally, as are pseudogenes.[32] Furthermore, you might recall that the back-of-the-envelope estimation of phylogenetic mutation rate agrees with the biochemical mutation rate and the direct mutation rate, and that agreement can only be true if most of the human and chimpanzee genomes are evolving at the neutral rate. The argument promoted by Mattick and others looks very much like another example of *ad hoc rescue* in an attempt to explain away the evidence that most of our genome is not conserved.

DELETING DNA TO PROVE THAT IT IS JUNK

Junk DNA is defined as DNA that is dispensable, so in order to prove that a given stretch of DNA is junk, it should be possible to remove it without having any effect on the survival of the organism. Unfortunately, such large-scale deletion experiments are difficult, so there's not a lot of data. One of the best-known experiments was completed in 2004 by a group at the Lawrence Berkeley Laboratory in Berkeley (California, USA).[33] They deleted two large blocks of DNA from the mouse genome – the largest one covered 1.511 Mb on chromosome 3, and the smaller one covered 0.845 Mb on chromosome 19 for a total of 2.3 Mb, or 0.07 percent of the genome. The deleted blocks were selected because they didn't contain any genes, but they did contain hundreds of short segments (less than 100 bp) that were at least 70 percent similar to the same sequences in the human genome.

Mice that were missing the 2.3 Mb of DNA showed no ill effects, including the ability to reproduce and pass on the smaller genomes to their offspring. This result demonstrates that the deleted regions were junk DNA and the short stretches of DNA that were similar in humans were probably just due to chance.

This is another example of an experiment that scientists can't do on humans but where nature has done the experiment for us. As more and more genomes are sequenced, it becomes apparent that some individuals are missing large blocks of DNA that are present in the reference genome or in other individuals. A typical genome has several thousand of these structural variants covering a total of about 20 Mb, or 0.6 percent of the genome.

If you add up all the blocks of DNA that can be missing without affecting survival, then the total amount of the genome that can be deleted is 224 Mb, or 7 percent of the genome. Keep in mind that no two individuals differ by more than 1 percent.[34] We can't prove conclusively that deleting this 7 percent of the genome in a single individual is harmless, but the data is consistent with the idea that it is junk DNA.

Bulk DNA hypotheses

This is as good a time as any to introduce another objection to junk DNA and the importance of sequence conservation. It's the idea that the total mass of DNA in the genome is important and not the specific function of any particular segment. This skirts around the genetic load argument and avoids dealing with the fact that less than 10 percent of our DNA sequences are conserved.

There are several different ideas, but they can all be grouped under the title of "bulk DNA hypotheses." These ideas are compatible with the presence of large amounts of defective transposons, pseudogenes, and repetitive sequences since these now become the mechanisms for bulking up the genome. Furthermore, they are potentially capable of explaining the C-Value Paradox if genome size correlates with the rationale for expecting large genomes. This latter point is important; it means that these explanations are no different than any other explanations of genome function. They still must pass the Onion Test, and they still have to explain why the pufferfish genome is so much smaller than ours.

Skeletal DNA hypotheses

One of the first bulk DNA explanations was advanced by Cavalier-Smith in 1978, although others had mentioned it earlier. He proposed the "skeletal DNA hypothesis," suggesting that an increase in genome size leads to an increase in nuclear volume, which, in turn, leads to an increase in cell size.[35] Thus, genome expansion by adding bulk DNA may be favored because it leads to an increase in the size of the nucleus. A larger nucleus means a larger surface area and more nuclear pores, and this could increase the efficiency of transport across the nuclear membrane. This might be important for importing and/or processing RNA.

It's difficult to test the skeletal DNA hypothesis except by looking for correlations between genome size, nucleus size, and cell size. Ryan Gregory reviewed the literature on cell size and genome size in animals and concluded that there's a rough correlation in many vertebrates. In fish and amphibians, for example, an increase in genome size is almost always associated with an increase in the size of the average cell. A similar correlation between genome size and cell size exists in plants, but the correlation is much less obvious in reptiles, birds, and mammals.[36]

One of the conceptual difficulties with Cavalier-Smith's hypothesis is that there could be selection for an expanded nuclear membrane (and more nuclear pores) that then allows for an increase in genome size.[37] In other words, it is not clear whether the increase in genome size is a cause or an effect.

Another problem is that there is no evidence that the number of nuclear pores impose a real limit on the growth rate of cells despite many additional functions of nuclear pores having been identified. Finally, the Cavalier-Smith hypothesis suffers from the same objection that makes us skeptical of all bulk DNA hypotheses – it does not pass the Onion Test. It lacks explanatory power because it does not explain why species with relatively small genomes are so successful and it does not explain why some species have genomes that are severalfold larger than closely related, phenotypically similar, relatives. This point has been made by many critics; for example, here's what Sean Eddy wrote in 2013 in response to bulk DNA hypotheses:

> But to explain mutational load – and the more modern observations from comparative genomics, showing that only a small fraction of

most eukaryotic genome sequence is conserved and under selective pressure – you have to posit an adaptive role where only the bulk amount of the DNA matters, not its specific sequence. To explain the C-value paradox, you have to explain why this bulk amount would vary quite a bit even between similar species. Although some such adaptive explanations have been speculated, a rather different line of thinking, starting with Ohno and others in the 1970s, ultimately lead to a reasonably well-accepted explanation of the C-value paradox.[38]

In order to be useful, an adaptive hypothesis needs to help us make sense of biology and it needs to provide a better explanation of the data than the null, or nonadaptive, hypothesis. In my opinion, the idea that most excess DNA in large genomes is junk is more consistent with the evidence and with modern evolutionary theory than any of the proposals that bulk DNA has a function. For example, Michael Lynch notes that in multicellular organisms, there's a general tendency for increases in cell size to be associated with increases in the size of the organism, and larger organisms tend to have lower population sizes. Thus, the expansion of the genome isn't driven by direct selection for a larger genome but by a reduced ability to select against the detrimental effects of a large genome because of the smaller population. If this is true, then the correlation between genome size and cell size is an epiphenomenon, and the real cause of the effects is larger organisms and smaller populations.[39]

Furthermore, if there's a selective advantage to genome size, then it's far more likely that there's selection for smaller genomes, and smaller cells, in species that would benefit from this efficiency. This may explain why birds and bats seem to have smaller genomes than their relatives who don't expend a lot of energy flying.[40]

The bodyguard hypothesis

The bodyguard hypothesis is sometimes called the "insulation theory." That's the term used by Nessa Carey in her 2015 book *Junk DNA*, where she reports that the function of extra DNA in our genome is to protect the important parts from mutations caused by ultraviolet light and other environmental insults.[41] Here's how she explains it: "If we think of our genome as constantly under assault, the insulation theory of junk DNA has definite attractions. If only one in 50 or our bases is important for protein sequence because the other 49 base pairs are simply junk, then there's only a one in 50 chance that a damaging stimulus that hits a DNA molecule will actually strike an important region."[42]

This hypothesis doesn't make sense for several reasons. First, assuming that radiation is a serious problem in evolution (it isn't), then bulking up the genome is not the most efficient way of insulating functional DNA from damage. After all, the genes and other functional regions are already packaged around protecting proteins within a nuclear membrane. There are lots and lots of other chemicals that the cell could make to protect its genome – it could, for example, have evolved extra membranes around the nucleus and around the outside of the cell. It could have bulked up carbohydrates or proteins to protect the DNA. It could bury the germ line cells deep within the body of an animal as is the case with ovaries (but not testes) in mammals.

The bodyguard hypothesis only works if the extra DNA intercepts a significant amount of incoming radiation before it hits a functional part of the genome. That means that the extra DNA must somehow be arranged so that it encases the genes and other functional regions. Just having extra DNA in the nucleus won't work because the radiation that hits it wouldn't have damaged the important regions anyway.

But there's a more serious problem. Recall that the vast majority of mutations are caused by errors in DNA replication. Imagine a genome that's 100 percent functional, and every time it is replicated there are 10 new mutations. Now imagine that you increase the genome size tenfold by adding junk DNA to "protect" the genome from mutation. When the junky genome is replicated there will now be 100 new mutations in total, but there will still be the same 10 mutations in the functional part, so bulking up the genome hasn't done anything except add more neutral mutations to junk DNA.

Another version of the bodyguard hypothesis has been advanced by Claudiu Bandea.[43] He argues that the excess DNA serves as a sink for transposon insertions, thus protecting the functional DNA from mutation due to insertion of a transposon.

The logic behind this argument is as flawed as the logic for all the other versions of bodyguard hypotheses. Yes, it's true that a transposon is more likely to insert harmlessly in junk DNA, but it's also true that a huge amount of junk DNA acts as a reservoir for active transposons that are jumping around in the genome, so more junk DNA means more transposons. Furthermore, the idea of protecting against mutation is known as second-order selection because it selects against a possibility (future transposon insertion) and not an actual event. Second-order selection requires powerful natural selection, and that's only possible with huge population sizes of the sort found in bacteria. It will not work in most eukaryotes, and it certainly won't work in species with large genomes where natural selection is already weak because of small population sizes.[44]

As if this weren't enough, none of these bodyguard ideas passes the Onion Test. If protecting DNA is so important, then how can it explain the fact that the onion needs so much more protection than humans and that different onion species have quite different amounts of insulating DNA? I'm surprised that anyone who has thought about the problem takes any of the bodyguard hypotheses seriously.

Genetic diversity

The genetic diversity argument assumes that having the potential for genetic diversity is advantageous. This is not the same as regular diversity, where the population contains multiple alleles of the same gene. This argument postulates that excess DNA is present in the genome in order to evolve extra diversity in the future. Let's look at an example.

The genes for antibody proteins of various sorts must be capable of generating lots of diversity to cope with all of the challenges faced by modern individuals. In this case, there are specific mechanisms for creating diversity, and it requires large numbers of similar but slightly different genes in the genome. Some of the genes in those regions may be dead genes (pseudogenes), but according to the genetic diversity hypothesis, they may not be junk. Instead, they may serve as a reservoir of genetic diversity by recombining with functional genes to create new combinations. The idea is that these "junk DNA" sequences are essential for creating diversity even though they may be nonfunctional on their own.[45]

There may well be examples of pseudogenes that contribute to genetic diversity. The most likely examples are in the major histocompatibility complex (MHC) Class I and Class II gene clusters on chromosome 6. These genes encode the proteins responsible for histocompatibility and tissue rejection. The clusters of genes include pseudogenes, and there is frequent recombination and shuffling to produce new genes with slightly different amino acid sequences.

It's unlikely that there are very many other examples in the human genome. Even if every single pseudogene was capable of contributing to genetic diversity, it would only account for 1 percent of the potential junk DNA component of our genome.

Pseudogenes usually confer no immediate advantage to the organism that carries them. The postulated advantage is only realized in future generations, so this is another example of second-order selection. Everything we know about human evolution indicates that an adaptive explanation for the presence of pseudogenes is highly improbable.

Medical relevance

Proponents of function often argue that the search for causes of genetic diseases is shedding light on the role of all that extra DNA in our genome. They frequently quote papers saying that most genetic diseases are associated with mutations in "noncoding DNA," and they use this fact to argue that this DNA is not junk.

The most important problem with this claim is the confusion between noncoding DNA and junk DNA. There's a lot of function in noncoding DNA, and no knowledgeable scientist ever thought that all noncoding DNA was junk. It's not a surprise that mutations affecting genetic diseases are found outside of coding regions – many of them occur in regulatory sequences and nobody thinks these sequences are junk.

The other problems are more subtle. I discuss them here because they are related to genetic load. It's clear that mutations causing genetic diseases such as cystic fibrosis and phenylketonuria contribute to genetic load even if we now have effective treatments. However, the regions where the mutations occur are not necessarily functional parts of the genome in most individuals. They could still be junk DNA. How is that possible?

Consider a region within a large intron in a gene that encodes an important protein. That gene is transcribed, and the intron sequence is spliced out. Now, imagine that most of the intron sequence is junk* and suppose a mutation occurs that creates a new splice site in this otherwise useless DNA. This new mutation will lead to aberrant splicing and a severe reduction in the amount of protein being synthesized. Such a change could cause a genetic disease even though the mutation occurred in junk DNA.

There are many examples of this phenomenon, thus demonstrating that mutations affecting genetic diseases don't necessarily mean that the DNA where these mutations occur is functional. Let's think about the implications of that observation. If deleterious mutations can occur in junk DNA, then it means that the more junk DNA in your genome the greater the chance of a deleterious mutation. This is almost certainly true, and it's referred to as the mutational-hazard hypothesis.[46] It's one of the reasons why excess DNA (junk) is slightly

* I explain in chapter 6 why most introns sequences are junk.

deleterious and can only accumulate in small populations where drift overcomes selection.

There's another subtle objection to the argument for medical relevance, but it requires a change in the way we look at genetic diseases. We tend to think that genetic diseases are due to mutations in very important genes and that these mutations will help us figure out the really essential parts of the genome. That's not a good way to think of genetic diseases.

Mutations in important genes, such as the genes for DNA polymerase or the enzymes of major metabolic pathways, are rarely seen in the medical literature. That's because mutations in those genes are lethal and embryos never develop into a newborn baby. The mutations causing genetic diseases usually affect functions that are nonlethal in cells and embryos so that the developing fetus survives until birth. As a general rule, genetic diseases affect genes that are not very essential for survival and that's why we see them.

This view is not meant to diminish the tragedy of genetic diseases – it's only meant to put the data into perspective. Medicine is not going to tell us what part of the genome is functional and what part isn't because (1) genetic diseases can be caused by mutations in junk DNA and (2) mutations in the most important parts of the functional genome are lethal and never show up as genetic diseases.

Ignoring history

[T]hose ignorant of history are not condemned to repeat it; they are merely destined to be confused.

Stephen Jay Gould (1977)

By the end of the 1970s, we saw a remarkable convergence of opinion on the organization of genomes. The smartest and most knowledgeable men and women had come to the conclusion that most of our genome is junk. They based their conclusions on several lines of evidence including mutation load, the variation in genome sizes

(C-Value Paradox), the small number of genes, and an understanding of the nearly neutral theory and population genetics.

They knew that fluctuations in genome size were due to increases and decreases in repetitive DNA, and they knew that transposons were a major contributor. They knew that protein-coding regions made up only a small percentage of mammalian genomes, and they knew that noncoding genes, such as genes for ribosomal RNAs and transfer RNAs, were common but not abundant. They knew that noncoding DNA contained many functional elements, such as regulatory sequences, centromeres, and the like. Junk DNA was consistent with all the evidence and with evolutionary theory, and it explained the C-Value Paradox.

In most cases, the models developed by the experts in a field will eventually become the dominant view, but that didn't happen in this case. The concept of junk DNA was too radical for most biologists, so it was never accepted by the majority. By the time the human genome sequence was published in 2000 the views of those experts had been dismissed.

Even worse, a remarkable false counternarrative has become entrenched in the scientific literature and popular science writing. The current revisionist history goes something like this. Scientists in the 1970s were surprised to discover that only 2 percent of our genome codes for protein. They assumed that all the noncoding DNA (98 percent) had to be junk because they thought that protein-coding genes were the only functional parts of our genome. The discovery of regulatory DNA and noncoding genes demonstrates that some noncoding DNA is functional, thus making us realize that the original argument for junk DNA was largely based on ignorance. Furthermore, the discovery of genetic disease mutations in noncoding DNA is taken as evidence that much of our genome is not junk.

Here's an example of this false counternarrative from a book by Alan McHughen published in 2020:

> When it was first discovered, the nongenic DNA was sometimes called – somewhat derisively by people who didn't know better – "junk DNA"

> because it had no obvious utility, and they foolishly assumed that if it wasn't carrying coding information it must be useless trash.
>
> In evolutionary terms, a DNA sequence with no function is simply dead weight that gets carried along, at some cost to the organism, to be jettisoned at the first opportunity. If the sequences were not adaptively important, evolution would have kicked them out as expendable excess baggage. The fact that nonrecipe DNA continues to be part of the human and other eukaryotic genomes over millions of years indicates that there is some adaptive value to carrying the "junk baggage" along, even if that value remains unclear to us today.
>
> In addition to various putative regulatory and structural functions, recent evidence indicates that mutations in the intergernic noncoding DNA leads to a increase in susceptibility to various diseases. If confirmed, it would show a clear adaptive value to "junk" DNA.
>
> Today, we appreciate that this is not useless junk and now call it noncoding DNA.[47]

This is not an isolated example of the dominant counternarrative. This false version of history is so common that it is widely expressed in the scientific literature as well as in popular books.

Let me be very clear about three things that are mentioned above. First, no knowledgeable scientist in the 1970s ever believed that all noncoding DNA was junk – such a claim is absurd despite what the fake news might tell you. Second, no modern evolutionary biologist believes that natural selection is powerful enough to remove all useless junk from human genomes – the mere presence of excess DNA is not evidence of function. Third, mutations in junk DNA can cause genetic diseases, so the presence of these mutations does not prove that the DNA is functional.

The most annoying thing about this false history is that it is false, but the second-most annoying thing is that it insults the intelligence of some very smart scientists such as Thomas Jukes, Francis Crick, Sydney Brenner, Tomoko Ohta, Jacques Monod, Masatoshi Nei, Motoo Kimura, Ford Doolittle, and Susumu Ohno. They were not "fools."

Sloppy genomes are consistent with nearly neutral theory and the importance of random genetic drift and population size as major factors in evolution. The idea that most of our genome is junk was based largely on three lines of evidence:

1. observed differences in genome size in different species (C-Value Paradox)
2. about 50% of our genome is littered with fragments of defective transposons and viruses
3. mutation load/genetic load

I've devoted a chapter to each of these three topics. The explanation adopted by most experts in molecular evolution is that a large percentage of our genome is junk. This model is consistent with the data, and it explains the data.

In order to see why many scientists have ignored most of these points and rejected junk DNA, we have to restart the historical overview at the beginning of the new millennium when the first draft of the human genome was published.

CHAPTER 5

The Big Picture

Some have said to me that sequencing the human genome will diminish humanity by taking the mystery out of life. Nothing could be further from the truth. The complexities and wonder of how the inanimate chemicals that are our genetic code give rise to the imponderables of the human spirit should keep poets and philosophers inspired for the millenniums.

Craig Venter (2000)[1]

Picture a huge room with rows and rows of laboratory benches. There are DNA sequencing machines sitting side by side on those benches – each one is about 1 meter (3 ft) high and 1 meter wide. There are electric cables running upward from each machine to conduits in the ceiling, and all of them connect to the main power source on the wall. The machines are tended by a small army of technicians in lab coats who are constantly cleaning and refilling the machines with bits of DNA.

This is what a large sequencing lab looked like in 2001. You didn't see the direct result of all this activity because the DNA sequences are sent electronically over cables to another room that's full of computers. There's another group of technicians in that room who program and monitor the computers that collect the data, and their job is to organize the information and store it in a large database.

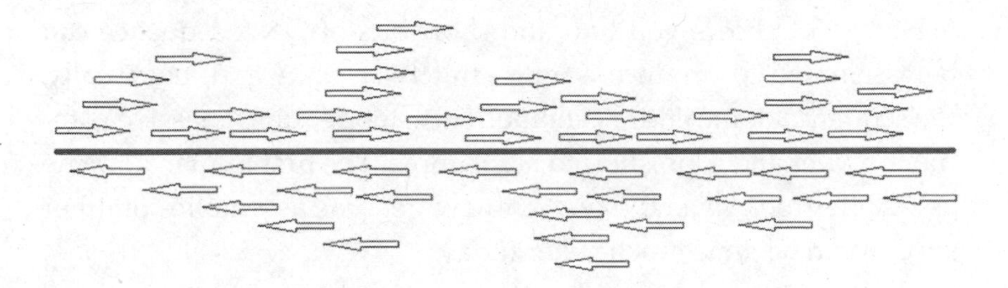

FIGURE 5.1 Assembling DNA sequences. Each arrow represents a short stretch of DNA sequence from a sequencing machine. The arrows indicate the direction of the single-stranded sequence from 5′ end to 3′ end. The overlapping fragments can be assembled to produce a long sequence of double-stranded DNA (solid gray line). The accuracy of the final genome sequence depends on the coverage of the individual fragments. The best genomic sequences are obtained when the average coverage is at least fivefold (5×), including at least one fragment from each of the two DNA strands.

Each individual bit of DNA sequence was only a few hundred nucleotides in length. Some of these bits overlap, and some of them are complementary to ones that are already in the database because they come from the opposite strand. The machines keep running, and the technicians keep feeding them until enough data has been collected to cover both strands of the entire genome. Usually this means that the average section of a genome will have been sequenced at least 5 times (5× coverage).

The assembly step is when you combine all these fragments into one contiguous sequence – it's where things can go horribly wrong if you don't have enough data.

Back in the olden days (i.e., 2001), the rate-limiting step was the generation of data. It was expensive to sequence DNA on a massive scale, and one reason why Celera and the International Human Genome Project had so much trouble with their first assemblies was because they didn't have enough data. Today, it's much cheaper to sequence DNA, so collecting data is not a problem – the cost has fallen from over US$1.00 to just a fraction of a cent for each nucleotide of the completed sequence.

Figure 5.1 shows you how the short bits of DNA sequence can be assembled to produce a long stretch of DNA and, eventually, the sequence of the entire genome. This step requires massive computer power and sophisticated algorithms. The first human genome assemblies took several days, but new genome assemblies of other species can be done much faster today.

A typical gene

The final result looks like the example in Figure 5.2. The figure shows the raw data after the genome has been assembled and the sequence has been deposited in GenBank, the public DNA database. This particular entry is from region p13.31 on chromosome 12.

The sequence is boring because it's just a large chemical expressed as a bunch of As, Ts, Gs, and Cs written in lowercase letters. We want to know what's in your genome besides nucleotides, and that's the job of the annotators who look at the sequence to see if they can recognize genes and other important features. One of the major developments over the past 20 years has been more and more sophisticated software for recognizing these features.

The easiest part of annotation is simply looking for genes that have already been cloned and sequenced. Those genes could be from the same organism – human in this case – or they could be from another species, in which case you're looking for similar, but not identical, genes in the human genome. One way to do this is to "decode" the DNA by searching for open reading frames that might encode a protein. Recall that open reading frames (ORFs) are stretches of DNA sequence that begin with an initiation codon (AUG in mRNA or ATG in DNA) followed by at least 100 codons and then a termination codon.

You can decode the DNA in Figure 5.2 by yourself. Start with the initiation codon, atg (ATG), at nucleotide 37 and decode each subsequent codon using the genetic code from chapter 1. If you do a good job, it should look like the result in Figure 5.3.

```
   1 gcagacactg accttcagcg cctcggctcc agcgccatgg cgccctccag gaagttcttc
  61 gttgggggaa actggaagat gaacgggcgg aagcagagtc tgggggagct catcggcact
 121 ctgaacgcgg ccaaggtgcc ggccgacacc gaggtggttt gtgctccccc tactgcctat
 181 atcgacttcg cccggcagaa gctagatccc aagattgctg tggctgcgca gaactgctac
 241 aaagtgacta atggggcttt tactggggag atcagccctg gcatgatcaa agactgcgga
 301 gccacgtggg tggtcctggg gcactcagag agaaggcatg tctttgggga gtcagatgag
 361 ctgattgggc agaaagtggc ccatgctctg gcagagggac tcggagtaat cgcctgcatt
 421 ggggagaagc tagatgaaag ggaagctggc atcactgaga aggttgtttt cgagcagaca
 481 aaggtcatcg cagataacgt gaaggactgg agcaaggtcg tcctggccta tgagcctgtg
 541 tgggccattg gtactggcaa gactgcaaca ccccaacagg cccaggaagt acacgagaag
 601 ctccgaggat ggctgaagtc caacgtctct gatgcggtgg ctcagagcac ccgtatcatt
 661 tatggaggct ctgtgactgg ggcaacctgc aaggagctgg ccagccagcc tgatgtggat
 721 ggcttccttg tgggtggtgc ttccctcaag cccgaattcg tggacatcat caatgccaaa
 781 caatgagccc catccatctt ccctaccctt cctgccaagc cagggactaa gcagcccaga
 841 agcccagtaa ctgccctttc cctgcatatg cttctgatgg tgtcatctgc tccttcctgt
 901 ggcctcatcc aaactgtatc ttcctttact gtttatatct tcaccctgta atggttggga
 961 ccaggccaat cccttctcca cttactataa tggttggaac taaacgtcac caaggtggct
1021 tctccttggc tgagagatgg aaggcgtggt gggatttgct cctgggttcc ctaggcccta
1081 gtgagggcag aagagaaacc atcctctccc ttcttacacc gtgaggccaa gatcccctca
1141 gaaggcagga gtgctgccct ctcccatggt gcccgtgcct ctgtgctgtg tatgtgaacc
1201 acccatgtga gggaataaac ctggcactag gtcttgtggt ttgtctgcct tcactggact
1261 tgcccagata atcttccttt ttgaggcagc tatataaatg atcatttgtg caagaaaaaa
1321 aaaaaaacaa gaacaggttt ctataacaac a
```

FIGURE 5.2. The sequence of part of the triose phosphate isomerase gene. This is a bit of DNA sequence from chromosome 12 in the region known as p13.31 where "p" indicates that it is on the short arm of the chromosome. Nucleotide 1 in this figure is actually nucleotide 6,867,527 on chromosome 12. The initiation codon of the open reading frame begins at nucleotide 37.

```
 M   A   P   S   R   K   F   F   V
Met Ala Pro Ser Arg Lys Phe Phe Val
atg gcg ccc tcc agg aag ttc ttc gtt ...
```

FIGURE 5.3 Decoding a DNA sequence. The open reading frame of part of the DNA sequence from Figure 5.2 beginning at nucleotide 37.

The potential protein-coding region has an ORF beginning with the amino acids MAPSRKFFV.... There are computer programs that will determine the sequences of all possible reading frames in both directions, and these sequences of potential protein-coding regions are then compared to all the protein-coding genes that have already been deposited in GenBank dating back to the early 1980s. In this case, the programs found a match. The amino acid sequence tells us that this is the gene for the enzyme *triose phosphate isomerase* (TPI) – one of the key enzymes of metabolism. It catalyzes one of the reactions in the pathway, leading to the synthesis of glucose (gluconeogenesis), and it's also used in the glucose degradation pathway (glycolysis), where glucose is broken down to produce energy. You may have had to memorize the name of this enzyme in school, but I bet you've forgotten it by now.

The Hugo Gene Nomenclature Committee (HGNC) is responsible for naming human genes and maintaining a database called genenames.org. The official name of the human gene for triose phosphate isomerase is *TPI1*. The National Center for Biotechnology Information (NCBI) in Bethesda (Maryland, USA) assigns a unique number to each gene and provides a freely available database of all genes. The Gene ID for the human *TPI1* gene is 7167.[2]

The gene for triose phosphate isomerase is present in almost all species, including plants and bacteria. It had already been cloned and sequenced several times, so it was easy for computer programs to recognize the function of this stretch of DNA on chromosome 12. Recognizing protein-coding genes in a newly sequenced genome becomes easier as more and more information gets deposited in the public databases.

The organization of the gene is shown in Figure 5.4. There are seven exons and six introns. The total length of the intron sequences is 2069 bp and the total length of the exons is 1350 bp. The coding region (colored black in the figure) is 750 bp – that's 250 codons – so the total length of the triose phosphate isomerase protein is 249 amino acids (aa). (The extra three nucleotides of the coding region are for the termination codon, TGA.)

The point of showing you one example of a gene is to let you see what the raw data looks like and how much data is out there on the web. I also want you to appreciate how many government agencies and scientists are required to maintain and organize all this data. We are paying for it with our taxes, and it's an important and worthwhile investment.

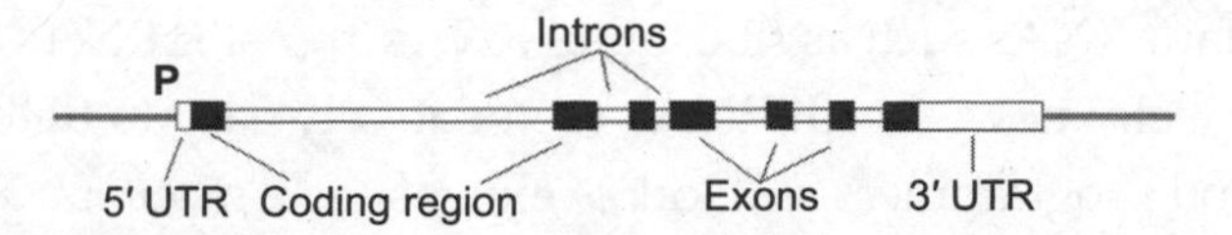

FIGURE 5.4 The human gene for triose phosphate isomerase (*TPI1*). The gene is 3418 bp long from the transcription start site (P) to the transcription termination site (t). Introns make up 2069 bp of this sequence and exon sequences total 1350 bp. The coding region (black) is 750 bp long, enough for 250 codons. The mature mRNA produced after processing contains untranslated regions (UTRs) at each end. The enzyme triose phosphate isomerase is made up of 249 amino acid residues with a total molecular weight of 26,669 daltons. This is a small protein-coding gene, about half the size of a typical human gene.

This is usually the time when we say that all this data is freely available and that anybody can use it, even high school students. The "free" part is true, but the rest is not true. You will have seen how complicated the data look if you visited the website, and you'll realize that it's not at all easy to interpret unless you've taken some advanced undergraduate courses in biochemistry or molecular biology. High school students can't do it without a great deal of guidance. Consider that I've been working with sequences for many decades and it still took me several hours to figure out the true sequence of the *TPI1* gene and calculate the amount of DNA in the coding and noncoding regions. Part of the problem is that the databases give you all kinds of information and speculation that are not accurate, so you really must know what you are doing to figure it out.

As I mentioned above, it's relatively easy to recognize protein-coding genes because we know how to detect a coding region, but it's harder to recognize genes for noncoding RNAs because we have to match up the nucleotide sequence and not the decoded amino acid sequence. This means comparing every possible nucleotide sequence to everything in the database – a task that used to take a long time on computers back in the 1990s but takes only seconds today.

Standard RNAs such as ribosomal RNAs, transfer RNAs (tRNAs), and the well-known small RNAs are relatively easy to detect if you have good programmers and lots of expensive computers. However, finding potential genes for some other RNAs is much harder because they may not be present in the reference database. This is the major source of controversy in genomics – how many genes for small RNAs are there in the human genome? Are there 1000, 10,000, or maybe even 100,000? We'll try to answer this question in chapter 8, but, for now, I'll just give you the numbers that most annotators agree on.

Annotators interpret the genome

If this was the language of God, then a holy editor would have been a great asset.

Adam Rutherford (2016)

There are other functional sequences in the human genome besides genes, and the annotators can often recognize them. The job of these genome editors is very important because they don't just run computer programs; they also make decisions about whether to accept or reject the output of those programs. Human interpretation is necessary because computer programs are deliberately set up to recognize anything that remotely resembles a functional DNA sequence, and they produce a large number of false positives in order to minimize the chances of missing something important. The job of checking all those hits falls to real, live humans who bring their knowledge and experience to the problem of annotation. These annotators are often young scientists with a PhD in genetics, biochemistry, or molecular biology, but they include people trained in computer science. This is the relatively new field of bioinformatics.

Interpreting a genome sequence is not straightforward and not without bias because humans are fallible. For example, some of the controversy over the amount of junk in the human genome involves differing opinions among sequence annotators about which part of the sequence should be included in a gene. When the correct interpretation is not obvious, the annotators sometimes give us several possible versions of a gene instead of making a decision about which one is the correct version. I encountered this problem with the data on the *TPI1* gene because there were two versions in the GenBank database and I had to figure out which one was correct.

In the case of the human genome, I'm going to rely mostly on the annotation of the Ensemble group at the Wellcome Genome Campus in Hinxton (UK), but I'll try to let you know when the Ensembl interpretation differs from that of other projects and from my own.

Many of the controversies in genome research, especially the amount of junk DNA, are reflected in differing opinions concerning genome annotation because some annotators err on the side of assigning function based merely on suggestive clues, whereas others demand more evidence.[3]

Armchair annotation of raw sequence data is not the end of the story. You also need to do experiments at the genome level and that's why hundreds of millions of dollars have been spent on analyzing the human genome to find which parts are truly functional and which parts are junk. The main effort has been carried out by a project called The Encyclopedia of DNA Elements, or ENCODE, which is largely funded by the American government. The ENCODE Consortium claims that most of the human genome is functional, but that view conflicts with the views of many other experts. We will discuss this controversy in the following chapters.

How much of the genome has been sequenced?

We learned in chapter 1 that the human genome is about 3.2 billion base pairs in size, and we learned that if all 24 pieces of this genomic DNA were joined end to end, it would be almost 2 meters (6 ft) long.

You would think that once the human genome sequence was published, we would know the exact size of each of the 22 autosomes plus the X and Y chromosomes, but that's not what happened. One of the problems is that there is no such thing as "the" human genome, as I mentioned in chapter 1, but a more serious problem is related to the fact that about 8 percent of our genome consists of stretches of highly repetitive sequences, and it was very difficult to sequence across them for a variety of technical reasons. Annotators know where these sequences are located, and they know what the beginnings and ends look like, but they often don't know the exact size of the repetitive sequence in between. What this means is that the standard version of the human genome reference sequence is only about 95 percent of the total genome size. Fortunately, the amount of the remaining bits of DNA could be estimated by measuring the approximate size of the gaps.

The latest assembly of the standard human reference genome at the time I'm writing this chapter is GRCh38.p14 from February 3, 2022,[4] and the total amount of sequenced DNA adds up to 2,948,318,359 bp. If you include the best estimates of the unsequenced part, then the size of the standard human genome is 3,099,441,038 bp or 3.1 billion bp (3 Gb). This is very close to the value that was estimated decades ago – the same size we calculated in chapter 1. (The estimated size of the standard human reference genome has been slowly dropping over the past few years.) There are still hundreds of gaps in the reference genome sequence, but most of them have been spanned, meaning that their size can be estimated even if the exact sequence is not known.

About 95 percent of the human genome has been sequenced in the latest version of the standard reference genome, and the rest is almost entirely repetitive DNA of various sorts. We know that the remaining 5 percent consists almost entirely of *highly repetitive* DNA sequences, and many scientists doubted that sequencing these regions would be worth the effort.[5] However, recent technical advances have enabled the sequencing of very long stretches of DNA that can cover the repetitive regions. The long-read sequences are about 20,000 nucleotides in length, and the less-accurate-but-still-useful ultra-long-read sequences are more than 100,000 nucleotides. This has led to the complete telomere-to-telomere (T2T) sequencing of an entire human genome from the CHM13 cell line. The new T2T-CHM13 sequence differs in many ways from the standard human reference genome (CRCh38), but most of those differences are due to polymorphisms in the human population.

The complete T2T-CHM13 genome is 3.054 billion base pairs long, but that doesn't include the Y chromosome because the CHM13 cell line was from a female. Adding the Y chromosome gives a size of about 3.1 Gb for the complete genome sequence.[6] This is the same as the predicted size of the standard reference genome and a bit smaller than historical estimates. It's likely that this value (3.1 Gb) will turn out to be a better estimate of the average size of the human genome, but I'll continue to use 3.2 Gb in this book until that slightly smaller average size is independently confirmed.

It will be very difficult to merge the T2T-CHM13 complete genome sequence with the current standard reference genome, so in the immediate future, we will have to deal with two different representations of the human genome.[7]

Whose genome was sequenced?

Let's stop for a minute and think about whose genome was sequenced before we get into more discussion about what's in your genome.

Celera Genomics choose to sequence DNA from five individuals when it entered the race to sequence the human genome. The samples came from two males and three females: one of the subjects was African American, one was Chinese, one was Hispanic, and two were Caucasian.[8] The final version was mostly Craig Venter's genome. (He was the male Caucasian; see "You Are Not Craig

Ventor" in chapter 4.) These five individuals actually contribute ten different genomes because each of our genomes contains one set of chromosomes from our father and one set from our mother.

The International Human Genome Project initially sequenced DNA from several different anonymous donors mostly from the area around Buffalo (New York, USA). They combined the sequenced genomes to create a standard reference genome with variations. The standard reference genome is not the genome of any one individual; instead, it's a consensus sequence based mostly on several of the original donors, with a large fraction coming from a single male donor with the code name RP11.

How many genes?

There are conflicting views on the total number of genes in our genome, but there's general agreement on the subset of genes that encode proteins. There are about 20,000 protein-coding genes. This is less than the number predicted in the draft sequences (30,000–40,000) because the annotators have eliminated most of the false predictions. I describe the evidence in the next chapter so you can judge for yourself whether this number is reliable. (Hint: it is!)

These genes are very large because they consist of protein-coding regions in the exons and large noncoding regions (introns). If you add up all the coding regions in the exons, it accounts for about 1.0 percent of the entire genome, but if you add up all the introns, you get about 37 percent of the genome, depending on how annotators define introns.[9] Since we define a gene as a DNA sequence that's transcribed (chapter 1), this means that something like 38 percent of our genome is devoted to protein-coding genes.[10] I will argue in chapter 6 that most of the intron sequences are junk DNA.

There are many noncoding genes and often there are multiple copies of each of these genes in our genome. If you add them all up, you get about 5000 more genes, for a total of 25,000 genes in the human genome. As I described in the first few chapters, this is close to the total number predicted back in the late 1960s and early 1970s by the experts who studied genomes, genetics, and evolution. The human genome project confirmed those predictions, so those experts were right.

The genes for tRNAs account for less than 0.1 percent of the human genome, and the genes for all the other small RNAs make up less than 0.1 percent of the genome. There are about 300 copies of each of the ribosomal RNA genes scattered over several chromosomes in five clusters of about 60 genes each, accounting for about 0.4 percent of the genome.[11] The total for all these well-characterized non-protein-coding genes is no more than 0.6 percent of your genome.

The main controversy over the number of genes is over how to count those parts of the genome that are transcribed to produce RNA (potential genes) but where there's no known function for those RNAs. The latest estimate from the Ensembl website (December 2022) lists an additional 26,000 such "genes," but note that the Ensembl annotators use a different definition of a gene than the one I'm using. They don't really care if the RNA product has a proven function or not, so they include many regions of DNA that may not be real genes. That's not going to work because the correct definition of a gene requires that it produce a *functional* product; otherwise, it's not a gene.

For now, let's assume there are about 5000 noncoding RNA genes in total. The functional parts of these extra noncoding genes might only cover about 0.2 percent more of the genome for a total of 0.8 percent devoted to all types of functional noncoding regions. However, many of the noncoding genes have introns, so they may cover about 6.8 percent of the genome if they contain lots of large introns. (This is a generous estimate.) We'll discuss whether all these noncoding RNA genes are real genes or not in the next few chapters, but for now you need to keep in mind that this controversy won't have a serious impact on the amount of functional DNA in our genome.

With respect to known genes, the totals are as follows:

Protein-Coding Genes
1.0% functional coding regions
37% introns, mostly junk DNA

Noncoding Genes
0.6% genes for well-characterized functional RNAs
0.2% genes for additional functional RNAs
6% introns (junk) in all noncoding genes

The total amount of the genome devoted to genes is close to 45 percent. Of this total, less than 2 percent is functional, and the rest is junk DNA in introns. What else is in your genome in addition to genes? The following sections briefly cover the other parts of the genome, including functional regions that are *not* genes.

Pseudogenes

Pseudogenes are sequences that look like functional genes, but they aren't functional because they have acquired inactivating mutations. They are broken genes. The latest Ensembl counts of pseudogenes fluctuate around 15,000, but this count only includes pseudogenes that once encoded proteins because they are the easiest ones to identify. There are probably thousands of other pseudogenes derived from noncoding genes. Broken genes are junk, but even if you add up all the pseudogenes, they account for only 5 percent of the junk in our genome. (Many of the pseudogenes have remnants of large introns, which is why they take up so much space.) We will see in the next chapter that a tiny number of pseudogenes have secondarily acquired an additional function, but this doesn't affect the big picture.

Regulatory sequences

We learned in chapter 1 that every gene is associated with a certain amount of extra DNA that controls its expression. These regulatory sequences include the promoter, where transcription begins, as well as the DNA sequences that bind various transcription factors controlling when and where the promoter becomes active. Regulatory sequences are not new – they've been in the textbooks since the 1960s thanks to the work of Jacques Monod and François Jacob on the genes for lactose metabolism in *E. coli* (Nobel Prize, 1965). (This is the famous *lac* operon where an "operon" is a cluster of genes.)

The amount of DNA devoted to regulation is well known in bacteria. It accounts for about 20 to 30 bp of DNA sequence per gene. The amount in well-studied eukaryotes, such as yeast, nematodes, and fruit flies, is about 200 bp or enough for binding several proteins that each recognize binding sites of 6 bp each, plus some extra DNA for spacing. If human genes are similar, then this extrapolates to 200 bp × 25,000 genes, or 5 million bp, of functional regulatory sequence, which is less than 0.2 percent of your genome.

There's a raging controversy about the total amount of your genome devoted to regulation, but we need to focus on the big picture for now. Even the most extreme claims don't envisage more than 400 regulatory binding sites per gene, so if a typical binding site is 6 bp, then this is 400 × 6 = 2,400 bp per gene, and it still won't account for more than 1.8 percent of your genome.

We discuss this controversy in chapter 10.

Origins of replication

Your entire genome must be copied every time your cells divide. DNA replication begins when a DNA replication complex assembles at a specific site called an *origin of replication*, and we've known about them for more than fifty years.

This is one class of functional DNA sequence that's noncoding and not associated with genes. The size of each origin is hotly debated, but we can make a reasonable estimate based on the amount of sequence required for the actual binding of DNA replication proteins (about 30 bp) and the amount required for the overall organization of the sites (about 70 bp) for a total of 100 bp per origin.

There may be as many as 100,000 origins of replication in the human genome, so the maximum amount of functional DNA needed for DNA replication origins is about 10 million bp, or 0.3 percent of your genome. That's a significant fraction, but it's very unlikely that all the putative origins are functional; instead, it's more likely that only 30,000 to 50,000 are real origins and that the others are either old, defective origins that no longer work or misidentified origins. There appear to be many more origins than necessary for DNA replication, giving rise to the "Jesuit model" for functional origins – "Many are called, but few are chosen." The best estimate is that origins of DNA replication account for less than 0.3 percent of your genome.[12]

Centromeres

There are 24 different centromeres in humans and each one consists of millions of base pairs consisting mostly of repetitive DNA (chapter 3). Centromeres make up about 6 percent of our genome, and some of that sequence is functional, although some of it can be deleted without doing any serious harm. We know that a lot of centromere sequences aren't essential because the amount of centromeric DNA in various individuals ranges from a low of 2 percent of the genome to as much as 10 percent.[13] Let's assume that the truly essential fraction of centromere sequences occupies about 1 percent of the genome and the rest is redundant. This is a pretty good estimate.

Telomeres

Telomeres consist mostly of repetitive DNA (chapter 3). They are essential, functional, DNA sequences that are required for proper DNA replication because the DNA replication complexes need help in replicating to the very ends of chromosomes. There are two telomeres per chromosome – one on each end. About 0.1 percent of the genome is devoted to these important structures.

Scaffold attachment regions

DNA binds to histones to make chromatin, and these chromatin fibers are organized into large loops that are attached to a protein-RNA scaffold. Particular DNA sequences called *scaffold attachment regions*, or SARs, are required to maintain this organization. Chromatin loops were first observed more than 100 years ago, but SARs were only identified by Uli Laemmli in the early 1980s.[14] There are about 100,000 loops in the human genome (chapter 1), and each one needs about 100 bp of SAR DNA for a total of 10 million bp. This is about 0.3 percent of your genome.

These sites are also called loop anchor sequences, and the loops are sometimes called topologically associated domains (TADs), although that term also refers to clusters of loops.[15]

The organization of chromosome loops is an active area of investigation, and it's not clear whether all the specific sites like SARs are essential or sufficient. Nevertheless, it seems reasonable to conclude that only a small fraction of the genome contains functional chromatin organization sites.

Viruses

Defective viruses make up about 9 percent of our genome whereas functional dormant virus DNA accounts for less than 0.1 percent (chapter 3).

Transposons

As I explained in chapter 3, the total amount of DNA containing bits and pieces of ancient transposons has not been nailed down precisely, but it's certainly a substantial part of the entire genome. Using fairly stringent criteria, the analysis of the first complete genome sequence revealed 13 percent SINEs, 21 percent LINEs, 9 percent LTRs, 0.15 percent intact retrotransposons, and 4 percent

DNA transposons for a total of only 47 percent.[16] This is a bit lower than other estimates of transposon-related sequences using less stringent criteria, but I'm going to assume that it is a more accurate measurement.

Some of the transposon and virus junk DNA is located within introns, but since I've already counted introns as junk and I don't want to count this DNA twice, I will assume that the roughly 55 percent is split 25:30 between introns and the rest of the genome, meaning that 30 percent of the genome outside of introns is occupied by fragments of defective transposons and viruses. With few exceptions, all these sequences are junk according to the evidence that I described in chapter 3.

Mitochondrial DNA

Mitochondrial DNA accounts for a tiny percentage of the entire genome (<0.01 percent). All of it is junk DNA because mitochondrial DNA is not supposed to be in the nuclear genome, and once inserted, it rapidly acquires mutations (chapter 3).

How much of our genome is functional?

If you add up all the sequences that have a known biological function, you get to about 4 percent of the entire genome (Table 5.1). I don't think anyone is going to quibble about those sequences that I've identified as functional, so it's safe to say that at least 4 percent of the genome has a known biological function. The real value is probably closer to 8 percent based on sequence conservation.

Recall that the protein-coding part of the genome accounts for only 1.0 percent of the total, so *noncoding* DNA accounts for much more functional DNA than *coding* DNA.

The data in Table 5.1 suggest that about 89 percent of the genome is junk, but keep in mind that the absolute values of each category are somewhat deceptive because the categories overlap. For example,

TABLE 5.1 The amount of functional DNA and junk DNA in the human genome.

	Functional	Junk
	~1.5% genes	~5% pseudogenes
	<0.2% regulatory sequences	43% introns (including 25% transposons)
	0.3% SARs	30% additional defective transposons
	<0.1% virus	9% defective viruses
	~0.1% active transposons	<0.01% mitochondrial DNA
	<0.3% origins of replication	2% extra repetitive DNA
	~1% centromeres	
	~0.1% telomeres	
Totals	**<4%**	**~89%**

many transposon and viral fragments are found in intron sequences or in pseudogenes, so they may be counted twice. We'll try to sort out the details later on, but for now, it looks like almost 90 percent of our genome is junk by any reasonable definition.

What about the rest of the genome – the missing 7 percent from Table 5.1? There are two main possibilities: (1) it's mostly DNA whose function remains to be discovered, or (2) it's junk DNA with no function. The rest of this book is about which of those possibilities is most likely. I'm certain that the values in the table are going to change from time to time as more information is published, but the big picture is not going to be altered in any serious way. You can keep tabs on the updates by checking my blog (*What's in Your Genome?*) where you can also ask questions and discuss the data and the conclusions. For now, let's assume that 10 percent of our genome is functional and 90 percent is junk.

About 90% of your genome is junk!

What is junk DNA?

I explained in the last chapter that the best operational definition of functional DNA is any stretch of DNA that *cannot* be deleted from the genome without reducing the fitness of the individual.

This is another way of saying that functional DNA is constrained by purifying selection. From this definition, it follows that junk DNA is any stretch of DNA that *can* be deleted without reducing fitness.

From the very beginning, the debate was always about the *amount* of nonfunctional DNA in various genomes, not about the *existence* of nonfunctional DNA. Nobody doubts there is nonfunctional DNA in every genome (i.e., pseudogenes) – the key question is whether or not the *majority* of DNA in large genomes is junk. Proponents of junk DNA claim that 80 to 90 percent of the human genome is junk, while opponents of junk DNA restrict the amount of junk DNA to much less than that. Proponents of junk DNA are comfortable with the idea that large genomes full of junk could have evolved by chance, while opponents of junk DNA believe that large genomes demand an adaptive explanation and that junk should have been eliminated by natural selection because it is harmful.

It's easy to be fooled by misleading press releases and articles written by science writers who haven't done their homework, leading to a situation in which the dominant themes in the popular press these days are not true representations of the state of scientific knowledge. Most people think we know very little about the human genome, and they believe that the evidence will eventually reveal a genome full of sophisticated software orchestrating an exquisitely fine-tuned control over our genes. They think our genome is well designed, although they may differ on whether the designer is God or natural selection.

The truth, in my opinion, is very different. Our genome is a messy product of evolution, and most of evolution has nothing to do with natural selection. Our genome is full of junk DNA that does absolutely nothing but just sit there wasting space, and even the functional part is not finely tuned but prone to mistakes and errors. This picture is very different from the one painted by the popular press.

Why is there such a discrepancy between what knowledgeable scientists know and the message conveyed to the public? There are many reasons – some of which I have already discussed – but now we must examine the results of experiments carried out after the completion of the human genome project. As I indicated in the previous chapter, we entered a new era in 2001, when all the previous work on genome organization was ignored and the data were reinterpreted in light of the human genome sequence.

CHAPTER 6

How Many Genes? How Many Proteins?

Therefore, in this book I will consider genes as DNA sequences encoding information for functional products, be it proteins or RNA molecules. With "encoding information," I mean that the DNA sequence is used as a template for the production of an RNA molecule or a protein that performs some function.

Kostas Kampourakis (2017)[1]

Let's talk some more about genes. Everybody wants to know how many genes we have, so I gave you a "big-picture" answer in chapter 5 – I think we have about 25,000 genes. About 20,000 of these genes are protein-coding genes, and approximately 5,000 are genes for noncoding RNAs. That's my personal, hopefully informed, opinion based on all the evidence I've seen so far, but for another opinion, you can look at one of the main websites for information on the human genome (Ensembl), where it says there are more than 45,000 genes.[2] That's a significant difference, but the number on the Ensembl website is dropping and will probably be lower by the time you read this. Much of this chapter is about why there's so much disagreement over the number of genes and why many scientists are upset that we appear to have so few genes. I'll also cover the data on how many proteins we have and how it relates to the number of genes.

You might think that publication of the human genome sequence would have answered the question once and for all, but this didn't happen for a variety of reasons. Some of the reasons are technical, and some of them are more theoretical. The technical reasons are complex, but they boil down to the fact that it's difficult for computers to identify genes by scanning 3 billion base pairs of DNA. As I explained earlier, protein-coding genes are relatively easy to identify, and genome annotators have concluded that there are about 20,000 of these genes. The number of genes specifying functional noncoding RNAs is still up in the air because it's much harder to identify those genes.

The theoretical reasons concern the definition of a gene. You can't really count the number of genes if you don't know what a gene is. I believe this problem was solved a long time ago but, unfortunately, there's still a lot of controversy, so we need to look at how to define a gene.

Defining a gene

In chapter 1 I defined a gene as a DNA sequence that is transcribed to produce a functional product, and this is the way genes have been described since the 1960s by many scientists who have thought seriously about the subject. There are lots of textbooks that use a similar definition.[3] Kostas Kampourakis in *Making Sense of Genes* concludes that the most acceptable definition of a gene is the one that includes the production of both proteins and noncoding RNAs (see the quotation at the beginning of this chapter).

The idea here is that the primary gene product is RNA and that any segment of DNA that's copied into RNA (transcription) qualifies as a gene as long as the RNA has a *biological function*. That last bit is important because there are sequences of DNA that are transcribed but the RNA doesn't do anything. Transcribed pseudogenes are an example, but so are many spurious or accidental transcripts that happen quite frequently when the transcription machinery makes a mistake. Those bits of DNA are not genes even though they are transcribed.

Very few definitions in biology are absolute. In fact, one of the main reasons why biology is so difficult is because it is full of exceptions to the rule. In the case of genes, the exceptions are well known; they include the fact that some genes are made of RNA, not DNA – the genes of coronavirus are an example. There are also more complicated exceptions, such as cases in which several different protein-coding regions are transcribed together to produce a long RNA where several different regions can be translated simultaneously. These complexes are called *operons*, and they are common in bacteria. The problem here is that we will often want to refer to each protein-coding region as a separate gene. There are also rare examples of two separate DNA sequences that are transcribed to produce RNAs that are then joined together to make a functional product; strictly speaking, neither of the separated DNA sequences qualifies as a gene.

What this means is that the best definition of a gene is only a "working definition" because there is no definition that covers all cases. This vagueness drives philosophers crazy, but biologists are used to such ambiguity.

The molecular gene and the Mendelian gene

The gene I'm talking about is the "molecular gene" (a gene that corresponds to a particular DNA sequence in the genome), but geneticists often talk about another kind of gene called the "Mendelian gene." Both definitions are used in the scientific literature, and they do not overlap.[4] I will use the molecular definition in this book, but keep in mind that there are lots of functional DNA sequences that are not genes by this definition.

The Mendelian gene concept is frequently used by geneticists and evolutionary biologists. It's a nebulous, ill-defined unit as seen in the definition used by Richard Dawkins in his book *The Selfish Gene*. To him, a selfish gene is any bit of chromosome that's subject to natural selection. This would include centromeres, origins of replication, and regulatory sequences that are not genes by the molecular definition. When Dawkins talks about the "selfish gene," he's talking about a unit of selection and not the molecular gene. The "selfish gene" is just a convenient metaphor, as he makes abundantly clear in his book: "To be strict, this book should be called not 'The Selfish Cistron' nor 'The Selfish Chromosome,' but 'The slightly selfish big bits of chromosome and the even more selfish little bit of chromosome.' To say the least this is not a catchy title so, defining a gene as a little bit of chromosome which potentially lasts for many generation, I call this book 'The Selfish Gene.'"[5]

The Selfish Gene was voted the most inspiring science book of all time in a 2017 poll by the Royal Society of London.[6] This is a problem because the book only deals with the Mendelian gene, so the general public (and many scientists) are not familiar with the molecular definition, and the distinction is crucial. It's also a problem because the concept of the selfish gene is highly adaptationist – it's all about natural selection – and you will never understand genomes if that's all you know about evolution.[7]

In the debate over the definition of a molecular gene, I favor a definition that focuses on RNA as the functional product because I think this definition captures the essence of what most scientists believe when they talk about genes as functional entities. This includes scientists who may slip up from time to time by restricting their gene talk to protein-coding genes. When you ask them in private, it turns out they are well aware of ribosomal RNA genes and a host of other noncoding genes. Sometimes scientists just forget to be precise when talking about genes and they give the impression that there's only one kind of gene, namely, the ones that encode proteins.

Counting genes

When a draft genome sequence is first published, the number of genes is estimated using software that looks for coding regions and for the standard noncoding RNA genes. As I described earlier, these programs are deliberately set up to reduce the chance of missing any genes, meaning that they will probably identify many false positives – sequences that resemble genes but aren't functional genes by any reasonable definition.

Careful examination by professional annotators usually eliminates some of these false positives, and this is why the number of predicted genes always drops when annotators get working on the draft sequence; for example, initial estimates of the number of protein-coding genes in the human genome were on the order of 30,000 to 35,000, but the latest estimates are less than 20,000, and the difference is due mostly to annotators who looked closely at the predictions and decided that many of them weren't actually bona fide genes. Many of the predicted genes were eliminated because they don't look like real genes when they are examined more closely. In other cases, the sequences of additional genomes from other individuals don't indicate the presence of a gene, suggesting that there was a mistake in the original sequence data. In many cases, the putative genes were very small, which is consistent with accidental open reading frames that don't represent true genes.

It's time to review the evidence for the current number of real protein-coding genes in the human genome. Many protein-coding genes are uncontroversial, and we can say with confidence that there are more than 17,000 such well-characterized genes producing known proteins. All these genes are found in other mammals, and most are present in more distantly related species like birds and fish and even bacteria.

We are less certain about the other 3000 potential genes. Some of them are simply small open reading frames that have start codons and stop codons. Are these genuine functional genes, or are they just chance occurrences of an open reading frame?

If a candidate gene also exists in other species, then that's a pretty good indication that it's a real functional gene. Conversely, if the candidate gene doesn't exist in other species – especially in closely related species like chimpanzees and gorillas – then it might not be a real gene. A lack of conservation doesn't rule out function entirely because there's always the possibility that a new gene has arisen in the human lineage, but although this is a theoretical possibility there are very few well-established examples of such recently evolved genes. Thus, sequence conservation is a very good way to identify real genes.

Scientists have looked closely at all the potential genes predicted by computer programs to see if they are conserved in other species and many of them are conserved – especially protein-coding genes that produce large proteins.[8] Thus, in addition to the 17,000 well-characterized genes, we can add about 2000 additional genes just because they are also present in other species. We often don't know the function of those genes, but they certainly have all the characteristics of real genes.

The latest version of the Consensus Coding Sequence Database (CCDS)[9] has 19,030 protein-coding genes. About 14,000 potential coding regions from the initial draft sequence have been rejected for various reasons. This estimate is very conservative and probably is too low. The three most popular genome annotations (RefSeq, GENCODE, UniProtKB/SwissProt) all agree on a larger core of 19,500 protein-coding genes.[10]

Many candidate genes with short open reading frames don't pass the sequence conservation test. Often when you look closely at the same position in the genomes of chimpanzees or other apes you see a very similar sequence except that it contains a stop codon or a small insertion or deletion. There is no open reading frame in these related species, but it's obvious that a change in one or two base pairs could create a spurious open reading frame by accident. It looks very much like the candidate gene is a fluke, and the computer programs came up with a false positive.

In order to test the gene prediction programs, a group in Seattle (Washington, USA), created a realistic artificial DNA sequence that had all the characteristics of the human genome but no genes. They tested all the gene prediction programs, and all of them came up with gene predictions when they analyzed this artificial genome.[11] This result was not a surprise since the programs are designed to pick out anything that might possibly be a gene. The study serves to remind us that gene prediction programs give us "predictions," and not real genes.

Many of the non-conserved predictions are simply *o*pen *r*eading *f*rames (ORFs), and they are called ORFans ("orphans") because if they are real genes, then they are restricted to only one species, namely, *Homo sapiens*. Some people make a big deal out of ORFan genes because they think they represent real genes that are only found in humans, and this finding confirms their belief that evolution must have favored new genes that make us human. However, it's important to keep in mind that most so-called ORFan genes are *putative* genes, not *bona fide* genes, and every new genome sequence has hundreds or thousands of ORFans when the sequence is first published. This number falls precipitously as the annotators get to work until, in the end, there are hardly any.[12]

Counting proteins

As I mentioned above, there are about 17,000 protein-coding genes with unimpeachable credentials, but there is considerable controversy over several thousand potential genes that look like real genes but where there's no solid evidence that they encode a functional protein. Are they real genes or not? Several groups have attempted to answer that question by looking for all the proteins made in human cells. The idea is to confirm the existence of potential genes by identifying the protein product using a sensitive technique called mass spectrometry. One study found only 11,840 proteins, but other studies have detected 19,610 proteins, 18,097 proteins, and 17,294 proteins.[13] The Human Proteome Project tries to sort out all the data to create a high-confidence database of human proteins. The 2022 version has 18,407 entries, and they estimate that there are probably an additional 1343 unconfirmed genes for a total of 19,750 genes.[14]

What do we know about the 1343 genes that are missing a protein? There are technical reasons why some proteins can't be detected, but there are also biological reasons; for example, the protein might be expressed in a tissue that wasn't examined.[15] However, it's also possible that some of the predictions aren't real genes so they never make a protein.

It's not easy to keep up with the latest studies in this field, and it takes a lot of time for annotators to review all the published papers and make changes to the standard human genome reference sequence. At this point, there probably aren't more than a handful of undiscovered protein-coding genes in the human genome. It's clear that scientists have reached the point where further improvements to the protein-coding gene count are going to be minor, so we can be confident that the total number of protein-coding genes is going to be somewhere between 19,000 and 20,000 genes – probably closer to 20,000. I'll often use 20,000 because it's a nice round number. Don't believe anyone who gives you an exact number like 20,807, because the data just isn't good enough to support such a specific claim.

The functions of protein-coding genes

What are all these protein-coding genes doing? According to the Human Protein Atlas, about 10,000 of them are expressed in most cell types, and 7500 are expressed in all cells. These are the well-characterized genes present in all other mammals, and they include "housekeeping" genes that encode all the normal metabolic enzymes plus all the genes necessary for transcription, translation, DNA replication, and gene regulation.[16] The other genes code for proteins that may not be required in most cells but may be essential in some specialized cells; an example might be hemoglobin in red blood cells. There are also hundreds of genes that control and regulate gene expression, including those that control the growth and development of a fertilized egg into a fully formed adult.

The set of all proteins in a cell is called the *proteome*. Our proteome of 20,000 or so proteins may seem like a lot, but the combined proteome of all species is very much larger. So far, the universal proteome contains more than 800,000 different proteins, and as is the case with the human proteome, a large fraction of these proteins are involved in basic metabolism. More than one-third of these proteins have no known function.[17]

Eukaryotes are complex organisms with lots of genes, but many scientists are interested in the minimum number of genes required to sustain the first free-living cell. Presumably, this subset of genes represents the earliest genes in the history of life. The smallest modern genomes are those of various strains of *Mycoplasma* with about 500 genes, and if you look closely at just the highly conserved genes shared by the simplest species, it looks like the minimal set is about 250 genes. However, this estimate is known to be too low since there are clearly less well-conserved genes that are essential in modern species. Recent work has concentrated on creating a synthetic organism by taking just the truly essential genes from simple bacteria and putting them on an artificial chromosome. The latest results produced a viable bacterium with a genome of only 531,000 bp (531 kb) and 473 genes.[18] The cells can grow and divide every 3 hours. The most amazing part of this work is that 149 of those essential genes have unknown biological functions! Nobody knows what these mystery genes are doing.

Historical estimates of the number of genes

Recall that estimates of about 30,000 genes were derived from genetic load arguments dating back to the 1940s (see chapter 4). This number of genes was also consistent with counts of mRNAs in the 1970s based on the data from experiments that looked at the total number of different mRNAs expressed in a wide variety of tissues.

This is exactly the estimate published by Susumu Ohno in 1972 in the famous paper that kicked off the junk DNA debate ("So Much 'Junk' DNA in Our Genome").[19] The combination of those two powerful estimates led to the widespread belief that humans must have about 30,000 genes and that most of our genome is junk – a conclusion that was reported in many textbooks throughout the 1980s.

For some inexplicable reason, most of this information was ignored when the U.S. Human Genome Project started; for example, in their 1990 report on the U.S. Human Genome Project the National Institutes of Health (USA) and the Department of Energy (USA) guessed that there were going to be about 100,000 genes, but this guesstimate was not based on real data – it was a crude back-of-the-envelope calculation by Walter Gilbert, who assumed that each gene was 30,000 bp long and that the genome was full of genes.[20] Both of these assumptions are incorrect, and furthermore, the scientists who had actually studied the problem knew they were wrong in 1990. Unfortunately, these experts were not consulted, and this incorrect guesstimate stuck in the minds of many scientists and science writers who assumed that it was an informed opinion because it was in an official report of the U.S. Human Genome Project.

The 1990s saw a different attempt to determine the number of genes. The idea was to isolate all the mRNAs from a wide variety of tissues and convert them into complementary DNA (cDNA) using reverse transcriptase to copy RNA into DNA. The resulting bits of DNA, called *expressed sequence tags* or ESTs, are then cloned and sequenced. By 1996 there were 600,000 ESTs in the GenBank database, and many scientists thought you could avoid sequencing the entire genome – and all that junk DNA – by concentrating on the protein-coding genes, where all the function was presumably located. That's why so much effort was expended on cloning and sequencing ESTs.

Individual ESTs are quite small, but they can be clustered into longer stretches that presumably represent a complete functional messenger RNA molecule. The technique is similar to that used to create a genome sequence from smaller pieces of DNA sequence (Figure 2.1).

Workers compiled a database of assembled clusters of ESTs called UniGene, and it was supposed to contain one copy of every protein-coding gene sequence. There were 49,625 putative genes in 1996.[21]

There are several problems with this approach. The most serious problem was that only 4563 of these clusters matched known genes (9 percent in 1996). This may not seem like a problem if you believe the human genome must contain vast numbers of previously unknown genes that had never been detected in any other species, but it was a problem if you believed that most of these ESTs represented spurious transcripts that had nothing to do with mRNA. Many scientists discounted the EST data for this reason – they didn't believe there could be more than 30,000 genes, so there must be something wrong with the UniGene data.[22]

The skeptics were right because most of those EST clusters turned out to be artifacts of some form or other. They do not contain open reading frames, although they are supposed to be derived from mRNA. They clearly are not copies of genuine mRNAs. The latest version of UniGene dates from 2012, and it contains 221,081 "mRNAs" – a ridiculous number.[23] Nobody believes that there are that many genes.

By the end of the twentieth century, the knowledgeable experts were still predicting about 30,000 genes based on 30 years of data, and the available sequence information at that time was in line with this prediction.

In June 2000, *Nature* published three papers on the number of genes in the human genome. One of them, from Craig Venter's The Institute for Genomic Research, speculated that there were 120,000 genes based on EST data. As I explained earlier, there was good reason to doubt that estimate. The other two papers presented solid evidence to back up predictions of 35,000 genes and of about 30,000 genes.[24] When you take into account the 1970s' data on the number of mRNAs and the 1960s' data on genetic load – both of which predicted about 30,000 genes – you can see that a great many experts weren't at all surprised when the actual number of genes turned out to be very close to those values.[25]

Confusion about the number of genes

What were the people working on the human genome sequencing project predicting? Most of these workers were technical types working on specific problems with sequencing, assembly, and data analysis. They weren't necessarily experts in biology or evolution, and they weren't experts on the history of gene number predictions. Ewan Birney organized a competition among this group to guess the number of protein-coding genes in the human genome and the results of "Gene Sweepstake 2000" were reported in *Nature* – a year before the draft sequences were published.[26] The gene sweepstake attracted lots of attention because it suggested that biologists were expecting many more genes than were actually found. In the year 2000, the estimates from these workers ranged from about 27,000 to more than 150,000. The average guess was 62,000 genes, but most were expecting fewer than 60,000 genes. Only a handful of guesses were above 100,000.

Bets continued after the draft sequence was published, and subsequent guesses were much lower than the numbers in 2000. (It cost US$1 to place a bet in 2000, $5 in 2001, and $20 in 2002.) The eventual winners included Lee Rowen of the Institute for Systems Biology in Seattle (Washington, USA), who bet that there were 25,947 genes in 2001 and won half of the $1200 pool. The other half was split between Paul Dear (Medical Research Council, UK), who guessed 27,462 genes in 2000, and Olivier Jaillon (Genoscope, Evry, France), who guessed 26,500 genes in 2002.[27]

The standard mythology in the popular press is that scientists were estimating 100,000 genes in 2000, and they were shocked when the actual number turned out to be a lot less. The facts are quite different, as we have just seen.

The Deflated Ego Problem

I am ... an advocate of the position that science is not an objective, truth-directed machine, but a quintessentially human activity, affected by passions, hopes and cultural biases. Cultural traditions of thought strongly influence scientific theories, often directing lines of speculation, especially ... when virtually no data exist to constrain either imagination or prejudice.

Stephen Jay Gould (1980)

The problem here is not the false history about estimating the number of genes. The real problem is why *some* people were genuinely surprised when the human genome sequence was published. As we have seen, the knowledgeable experts turned out to be correct, but there were many scientists who had not kept up with the literature and were genuinely "shocked" to learn that humans had only 30,000 genes (since revised downward to 25,000). Some of them still don't believe it, and there must be a reason for their skepticism – why don't they want to believe that humans have so few genes despite all the evidence?

In 2005 the journal *Science* published a list of important things we don't know, and science writer Elizabeth Pennisi claimed that one of the great mysteries was "Why Do Humans Have So Few Genes?" Naturally, she begins with the standard mythology: "When leading biologists were unraveling the sequence of the human genome in the late 1990s, they ran a pool on the number of genes contained in the 3 billion base pairs that make up our DNA. Few bets came close. The conventional wisdom a decade or so ago was that we need about 100,000 genes to carry out the myriad cellular processes that keep us functioning. But it turns out we have only 25,000 genes – about the same number as a tiny flowering plant called *Arabidopsis* and barely more than the worm *Caenorhabditis elegans*."[28]

Pennisi is echoing a popular view among scientists, namely, that we should be a lot more complex at the molecular level than a flowering plant or a nematode worm. Many scientists believe that species complexity should be correlated with the number of genes, so humans should have a lot more genes than a worm. It was a surprise for them to learn that we don't have more genes. I call this the *Deflated Ego Problem* for obvious reasons.

Pennisi and others should not have been surprised. The reason we have about the same number of genes as other animals is mostly because all animals need the same basic genes for metabolism and building cells – these are the 10,000 or so standard housekeeping genes. A much smaller subset of common genes is required for controlling and regulating other functions, such as cell division. Humans have genes for things like making bone and hair and other structures not found in worms, but there aren't as many of these genes as you might expect. Besides, worms also have genes that are unique to their class, so the numbers balance out to some extent. Many of the obvious differences in complexity are not due to extra genes but to when genes are expressed during development; for example, humans and whales have pretty much the same genes but their genes are regulated differently during development to give very distinct bodies. I discuss the evidence more thoroughly in chapter 10.

Scientists who disagree with what I just wrote will continue to believe that the low number of genes in humans is a problem requiring an explanation. They have created several narratives to explain how we can be so complex with "only" 25,000 genes. Almost all of these rationales involve attributing function to the excess DNA that I refer to as junk DNA, and this is why there's an ongoing debate over the amount of our genome that's functional. One of these attempts to solve the Deflated Ego Problem involves introns, but in order to understand it, we must review our knowledge of eukaryotic gene organization.

Introns and the size of genes

We have seen that most eukaryotic genes consist of some combination of exons and introns and that the primary product of transcription is a large precursor RNA that's subsequently processed by cutting out the intron sequences to make a mature RNA. The common term for this is *splicing*.

Recall that splicing is catalyzed by a large enzyme complex called a *spliceosome*. The spliceosome brings together the two ends of an

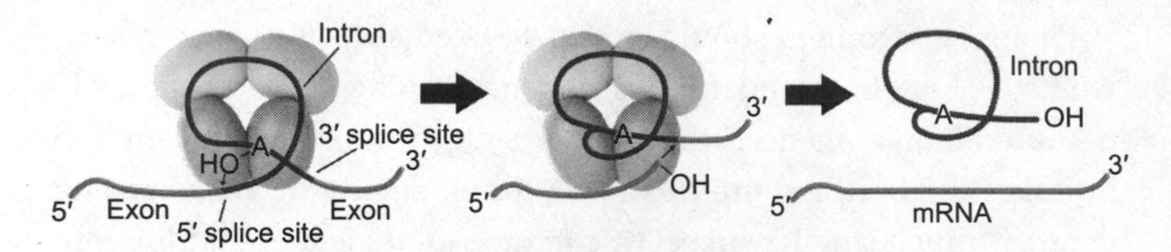

FIGURE 6.1 The splicing reaction. (Left) The spliceosome (light gray) brings together the three signal sequences in introns. The branch site with its adenosine (A) residue is positioned near the 5′ splice site. The spliceosome catalyzes a reaction between adenosine (A) at the branch site and the bond joining the nucleotides at the splice site, resulting in cleavage of the precursor RNA between the exon (gray) and the intron (black). The 5′ end of the intron is then joined to the branch site to form a circular RNA. (Middle) The spliceosome complex then positions the end of the left exon at the 3′ splice site and catalyzes a joining reaction between this end and the right-hand exon. (Right) The mature RNA with exons joined is released along with the excised intron in the form of a lariat structure.

intron to form a large loop that is then cut out of the precursor RNA and discarded as shown in Figure 6.1. The chemical reaction involves joining the tail end of one exon to the front end of the next exon, and this joining reaction is very efficient because the spliceosome binds to the precursor RNA and brings the two ends of the exons close to each other in the active site of the enzyme.

How does the spliceosome know where to cut? You can probably guess the correct answer since there aren't many choices. The cut occurs at specific junction sequences called *splice sites* located at each end of the intron; the one at the beginning of the intron is called the *5′ splice site*, and the one at the end of the intron is called the *3′ splice site*. There's also a *branch site* containing an adenosine (A) residue located in the middle of the intron.

Both protein-coding genes and noncoding genes can have introns. The average protein-coding gene in humans has 9 exons and 10 introns. The number of introns ranges from zero to more than 30, but the vast majority of protein-coding genes have fewer than 10. In contrast, those genes that produce noncoding RNAs often don't have introns, and those that do usually have only one intron (two exons).[29]

If the average number of introns in a human protein-coding gene is 10 and the average length of introns is 6 kb, then introns account for 60,000 bp in a gene.[30] The average exon is only 160 bp, and since there are 11 of these in a gene, this takes up 1760 bp. That's enough to code for 585 amino acids and a protein with a molecular weight of about 70,000.

If you add together the exons and introns, you get 61,760 base pairs, and we'll assume that this is the size of the average protein-coding gene (transcribed region).[31]

Let's assume there are roughly 19,500 of these protein-coding genes, so using these average values, they should occupy 61.7 kb × 19,500 = 1.20 Gb, or 38 percent of the genome (see chapter 5). The total amount of coding region is 1760 bp × 19,500 = 34 Mb, or 1.0 percent of the genome. When you add in 5000 noncoding genes, the total amount of the genome devoted to genes comes to about 45 percent.[32]

This may seem like a lot – I know I was a bit skeptical when I did the calculations – but it seems to be roughly correct, and it's consistent with the data from other mammals where genes seem to occupy about 50 percent of the genome.[33] We can quibble a bit about the exact number, but if you look at different sections of the human genome using the University of California at Santa Cruz Genome Browser, then it's obvious that genes take up about half of the sequence.[34] This is a fact that's not often mentioned in the popular press or even in the scientific literature. It's much more common to hear that genes only take up 1 or 2 percent of our genome because writers are assuming that genes only consist of coding regions.

Introns are mostly junk

I concluded that introns were mostly junk when I was describing the big picture in the last chapter. There are several reasons for thinking this way. First, intron sequences are not conserved, and the lengths of introns are not conserved. Second, homologous genes in different species can have different numbers of introns, and homologous bacterial genes get along quite nicely without introns. Third, researchers routinely construct intronless versions of eukaryotic genes, and they function normally when reinserted into the genome. Fourth, intron sequences are often littered with transposon and viral sequences that have inserted into the intron, and that's not consistent with the idea that intron sequences are important. Thus, we can conclude with confidence that neither the sequence nor the size of introns is under selection except for the splice sites and a minimum length required to form the loop during splicing. That minimum length is about 30 bp in most eukaryotes, and the splice sites account for about 20 bp – all the rest seems to be junk.[35]

Not everyone agrees that introns are junk. Some scientists argue that introns may be a subtle way of influencing gene expression by lengthening the time it takes to transcribe a gene, but let's think about that for a minute. The rate of transcription is about 50 nucleotides per second, so transcribing an intronless gene of 2000 bp would take 40 seconds. The same gene with introns might be 60,000 bp long and would take 1200 seconds (more than 20 minutes) to transcribe. While it's conceivable that there might be situations in which such a delay between transcription initiation and protein synthesis would be beneficial, I don't know of any proven examples. In any case, it's unlikely that this would apply to more than a handful of genes. There's no evidence that long introns are being selected, and besides, there are much easier ways to evolve mechanisms for controlling gene expression, so this does not seem to be a good argument for introns.

By way of contrast, we know that there *is* selection for *deleting* introns when rapid gene expression is necessary because there are several examples of genes involved in a rapid response to environmental change that have lost their introns in order to speed up expression. There is also some evidence indicating that highly expressed genes in humans have smaller introns than other genes.[36] This evidence suggests that there has been selection for removing excess junk DNA from these introns in order to speed up gene expression.

I think it's reasonable to conclude that introns are mostly junk as the evidence suggests.

YEAST LOSES ITS INTRONS

Baker's yeast (*Saccharomyces cerevisiae*) is one of the best-studied eukaryotes. Its genome is just slightly larger than the largest bacterial genome, and in 1997, it was the first eukaryotic genome to be sequenced.[37] It has about 7000 genes in total, and 6604 of these genes are protein-coding genes. Only 280 of these genes contain introns because the rest have lost their introns over the course of several hundred million years of evolution.[38]

We know that introns have been lost in yeast because the genes of related species have lots of introns, meaning that the common ancestor of all fungi undoubtedly had genes with multiple introns. This makes sense since the available evidence indicates that introns invaded eukaryotic genes very early in the evolution of eukaryotes so that the ancestors of all modern eukaryotes had lots of introns. The fact that most introns have been purged from the yeast genome suggests that introns are not essential for gene function. In other words, introns are mostly junk.

What about the remaining introns? There's an ongoing attempt to build a completely synthetic yeast genome by synthesizing artificial chromosomes, and part of the process is to eliminate all the junk DNA by cutting out all unnecessary introns. So far it looks like most of the remaining introns can be removed without any obvious effect on the viability of the cells. However, there are a few introns that are essential and others that are nonessential but affect growth under certain conditions. Some of the essential introns contain genes for small noncoding RNAs, and that's why they can't be deleted, but the role of other essential introns is still a mystery. The nonessential introns that affect growth appear to be required to regulate growth under starvation conditions.[39]

(*continued*)

YEAST LOSES ITS INTRONS (continued)

If we assume that the common ancestor of yeast and other fungi had three introns in every protein-coding gene, then there would have been about 18,000 introns in the common ancestor. Most of them (>98 percent) have been eliminated by evolution without serious consequences, but these results indicate that a few of them have secondarily acquired an important role in gene expression. The evidence is compatible with the idea that most introns are junk and incompatible with the idea that most introns have a function.

Alternative splicing: common or rare?

Alternative splicing refers to the production of different functional RNAs due to differences in the way the primary transcript is spliced. It's a mechanism whereby a single gene can be responsible for producing two different products with different functions.

Let's look at a simple example in order to grasp the concept. Figure 6.2 shows a hypothetical gene with just two introns. The normal splicing pattern is shown at the top, where three exons are joined to produce the mature mRNA. A different splicing pattern is shown at the bottom of the figure, where exon 1 is joined to exon 3 and exon 2 is skipped. In theory, this single gene could produce two different proteins by alternative splicing, and for typical genes with nine exons, the number of possible protein variants could be closer to 10 to 20. If it were true that most genes produce many different proteins by alternative splicing, then this could potentially solve the Deflated Ego Problem for those who are still upset by the fact that we have the same number of genes as worms. The idea is that the evolution of complexity in humans would be explained by making different proteins from a single gene rather than by increasing the total number of genes.

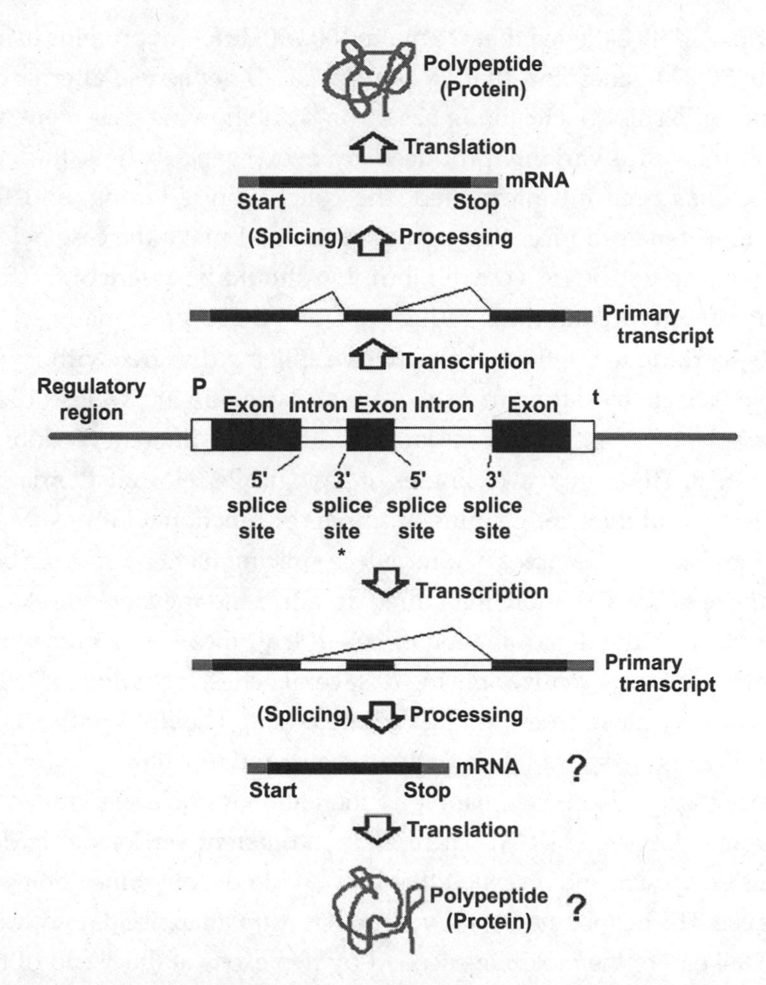

FIGURE 6.2 Exon skipping. The top part of the figure shows the normal splicing pattern following transcription of a gene with two introns. Both introns are removed by the spliceosome, which recognizes the 5′ and 3′ splice sites at the ends of each intron. The bottom part of the figure shows the splicing pattern when the first 3′ splice site (asterisk) is not recognized and splicing removes both introns plus the middle exon sequence. This pattern is called "exon skipping" since one of the exons is excluded from the final mRNA sequence. If this were an example of true alternative splicing, then both polypeptides would be functional even though the bottom one is missing some internal sequence. Conversely, if this were an example of aberrant splicing then the bottom polypeptide would not be functional and probably would never be made because the incorrect RNA would be rapidly degraded.

It is widely believed that we make 100,000 different proteins using only 20,000 genes and that 95 percent of all genes use alternative splicing.[40] This conclusion is based on data showing that there are multiple splice variants produced by every gene. I think the evidence has been misinterpreted, the conclusion is wrong, and the average gene produces just a single protein. I make the case below so you can decide for yourself, but you should be aware of the fact that my interpretation is a radical position that a great many scientists working in the field of alternative splicing disagree with.

Let's begin by stating that there are many legitimate, well-studied, examples of different splice variants, giving rise to different versions of a protein. These are real examples of biologically relevant alternative splicing, and they are genuine examples of functional introns. Most of these classic examples of alternative splicing have been described in the textbooks for more than three decades, and they include examples such as the determination of sex in fruit flies – a process that's controlled by alternative splicing of several genes, including *Sxl* (sex–lethal), *tra* (transformer), and *dsx* (double–sex). The RNA products of these genes are spliced differently in males and females.

Another well-known example is the antibody molecule known as immunoglobulin M (IgM). There are two different versions of IgM: a secreted version and a version that is bound to the outer membrane of the cell. The membrane-bound version has extra amino acid residues at the tail end of the molecule encoded by two exons at the 3′ end of the gene. These exons are not included in the mature mRNA of the secreted version, so it lacks the membrane binding signal. The secreted form of the IgM protein is produced first in antibody cells (B cells), where the precursor RNA is processed in a way that eliminates the last two exons, but as the cells mature, the processing is switched to include the last two exons.

How does alternative splicing work?

There are many other well-studied examples of true, biologically relevant, alternative splicing. The point is not whether true alternative splicing occurs; it's whether alternative splicing is rare or ubiquitous. The well-known examples show a wide variety of different strategies involving promoters, splice sites, termination sites, and reading frames. This sounds complicated, but in reality, we need to learn only a few basic principles in order to understand alternative splicing. The key to understanding the concept behind alternative splicing is to focus on the 3′ splice site of the first intron in Figure 6.2 – I marked it with an asterisk.

Not all splice sites are identical. The sequences of nucleotides at a 3′ splice site, for example, resemble a certain pattern that corresponds to the ideal splice site. This ideal sequence is known to be the best, or strongest, site for spliceosome binding. These strong binding sites will be successfully recognized and cut 99.999 percent of the time, and the intron will be removed correctly in the vast majority of cases.[41] That process may seem pretty accurate, but the splicing error rate (10^{-5}) is 10 times higher than the error rate for transcription and many orders of magnitude higher than the error rate of DNA replication. That's because the spliceosome is a very large and messy complex carrying out a very complicated reaction involving many components. Furthermore, the ideal splice site is less common than you might think because both 5′ and 3′ splice sites can deviate from the ideal sequence by one or two nucleotides and still work effectively in spliceosome assembly. They may only be recognized 99.99 percent of the time or 99.9 percent of the time or even 99 percent of the time.[42]

What happens if the 3′ splice site isn't recognized by the spliceosome? Recall that transcription begins at the 5′ end of the gene, so the first sequence to appear in the RNA precursor is the 5′ splice site. Spliceosome components bind first to the 5′ splice site as transcription proceeds, and then these components wait until a 3′ splice site appears in order to complete the assembly of the spliceosome complex. Imagine, in our hypothetical example, that the first 3′ splice site is a relatively weak binding site and the spliceosome fails to recognize it from time to time. It may skip that site 0.1 percent of the time or even 1 percent of the time. What happens in such cases is that the spliceosome continues scanning the RNA looking for the next match to the 3′ splice site, and it finds it at the end of the second intron. The splicing reaction is then completed by splicing together exons 1 and 3 and eliminating exon 2, as shown in Figure 6.2. This produces a splicing error triggered by a weak 3′ splice site. This low level of splicing mistake can easily be tolerated in cells that are churning out dozens of transcripts because one or two bad RNAs aren't going to make much difference. That's why natural selection hasn't led to perfect splice sites at every location – they only have to be good enough to do the job most of the time.

Imagine that our hypothetical gene screws up splicing about 0.1 percent of the time. If you isolate RNA from a few million cells, you will find that 99.9 percent of them are correctly spliced. About 0.1 percent of processing events will be errors, but the total number of incorrectly spliced RNAs will be much less than that in your sample because the incorrectly spliced RNAs are usually less stable – they are rapidly degraded by enzymes that clean up mistakes. There might be 100 transcripts from this gene in every cell, so if you extract RNA from 1 million cells, you will get 100 million RNA molecules. If the noncanonical mistakes represent only 0.001 percent of the total, that's still 1000 bad RNAs in your sample, and modern methods are quite capable of detecting those RNAs and putting their sequences into the databases. Thus, we end up with a database that contains the correct high-abundance mRNA sequence plus all the errors and mistakes that have been detected in hundreds of experiments over the past 30 years. The noncanonical mistakes are not functional molecules – they are junk RNA. They are not real examples of alternative splicing since the term should only refer to different *functional* RNAs produced from the same gene. These junk RNAs are just transcripts, but in the scientific literature, they are frequently counted as true alternative splicing events.[43]

In order to understand whether alternative splicing is a real phenomenon we need to distinguish between junk RNA transcripts and biologically functional alternative splicing variants. We'll see how to do this shortly, but meanwhile, let's look at how real biologically relevant alternative splicing occurs.

Our hypothetical gene produces two different spliced mRNAs, and if this were biologically relevant, then the two different proteins produced by alternative splicing would have different functions. In theory, this deliberate production of different RNAs from the same gene could be achieved by having a weak 3′ splice site at the end of the first intron as in the hypothetical example described above. However, in this case, the weak binding site would be selected by natural selection to allow for alternative splicing. In order to be biologically relevant, the amount of each transcript would have to be very much greater than the rare transcripts produced by splicing errors.

There aren't very many proven examples of this particular form of true alternative splicing. Almost all well-studied examples rely on additional proteins that regulate splicing, and there are two different types of regulation: repression and activation. Repression requires a splicing factor that binds to a splice site and blocks its utilization by the spliceosome; an example would be a repressor that binds to the 3′ splice site in Figure 6.2 – the one marked with an asterisk – causing exon 2 to be skipped. This is a way of blocking the activity of a strong splice site. Activation is when the splicing factor binds to a weak splice site, causing it to be recognized more efficiently by the spliceosome; in this case the default event is to skip the weak splice site, but when the activator factor is present, the splice site is used.

In both cases, splicing is regulated so that different RNAs are produced depending on whether the activator or repressor factors are present. There are many real examples of such regulated alternative splicing, including the classic textbook examples, so one of the ways of recognizing real biologically significant alternative splicing is to look for regulation. If one spliced version is present in one type of cell while an alternative version is present in another cell, this may indicate regulation. However, it's easy to be fooled by cell or tissue-specific splicing because the various activators and repressors themselves can cause error-prone splicing when they bind inappropriately. Thus, tissue specificity, by itself, does not prove regulation and does not distinguish between real alternative splicing and spurious events due to incorrect splicing.

You can often tell the difference by looking at the abundance of various processed RNAs because, as I mentioned earlier, splicing errors are rare so nonfunctional RNAs resulting from errors will be usually present at very low concentrations relative to the correct version.[44] A good rule of thumb is that any RNA present at less than one copy per cell is probably a mistake, and this criterion alone rules out the vast majority of splice variants where relative concentrations have been reported. Unfortunately, most new reports of splice variants don't include concentration data, so there's no way to tell if they are real examples of alternative splicing.

The same techniques used to count the total number of proteins can be used to determine whether different proteins can be produced from the same gene by alternative splicing. This number is important because many scientists claim that alternative splicing produces protein diversity. Unfortunately, the evidence does not support the claim because less than 5 percent of genes produce different proteins that have actually been detected.[45]

We don't know for certain how many genes are affected by true alternative splicing, but the number is undoubtedly much lower than the 95 percent claimed in the literature. A thorough and systematic review of the literature discovered only a few dozen examples.[46]

Splicing errors are the best explanation

Another way to tell the difference between splicing errors and real alternative splicing is to look at the transcripts in other species. Real, biologically significant alternative splicing is usually conserved in multiple lineages, and this is one of the most important criteria when you are trying to decide whether something is truly functional or not. It turns out that other species have only a small fraction of the

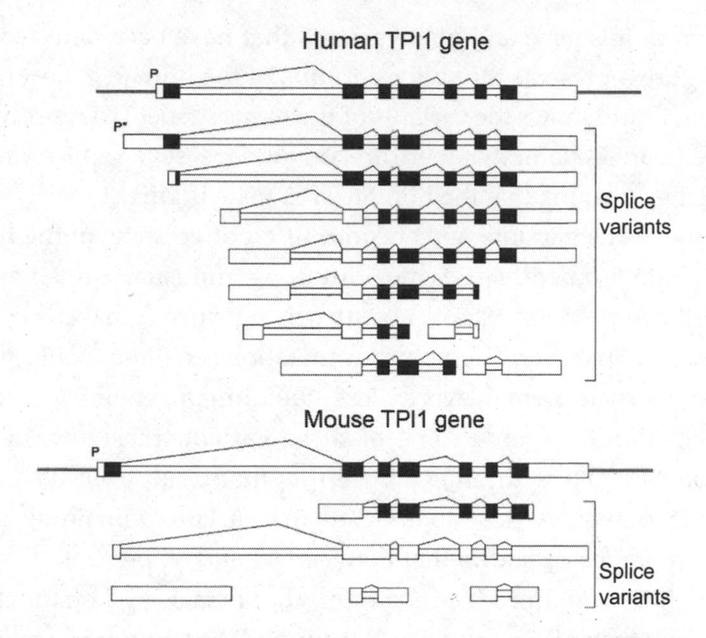

FIGURE 6.3 Splice variants of the human (top) and mouse (bottom) TPI1 gene from the Ensembl gene database. Several of the variants will encode a different version of the triose phosphate isomerase enzyme (solid boxes) while others don't have an open reading frame (open boxes).

transcripts seen in humans, a fact that's consistent with splicing errors and not consistent with something that's biologically relevant.[47]

This is an important observation since, as we will see throughout this book, evolutionary conservation is a good test of function. If the presence of multiple splice variants were conserved in multiple species, then that would be strong evidence for function. However, the evidence shows that only about 5 percent of genes have conserved alternative splicing,[48] and that's nowhere near sufficient to justify the claims made by alternative splicing advocates.

One of the other arguments against alternative splicing is just plain common sense. Let's look at the splice variants of the *TPI1* gene from the Ensembl database (Figure 6.3). Recall that the gene has seven exons and six introns, and the figure shows the normal splicing pattern of this gene and nine variants that have been detected. The top figure represents the splice variants of the human gene, and the bottom figure shows the variants of the mouse gene. The most widely quoted papers claim that there are, on average, seven splice variants per gene, meaning that the human TPI1 gene is fairly typical.

Some of the variants will encode different versions of the trisose phosphate isomerase enzyme, but these unusual proteins have never been detected. That's not surprising since it makes no sense to imagine that such an important metabolic enzyme would have a weird alternate form. Nevertheless, the human genome annotators have chosen to include some of these variant transcripts in most databases along with other transcripts that don't even have open reading frames. A close look at all the variants remaining in the curated databases shows that between 45 and 80 percent of variant transcripts would produce a protein that is missing key functional and/or structural information.[49] Informed common sense tells you that these variants are probably splicing errors.[50]

It's apparent from looking at the figure that the splice variants produced by the mouse and human genes are quite different. What this means is that the production of splice variants is not conserved in mammals. If proponents of alternative splicing are going to claim that production of all these variants is biologically significant, then they must claim that different species of mammals, such as humans and mice, have evolved about seven different RNAs per gene within the past 100 million years or so. Furthermore, many of these versions arise from thousands of highly conserved genes encoding highly conserved proteins for standard metabolic functions. If true, that process would be inconsistent with everything we know about evolution.[51]

It's interesting that the number of splice variants for a given gene can often be predicted by looking at the number of introns in a gene and the level of expression. Genes with lots of introns that produce lots of RNA will have many splice variants in the databases, but there's no reason why this correlation should hold if all those splice variants have a biological function. However, the observed correlation is exactly what you expect if these splice variants are due to splicing errors. This logic compelled the authors of a 2009 paper to reach the following conclusion: "The results strongly support the hypothesis that most alternative splicing is a consequence of stochastic noise in the splicing machinery and has no functional significance."[52]

This view is consistent with all available evidence, and it is almost certainly correct. Most splice variants are due to errors in processing RNA – they are not biologically relevant, and many of them have already been dismissed as junk RNA by annotators of the human reference genome.

The case for splicing errors

Here are what I think are the important points about splice variants:

1. The number of spice variants can be accounted for by the known error rate of splicing, suggesting that the vast majority of variants are due to splicing errors. They are not included in the human genome sequence databases because they are rare, they are not conserved, and they make no sense.
2. The annotation of most human genes shows an average of seven variants in the best-curated databases. These *may* be biologically relevant, but very few have been proved.
3. Only a few hundred genes, at most, have been shown to produce real variants due to alternative splicing. In the absence of evidence of function, the null hypothesis is that the transcripts are due to splicing errors.
4. The Deflated Ego Problem cannot be solved by assuming that each human gene makes many different proteins.

In my opinion, the scientifically correct statement about alternative splicing is that less than 5 percent of human genes are regulated by alternative splicing and that almost all of these have only one additional functional variant.

The controversy and how it's reported

Why is the role of alternative splicing so important? It's important because there's a vigorous debate going on in the scientific community about its frequency. Many scientists believe that the majority of human genes exhibit some form of alternative splicing, meaning that most human protein-coding genes produce several different functional proteins. This view, which is probably the majority view among workers in the field of alternative splicing, has dominated the scientific literature and spilled over into the popular press. The average science journalist doesn't even know that there's a debate, so neither does the general public, who rely on information from articles written by science journalists.

Let's look at an example. The quotation below was from *Scitable*, a website run by the *Nature* publishing group as an education service:

> Alternative splicing was the first phenomenon scientists discovered that made them realize that genomic complexity cannot be judged by the number of protein-coding genes. During alternative splicing, which occurs after transcription and before translation, introns are removed and exons are spliced together to make an mRNA molecule. However, the exons are not necessarily all spliced back together in the same way. Thus, a single gene, or transcription unit, can code for multiple proteins or other gene products, depending on how the exons are spliced back together. In fact, scientists have estimated that there may be as many as 500,000 or more different human proteins, all coded by a mere 20,000 protein-coding genes.[53]

This is the common view among scientists today, but in my opinion it is almost certainly wrong. I agree with many experts in the field who think that the existence of abundant splice variants is a real phenomenon but that biologically relevant alternative splicing is restricted to a small number of genes. I agree with them that most predictions about widespread alternative splicing are based on errors in splicing that are mistakenly assumed to be functional alternatives.[54] It's probably not true that "genomic complexity" can be explained by making as many as 100,000 proteins from only 20,000 genes. This minority view – that the importance of alternative splicing is vastly overrated – is shared by a great many skeptical scientists, so let's be clear about what that position means. It means that statements backed by leading science journals, such as *Nature*, may not be true.

Papers critical of alternative splicing are rare in the scientific literature, so it's no wonder that many science journalists have been deceived into thinking that real alternative splicing is the only game in town. However, it's their job to be skeptical. They should be doing the hard work necessary to dig out the truth or, at the very least, mention the existence of opposing views.

This is a genuine scientific controversy that has not been definitively settled one way or the other, although the evidence is gradually tilting strongly in favor of splicing errors, and that's the view that I'm promoting. As you might imagine, I've had some interesting conversations with some of my colleagues who strongly support alternative splicing. Controversy is what makes science so much fun.

It's important to keep in mind that the main problem with the scientific literature is not that alternative splicing is widely assumed to be correct, but that most scientists ignore the controversy and don't even mention the possibility that there might be other explanations for the data (e.g., splicing errors).

I think the consensus is shifting toward splicing errors as the main cause of splice variants but it's going to take some time before the majority of researchers in the field change their views. They have a lot invested in their previous publications and grant applications, and it's going to be very difficult for them to concede that they may have been wrong about the frequency of alternative splicing.

THE FALSE LOGIC OF THE ARGUMENT FOR COMPLEXITY

If you think that our low number of genes compared to other species is puzzling, then alternative splicing looks like a way to solve the Deflated Ego Problem. The argument goes like this: we make up for our small number of genes by producing lots of different proteins from the same gene. However, the argument carries an implicit assumption, namely, that other, less complex species don't resort to alternative splicing, but, as it turns out, that's just not true. Splice variants are common in all eukaryotes, so the assumption (premise) is wrong. In fact, the number of splice variants correlates with the number of introns in a gene and not with the complexity of the organism; for example, flowering plants have large, complex genomes with lots of introns and just as many splice variants as humans.

The implicit part of the argument is often omitted, but in order to be logical, it must be accompanied by evidence that alternative splicing distinguishes humans from other "lower" species with the same number of genes. We may have the same number of genes as a plant, but if both humans and maple trees produce the same number of splice variants from this number of genes, then you can't use alternative splicing as an argument for being more complex than a tree.

The claim that the production of splice variants increases with complexity only applies if you are very selective in your comparisons. In addition to the problem with plants – as complex as humans by this criterion – you need to consider all other animals. They all produce splice variants that are as abundant as those in humans. For example, the nematode worm *Caenorhabditus elegans* is often touted as the prime example of a very simple organism that has the same number of genes as humans but little or no alternative splicing, but the latest data show that *C. elegans* has just as many splice variants as we do.[62] Thus, abundant alternative splicing, even if it exists, can't be a valid logical argument for human exceptionalism.

Alternative splicing and disease

An additional factor needs to be taken into account when trying to decide whether a given splice variant is functional or just due to a splicing error. I've left discussing it to the end because it requires a lot of background knowledge.

A substantial proportion of mutations that cause human disease is due to mutations affecting splicing. Michael Lynch has estimated that up to 8 percent of all human deaths by natural causes are due to mutations affecting splice sites and that this would include two kinds of mutations: those that make a natural splice site inactive (or defective) and those that create new spurious splice sites.[55] This may seem like a lot (8 percent), but it doesn't look like splice mutants are any more common than mutations affecting other sites controlling gene expression.[56] Many of these mutations occur after birth, and they affect somatic cells or cells that are not part of the germ line. Such somatic cell mutations may cause great harm or even death to the individual, but they will not be passed on to the next generation; mutations that cause cancer are a good example.

In my view, most of those splicing mutations occur within introns and other junk DNA regions of the genome. They may accidentally create new binding sites for the spliceosome or for the activators and repressors that regulate splicing. I believe that disease-causing mutations are usually the consequence of a messy and error-prone system that also produces many spurious transcripts. The presence of large amounts of junk DNA presents a big target for mutations that create a new functional binding site (gain-of-function mutations).

The type of hemophilia found in Queen Victoria's descendants is a classic example of a mutation that affects splicing. In this case, a mutation occurred on the X chromosome in the germ line of one of Queen Victoria's parents. She inherited it and passed it on to her descendants, notably, Alexei, the son of Tsar Nicholas of Russia, and his wife, Alexandra, a granddaughter of Queen Victoria. The mutation occurred in one of the introns of the gene for a blood clotting factor (factor IX), and it created a new 3′ splice site, causing aberrant splicing. The result was an extra intron in the mature mRNA that interrupted the reading frame, so the correct protein couldn't be made and blood clotting was defective.[57]

I think gain-of-function mutations like this are common because there's a lot of excess junk DNA in introns, making it likely that a mutation will create a new aberrant splice site. This is the same objection that I raised in connection with the medical relevance of other mutations in chapter 4. Mutations causing genetic diseases can occur in junk DNA and not just in functional regions of the genome. You can't just assume that a stretch of DNA is functional because a mutation in that region causes a disease.

Proponents of alternative splicing see this from a different angle because they believe that the expression of a typical human gene is finely tuned and highly regulated. They believe that many different proteins are produced by alternative splicing and that the process is more prone to errors because of this complex fine-tuning. According to them, this complexity explains why so many genetic diseases and cancers are caused by errors in splicing because they are loss-of-function mutations that disrupt normal splicing. Here's an example of such a perspective from a 2016 review in the prestigious journal *Nature Genetics*: "This diversity [of multiple proteins from a single gene] is primarily generated by alternative splicing, with >90% of human protein-coding genes producing multiple mRNA isoforms. Given the complexity of the precursor RNA sequence elements and trans-acting splicing factors that control splicing, it is not surprising that this RNA processing step is particularly susceptible to both hereditary and somatic mutations that are implicated in disease. The central importance of splicing regulation is highlighted by the observation that many disease-associated single-nucleotide polymorphisms (SNPs) in protein-coding genes have been proposed to influence splicing."[58]

And here's another example from a recent review: "Disruption or disregulation of alternative splicing has also been associated with pathological states further supporting the functional contribution of this process."[59]

I argue that this conclusion – that disease-causing mutations are loss-of-function mutations and that this supports alternative splicing – is not correct because it's wrong to assume without evidence that pathological states are caused by disrupting normal alternative splicing. They could just as easily be caused by gain-of-function mutations that create a new aberrant splice site that has nothing to do with normal splicing. The observation that splicing mutations can cause disease is compatible with both views of splice variants, namely, the view that they have a biological function (alternative splicing) and the view that they are splicing errors. You shouldn't just assume that splicing mutations support one of these views and not the other.

I mention this only to show that an incorrect assumption – that more than 90 percent of human genes are alternatively spliced – can easily lead to an incorrect interpretation of the data on disease-causing mutations.[60] Many scientists implicitly assume that the frequency of disease-causing splicing errors is evidence for biologically relevant alternative splicing and ignore the possibility that it can also be due to junk DNA and accidental formation of new splice sites. However, recent data on cancer mutations suggest that one-third of mutations affecting splicing are due to alterations of natural splice sites while two-thirds enhance splicing at newly created aberrant splice sites.[61] Thus, the evidence tends to support the idea that disease can often be caused by mutations creating new splice sites rather than by mutations disrupting existing splice sites.

CHAPTER 7

Gene Families and the Birth and Death of Genes

> *[R]ecent studies indicate that most gene families which are concerned with genetic systems or phenotypic characters evolve following the model of birth-and-death evolution.*
>
> Masatoshi Nei (2005)

Remember the histones? Histones are the proteins that make up the nucleosome core particle that helps package DNA. I mentioned them in chapter 1 when I dealt with the problem of folding up those long strands of DNA in the nucleus. Recall that DNA wraps around the core particle to form the nucleosome and that there are about 200 bp of DNA in every nucleosome. Since there are 6.4 billion base pairs of DNA in your cells, this means there are about 32 million nucleosomes and each one contains two molecules of each of the four types of histone proteins.[1] Thus, whenever one of your cells divides, it has to make 64 million new proteins for each of the four histones. That's a lot of proteins. In fact, it's too much protein for the average gene because even though mRNA can be translated many times, a single gene cannot be transcribed fast enough to produce enough mRNA.[2]

Animals with large genomes solve this problem by having multiple copies of the histone genes. There are two clusters of histone genes in humans. The largest cluster is called *HIST1*, and it's located on chromosome 6 in region 6p21–6p22. Cluster *HIST1* has about 55 histone genes with between 6 and 16 copies of each of the five histone genes (H1, H2A, H2B, H3, and H4). The other cluster is *HIST2* + *HIST3* on chromosome 1 in region 1q21 and has about 16 genes. Together, the two large clusters account for about 3 Mb of DNA, or 0.1 percent of the genome.[3]

There are 16 copies of the gene for histone H2B plus five pseudogenes. The active copies are not identical – they differ slightly in their nucleotide sequence, and even the proteins are slightly different because of one or two differences in amino acid sequence. This is a typical example of a *gene family* composed of multiple copies of very similar genes.

The total number of genes in the human genome may be about 25,000, but that doesn't mean 25,000 *different* genes. There are many cases like the histone genes where an individual gene is a member of a gene family. Another example is the eight members of the globin gene family, two of which (the alpha-globin gene and the beta-globin gene) encode the proteins that combine to make adult hemoglobin. The other globin genes make fetal and embryonic versions of hemoglobin.

One of the main reasons for having multiple copies of the same gene is that more genes make more products – this explains the multiple copies of the histone genes. For another example, consider the ribosomal RNA gene family. A rapidly dividing human cell needs to make about 1 million new ribosomes every time a cell divides because each daughter cell needs to inherit a complete set of ribosomes. Ribosomal RNA genes are transcribed by a special RNA polymerase, and it takes 100 seconds, or almost 2 minutes, to make each ribosomal RNA molecule.[4] Even though many RNA polymerases are active on the gene at the same time – they are lined up like streetcars in a traffic jam – it still takes too long to make 1 million copies from a single gene. That's why we need 300 copies of these genes in order to make enough ribosomes.

The birth and death of genes

From an evolutionary perspective, the human genome is a dynamic, constantly changing structure. New genes are often created, or born, by accidental gene duplication events, and one copy of this redundant pair of genes usually dies by acquiring a mutation – it then becomes a pseudogene. But while it's correct to emphasize the birth and death of genes and the dynamic nature of our genome, it can also be misleading if you don't appreciate the time scale – we're talking about millions of years, not human life spans.

The idea of a dynamic genome where genes are constantly being born and dying is mostly due to the work of the population geneticist Masatoshi Nei, and he refers to it as "birth-and-death evolution."[5] The idea can be traced back to Susumo Ohno in 1970.

Newly duplicated genes can escape death by mutation if they diverge to provide a new function, and there are many clear examples including the globin genes. Each of these genes encodes a protein with a new function that was not present in the common ancestor with only a single gene. In this case, all members of the gene family are essential, so mutations in any one of them will be detrimental. Thus, all the genes will be conserved.

The large family of olfactory receptor genes is another example of this phenomenon. These genes encode proteins found on the surface of cells in your air passages, especially the lining of your nose, and they bind various molecules in the air causing a signal to be transmitted to nerve cells, resulting in the detection of an odor. The gene family was discovered and characterized by Linda Buck and Richard Axel, who won the Nobel Prize in 2004 for this work.

The olfactory receptor gene family is the largest gene family in mammals, and each olfactory receptor gene encodes a member that recognizes a particular chemical (odor). Most mammals have about 1000 different genes, giving rise to very sensitive odor detection, but humans don't have a very good sense of smell because we have only 400 functional olfactory receptor genes – the other 600 copies are pseudogenes.

Another well-studied example involves the opsin genes. Opsins are proteins found in the photoreceptors in your eye, and they are part of the vision pigment that absorbs light and determines what colors you can see. Most mammals have two different opsin genes: one that absorbs light in the violet range of the spectrum and the other in the yellow region of the spectrum, resulting in *di*chromatic color vision. The third member of the human opsin gene family – the one responsible for perceiving the color green – is located on the X chromosome right beside the gene for the yellow-absorbing opsin. The adjacent genes are the result of a gene duplication event that occurred about 40 million years ago in the ancestor of all Old World primates, including us. One copy of these duplicated genes diverged by acquiring several mutations that shifted its light-absorbing characteristics to the yellow–green part of the spectrum, and the new gene became fixed in all descendants, leading to *tri*chromatic color vision. We can see more colors than our mammalian cousins because of this gene duplication event where one of the genes diverged from the parent.[6]

The evolution of a new function for one of the genes following a gene duplication event is called *neofunctionalization*, and it's the main process giving rise to distinctive gene families. But there's one other fate of duplicated genes, and it's called *subfunctionalization*. It goes like this: immediately following the duplication event, you will have two copies of the gene, but only one is necessary. By chance, one of the genes is mutated in a way that reduces its activity – the mutation could affect transcription, RNA processing, or translation. How expression is reduced doesn't matter as long as a gene is still partially active. Normally this mutation would be the first step to complete inactivation on the way to becoming a pseudogene, but before this happens the other copy also acquires a mutation that reduces activity. Now you have a situation in which neither copy can make enough protein on its own and both copies are required in order to make sufficient protein. This is subfunctionalization.

To summarize, there are five fates of duplicated genes:[7]

1. death of one copy by acquiring inactivating mutations and becoming a pseudogene
2. death of one copy by deletion
3. birth because both copies are retained in order to make more product
4. birth by neofunctionalization
5. birth by subfunctionalization

THE SMELL OF SWEAT

Isovaleric acid is one of the chemicals that contribute to the smell of sweat in a well-used locker room. We can all detect this odor, but some of us are more sensitive than others. Part of the difference is due to the presence or absence of an active olfactory receptor (OR) gene (OR11H7P) that's part of the large cluster of OR genes on chromosome 14. People with an active copy of OR11H7P are quite sensitive to the smell of isovaleric acid.

The OR11H7P locus is identified as a pseudogene in the public databases, although we now know that some people have retained an active copy of the gene. People who are homozygous for the pseudogene are less sensitive to isovaleric acid.

There are other examples of olfactory gene differences; for example, about 12 percent of the population is relatively insensitive to musky odors, such as those found in many perfumes. I am not one of those people, but it might

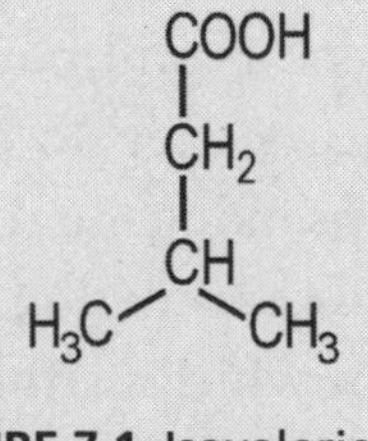

FIGURE 7.1 Isovaleric acid

explain why some people (men and women) don't seem to mind wearing a strong musky scent. Another example concerns the smell of butyl mercaptan, the main component of skunk spray. A small percentage of people can't detect this odor – I don't know whether this is a good thing or a bad thing.[8]

Gene duplication and mutationism

It appears that mutation (including gene duplications and other DNA changes) is the driving force of evolution at both the genic and phenotypic levels.

Masatoshi Nei (2005)

Gene duplication events are mutations that are mostly due to errors in recombination. A process called homologous recombination occurs when sister chromosomes align during meiosis.[9] Each of the chromosomes is then broken at the same spot and the ends are swapped. Homologous recombination results in the shuffling of different alleles on the same chromosome, and it's often thought of as one of the four main mechanisms of evolution along with mutation, random genetic drift, and natural selection.

An example of gene duplication is shown in Figure 7.2. In this case, a gene is flanked by identical repetitive sequences such as a transposon. Parts of the sister chromosomes misalign at the repetitive DNA elements and homologous recombination occurs at that site. This results in a chromosome with duplicated genes. (It also produces another chromosome with a gene deletion.) The duplication of the opsin genes in Old World primates is an example of this kind of event because the gene is flanked by *Alu* sequences.[10]

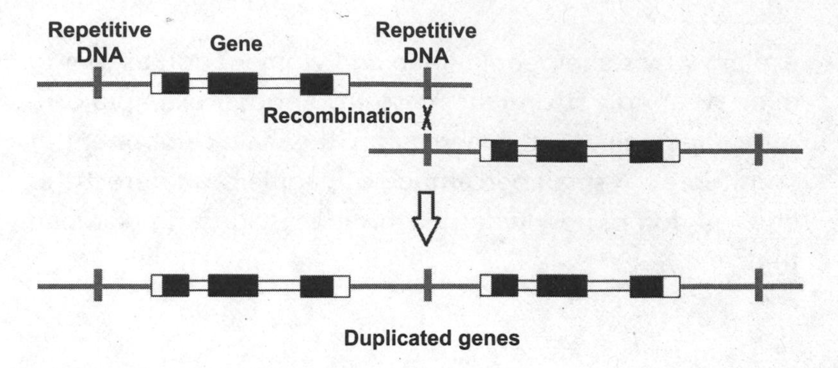

FIGURE 7.2 Gene duplication. A gene with three exons and two introns is flanked by identical repetitive sequences, such as a transposon. The two sister chromosomes misalign at the repetitive sequences and a homologous recombination event (crossover = X) occurs at the site of the repetitive DNA. This results in a new chromosome containing two copies of the gene.

Although some gene duplications are adaptive because extra copies of the gene are beneficial, it appears that 98 percent of all gene duplications are nonadaptive, meaning that one of the duplicated genes is going to die because it will acquire mutations rendering it nonfunctional.[11] The net effect of this birth and death is genome expansion by duplication. The process is not restricted to segments containing genes; in fact, most duplications of the sort shown in Figure 7.2 will occur in junk DNA where no genes are involved. The general term for duplication of large stretches of DNA is *segmental duplication*. Recall from earlier chapters that I attributed differences in C-value to expansions caused by insertions of transposons, and that's still true for most expansions. However, segmental duplications also play an important role in genome expansions.

Masatoshi Nei argues that much of the history of life is determined by accidental events, such as mutation, and one important class of mutation is gene duplication – a major source of new genes. The idea that mutation is a prime mover in evolution is called *mutationism*, and in addition to Nei, there are other strong proponents of this minority view (e.g., Arlin Stoltzfus).[12]

Nei emphasizes the importance of mutation by asking us to consider a parallel universe – one with natural selection but no mutation and one with mutations but no selection. There will be no evolution in the universe without mutations since there will be no variation for evolution to act on. In contrast, there will still be evolution in the world with mutations but no natural selection since new genes and new variants will appear and their frequency in the populations can increase by random genetic drift.

If the direction of evolution is largely determined by chance events, such as mutation, then this has consequences with respect to how we think about the future. It means we can't predict future evolution – including our own – because we don't know which mutations will appear and in what order they will appear. Similarly, if we look at the history of life, we must recognize that the pathways that were followed were more accident than design. This concept is a major theme of Stephen Jay Gould's view of evolution; he argues that if we rewind the clock and start over, the end result of millions of years of evolution would look very different, and intelligent life might not exist.[13]

Mutationism is just one of many ideas that emphasize the randomness and unpredictability of evolution, and it fits nicely into the view that genomes are sloppy and not exquisitely designed like a Swiss watch.

Whole genome duplications and the fate of genes

So far I've been discussing gene birth and death in the context of segmental duplication events in which only a small segment of DNA is duplicated, but there's another way to get extra copies of genes: whole genome duplication or polyploidization.* ("Haploid" is one copy of

* See "Instantaneous Genome Doubling" in chapter 2.

the genome, "diploid" is two copies, and "polyploid" is many copies.) Polyploidization is fairly common in plant lineages, but although it's quite rare in animals, there are a few excellent examples that are relevant to our discussion of the birth and death of genes.

The genome of modern salmon underwent a duplication event about 96 million years ago, and in another fish, carp, there was a polyploidization event about 8 million years ago. In the frog *Xenopus laevis*, the whole genome duplication event occurred about 18 million years ago.[14] All three genomes have been sequenced, so estimating the fate of duplicated genes at three different time points is possible. As expected, the carp genome – the one with the most recent polyploidization event – has about 57,000 protein-coding genes, which is twice as many as their close relatives that have not undergone polyploidization. This indicates that gene "death" is a slow process, taking much longer than 8 million years.

In the case of salmon, the genes were duplicated 96 million years ago, so they have evolved for 10 times longer than the carp genes since being duplicated. More than half of the duplicated genes in salmon have been lost by deletion or have become pseudogenes, giving a rate of gene loss of about 100 gene deaths per million years. It's hard to figure out whether the remaining duplicate genes have just survived by chance or whether they have acquired new and distinct functions but the evidence in the salmon genome suggests that subfunctionalization and neofunctionalization are not common. Thus, most of the extra genes will eventually die.

Many of the duplicates in the frog genome have been lost by deletion, which is consistent with an overall reduction in the size of the genome giving rise to a genome that's considerably smaller than it was immediately following the genome duplication. We see that more genes were lost by becoming nonfunctional pseudogenes in the frog genome; thus, it seems to be more common for genes to die by acquiring mutations than by deletion. Keep in mind that the accumulation of pseudogenes is one of the mechanisms that creates junk in the genome and the frog genome has a lot of junk DNA.

As is the case with the carp and salmon genomes, it's difficult to figure out what the remaining copies of the frog genes are doing, but it seems likely that neofunctionalization and subfunctionalization are not common. It looks like most of the duplicated genes just haven't had time to change significantly, but this does not mean that these duplicated genes have totally escaped mutation. Over the course of 18 million years, individual copies will have frequently been inactivated by mutation so that the population will then contain two alleles consisting of an active gene and an inactive pseudogene. The newly mutated gene (pseudogene) will initially be present at a very low frequency in the population, so it will likely be lost by chance before it becomes fixed. We don't see any trace of all those mutated versions in modern genomes, so we tend to underestimate the mutation rate.

Occasionally the pseudogene will survive by lucky accident, and its frequency in the population will increase by random genetic drift until it becomes fixed in the genome and the allele for one of the copies of the active version of the gene is eliminated because it was redundant. In this case, we see only the pseudogene when we examine the modern genome.

Don't forget that the composition of the modern genome is due to a combination of mutation and fixation by selection or drift. We're mostly talking here about nonadaptive fixation of neutral alleles, and this is due to the slow process of random genetic drift. The point is that the death of a gene requires a mutation *plus* the fixation of the pseudogene in the entire population. It's the combination of these two events that takes tens of millions of years or more.

One of the interesting consequences of slow death is that most genomes will contain a significant number of dispensable genes – genes that can be mutated or deleted without any noticeable effect on the organism.[15] We saw evidence of this earlier in chapter 4 in the box "Human Gene Knockouts," where I quoted a study showing that thousands of human genes could be knocked out without serious consequences.

REAL ORPHANS IN THE HUMAN GENOME

Recall that most ORFans are putative, or potential, genes that might be restricted to humans except that most of them are not real genes. The best available surveys suggest that only a small number of genuine new genes appear to have arisen in the human lineage. They are called *de novo* genes, and all of them are controversial.

However, there's another kind of ORFan gene that's quite real. These are genes that arise from a recent gene duplication event where "recent" means some time in the past 2 or 3 million years. Our genome may have two adjacent copies of a gene that's present in only one copy in our closest relatives such as chimpanzees and gorillas.

The birth of a new gene in this manner gives rise to a true ORFan gene – albeit one that's a copy of an existing gene. We expect that one copy of this pair will eventually die by mutation, but, for now, it's a real example of an orphan gene because other species have only one gene.

These new genes are usually polymorphic since fixation or loss of these copies takes millions of years – some people have them and some people don't. About 10 percent of our genome is present in different copies in various individuals, and our species has about 100 orphan genes of this type. Many of us have extra copies of gamma-globin genes, for example (see "Conserved Pseudogenes and Ken Miller's Argument Against Intelligent Design" later in this chapter).[16]

The ribosomal RNA genes illustrate an extreme example of copy number variation. The average person has about 300 genes, but the total number can vary between 200 and 400, and some individuals have as many as 600 copies.[17]

Different kinds of pseudogenes

Susumu Ohno argued in 1972 that a major factor in genome expansion is gene duplication caused by errors in DNA replication or recombination giving rise to two side-by-side copies of the same gene. As I described in the previous section, the genome will usually accumulate junk DNA by mutational inactivation of one of the duplicated copies, and after about 100 million years, these pseudogenes will have accumulated so many mutations that they will be almost indistinguishable from random sequence junk DNA. This is one way in which junk DNA accumulates in the genome, but it's probably not as important as transposon insertion.

It's easy to see why an accidental duplication event can lead to the formation of a pseudogene because the second copy of the gene is not essential, and it will not be conserved by natural selection. This is the first kind of pseudogene.

A second kind of pseudogene arises from an entirely different mechanism beginning with a gene transcript that is accidentally copied by reverse transcriptase to make complementary DNA (cDNA).* It can be subsequently integrated back into the genome becoming a pseudogene. Note that the cDNA resembles the mature RNA produced after the primary transcript is processed to remove introns, and that's why the result is called a *processed pseudogene*.

Although processed pseudogenes can arise in any cell type, the ones that become a heritable part of the genome must have been present in germ cells. Genes that are expressed in germ cells are much more likely to give rise to processed pseudogenes, so it's not surprising that many of these pseudogenes are derived from the genes for standard metabolic functions; for example, the gene for the important

* Recall that our genome is littered with multiple copies of endogenous retroviruses and retrotransposons and that many of them are transcribed to produce reverse transcriptase.

metabolic enzyme, triose phosphate isomerase, that I described in chapter 5 has spawned four processed pseudogenes on four different chromosomes.

A third type of pseudogene is called a *unitary pseudogene* because there's only one copy in the genome. They are remnants of genes that once were active but not essential. At some point the active gene acquired a mutation, rendering it inactive, and the inactive version became fixed in the genome by random genetic drift because there was no selection against losing the gene. A classic example in humans is the gene for the enzyme *L-glucono-γ-lactone oxidase* required for making vitamin C. Most mammals have an active gene and can make their own vitamin C, but humans lack this gene, meaning we can't make vitamin C and are prone to developing scurvy if we don't get enough vitamin C in our diet. We can still recognize the remnants of the original gene in our genome right in the same place where the active gene is found in most other species.

The fourth type of pseudogene is an inactive version of a gene found in some members of the species but not in others. These are called *polymorphic pseudogenes* because there are different alleles segregating in the human population – one of the alleles is an active gene, and the other is a pseudogene. This is not the same as when a deleterious mutation produces an allele causing a genetic disease. Such deleterious alleles are found in the human genome, but they are rare and the number of individuals carrying one copy of this defective allele is only a small percentage of the population. These deleterious (harmful) versions of the gene are usually lethal when homozygous.

Defective alleles are only called pseudogenes when their frequency in the population is quite high and individuals with two copies of the pseudogene are not at a selective disadvantage. They are like the third type of pseudogene that hasn't yet become fixed in the population by random genetic drift. They may eventually take over the entire population, or they could just as easily be lost before becoming fixed.

One of the best-known examples of a polymorphic pseudogene in humans is the gene for *N-acetylgalactosaminyltransferase*. This enzyme is responsible for attaching sugar groups to the proteins on the surface of red blood cells, giving rise to the ABO blood types. The "O" allele is a defective copy of the gene that doesn't make any of the enzyme, and it's a polymorphic pseudogene because there are still active copies of the gene in the human population. People with two copies of the defective gene cannot attach sugars to the proteins on blood cells, so they have blood type O, and they appear to be perfectly healthy.[18]

At some time in the distant future, the entire population of humans may have O-type blood in which case the gene will become a unitary pseudogene like the gene for making vitamin C. In fact, there are several groups of Native North Americans that have lost the A and B alleles, so everyone has O-type blood.[19] Alternatively, the O allele may be lost entirely, in which case the entire population will contain functional genes for the enzyme and everyone will have blood types A, B, or AB.

Another example involves cilantro (coriander). Many of us have a perfectly normal olfactory receptor gene called *OR6A2* that allows us to detect certain aldehydes in cilantro, giving it an obnoxious taste. It's obvious that normal people – I am one, and so is the famous chef Julia Childs – were not meant to eat cilantro! Some of you (about 90 percent of the population) are mutants. You have a mutated version of the *OR6A2* gene that doesn't produce a functional protein, and you like cilantro because you can't smell or taste the obnoxious aldehydes. The *OR6A2* pseudogene is a polymorphic pseudogene.[20]

The human genome contains approximately 15,000 pseudogenes, but different research groups use different criteria to describe pseudogenes, so there are several databases with differing numbers of pseudogenes.[21] The overlap between databases is not impressive, as shown in a 2012 study in which less than two-thirds of the pseudogenes in two large databases were the same.[22] What this result means is that there could be only 7500 pseudogenes or as many as 18,000, but the actual number isn't that important. What's important is that there are almost as many pseudogenes as there are genes.[23]

All these pseudogenes are junk, with rare exceptions that we'll discuss shortly. That includes a substantial proportion where the pseudogene is still transcribed but is incapable of making protein. This is a very good example of junk *RNA* – transcripts that are reproducibly detected but have no biological function, thus illustrating the important point that you cannot assume that all RNAs must be functional. This point will become important in the next chapter when we discuss noncoding RNAs.

Many of the pseudogenes in the human genome are small, processed pseudogenes that are "dead on arrival," but many are the result of gene duplications that have acquired various inactivating mutations. Since this latter set of pseudogenes has introns, it means that these pseudogenes can be quite large, so they represent a significant fraction of the human genome – perhaps as much as 5 percent.

Let's summarize what scientists have learned about the birth and death of genes. New genes can be born by duplications, and the fate of most of these genes is the inactivation of one copy since, in the vast majority of cases, the extra copy is redundant. In mammals, these loss-of-function alleles can be expected to appear within about 5 million years, and the pseudogene may start to lose most traces of its ancestry after about 100 million years.[24] This slow process of disappearing is one of the reasons why it's so hard to estimate the actual number of pseudogenes in the human genome because different researchers will often use different similarity criteria to identify potential pseudogenes.

Keep in mind that the formation of a nonfunctional pseudogene from a duplication event is just one of the steps required before it becomes part of the genome. The other step is fixation in the population by random genetic drift and fixation will depend on the effective population size (see chapter 2). On theoretical grounds, one expects that species with small population sizes should have more pseudogenes than species with large population sizes, and this is exactly what is observed.[25] Throughout most of our history the effective size of the human population has been less than 10,000 individuals, which explains why we have so many pseudogenes in our genome.

CONSERVED PSEUDOGENES AND KEN MILLER'S ARGUMENT AGAINST INTELLIGENT DESIGN

Ken Miller is a professor of biology at Brown University in Providence (Rhode Island, USA). He is the author of a very popular biology textbook and has been active in the fight against creationist attacks on evolution.

One of his arguments against the intelligent design version of creationism concerns pseudogenes.[26] He points out that humans and chimpanzees both have a nonfunctional pseudogene at the same relative position in the beta globin gene cluster as shown in Figure 7.3. The two pseudogenes share a number of mutations that would have inactivated the functional gene after it first arose by a gene duplication event. The presence of the pseudogene strongly suggests that it arose before the common ancestor of humans and chimpanzees more than 5 million years ago.

Miller's argument is not about the amount of junk DNA in our genome. It's about a specific bit of DNA that most people recognize as junk and the fact that two different species, chimpanzees and humans, have the same piece of junk DNA in the same position is powerful evidence of descent from a common ancestor. The fact that both species have the same inactivating mutations is additional support for a common ancestor. The evidence challenges explanations based on intelligent design because it's difficult to imagine why a designer would insert two similar broken genes at the same spot in both species.

The creationists counter with arguments that the pseudogene is actually functional.[27] The idea here is that, for some reason, an intelligent designer could have made something that looks like a pseudogene and inserted it in both species because it actually performs a function. Thus, scientists are mistaken when they claim that this is junk DNA; instead, it's evidence of a common design.

(*continued*)

CONSERVED PSEUDOGENES AND KEN MILLER'S ARGUMENT AGAINST INTELLIGENT DESIGN (continued)

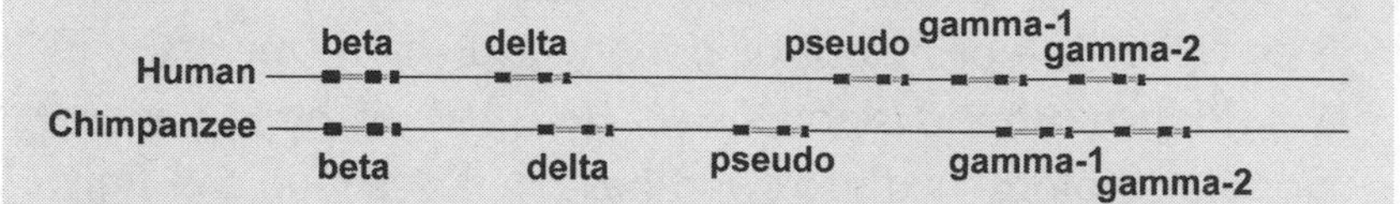

FIGURE 7.3 Pseudogenes in human and chimpanzee DNA. The human beta-globin gene family is located near the end of chromosome 11 at position p15.4. It starts about 5 million base pairs from the end to the left of this figure. There are four active genes (beta, delta, gamma-1, gamma-2) in the human and chimpanzee genomes, and they are expressed at different times during development. A pseudogene (pseudo) is located between the delta and gamma-1 genes. The chimpanzee genes are on chromosome 11, and they are also about 5 million base pairs from the end. A very similar pseudogene is located in the chimp cluster at the same position as the human pseudogene. The genes are transcribed from right to left in this orientation. Differences in spacing between the human and chimp genes are due to various insertions and deletions that have occurred in the past 5 million years.

The beta-globin pseudogene is still transcribed, but this is not evidence of functionality; it just means that this property of the original gene has not yet been lost to mutation. However, there's a complication in the case of this particular gene because the region encompassing the pseudogene happens to contain a scaffold attachment site that functions in organizing chromatin. Thus, a small bit of the region is conserved and functional, even though the original beta-globin gene has lost its original function of making a beta-globin protein.[28]

Are they really pseudogenes?

You would think that pseudogenes are clearly junk by any definition because they are broken genes, and when most things are broken, we think of them as junk. (See "If It Walks Like a Duck..." in chapter 3.)

Surprisingly, there are some scientists who believe that a majority of pseudogenes aren't "pseudo" at all. According to these scientists, they are new functional genes! A recent review, for example, declares that tens of thousands of pseudogenes are functional non-coding elements.[29] One of the justifications for this bizarre claim is simply the unfounded belief that most of our genome must be functional because natural selection would remove junk. If this extreme adaptationist view were true (it isn't), then it follows logically that most pseudogenes must be functional.

The claim of functionality also relies on the fact that many pseudogenes are transcribed and that some pseudogene mRNAs are translated to produce a defective protein. As we have seen, these observations are not proof of functionality – they are proof that the slow death of a gene can occur by mutations at all levels of gene expression. Some pseudogenes have mutations that disrupt the protein by mutating key amino acids or deleting part of the coding region, and those pseudogenes will still be transcribed and still produce a protein. This does not mean they have a biological function.[30]

All the scientists who claim that most pseudogenes are functional make the same logical error of cherry-picking a few special examples of pseudogenes that have secondarily acquired a new function while ignoring the vast majority of cases in which pseudogenes are just broken genes.[31]

How can you determine whether a pseudogene is functional? One way is to rely on sequence conservation – a criterion that has proved effective in determining whether a gene is real and whether alternative splicing is real. As we saw in the previous section, many human pseudogenes are ancient, as demonstrated by the evidence that they are present in the genomes of other apes. This means that the inactivation of the original gene to create a pseudogene occurred in the common ancestor of these apes.

Modern evolutionary theory predicts that these pseudogenes will accumulate mutations over time because they are not under any constraints. The rate of change (fixation of new mutations) is equal to the mutation rate as predicted by population genetics (chapter 4), but if the pseudogene has acquired a new function, it will be subject to natural selection in order to preserve that function, meaning it will not be changing as rapidly as other pseudogenes that are not constrained by selection. The differences can be detected by comparing the genomes of several species.

A careful examination of all 15,000 human pseudogenes reveals a small number that appear to be evolving more slowly, and this slow evolution is a strong indication that they have secondarily acquired a new function that is now subject to negative selection. In other words, they are partially conserved. There may be a dozen examples in the human genome corresponding to about 0.2 percent of all pseudogenes. That means 99.8 percent of all pseudogenes are really pseudogenes or junk DNA because they are evolving at the neutral rate.[32]

THE SHORT LEGS OF DACHSHUNDS

The short, curved-leg shape of several breeds of dogs, such as dachshunds, is known as chondrodysplasia. It is caused by the inappropriate expression in bone cells of a gene for fibroblast growth factor 4 (*fgf4*). Short-legged dogs contain a processed *fgf4* pseudogene that has fortuitously landed near a LINE element that's transcribed in bone cells. The region surrounding the processed pseudogene shows evidence of strong recent selection, as expected, since over the past few hundred years, dog breeders have been applying strong selection for the chondrodysplasia trait.[33]

Most processed pseudogenes are incapable of producing a functional protein because they lack a promoter, but in this case, it is transcribed by a nearby promoter. The fact that the promoter just happened to be active in the right cell types is a perfect illustration of the dictum that in evolution, just about anything is possible.

Humans also have an *FGF4* gene. Some alleles of this gene cause inappropriate expression leading to dwarfism with a trait that resembles chondrodysplasia.

How accurate is the genome sequence?

All this talk about counting genes and identifying pseudogenes would be pointless if the published human genome sequence wasn't highly accurate. Recall that "draft" sequences were published in February 2001, but it took several more years to "finish" the sequence. The initial standard of accuracy was to generate sequences that had only one error in 100,000 base pairs, and it took almost a decade to develop the technology for such accuracy. It's worth noting that even this level of accuracy would mean 30,000 errors in the genome sequence, and that doesn't count the potential errors introduced during assembly.

The error frequency can be reduced by sequencing each region multiple times and combining the results. For most genome sequences, the coverage is about fivefold (5× coverage), and this coverage should include sequences from both strands of DNA. However, in many cases that's not good enough to distinguish genes from pseudogenes and identify regulatory DNA motifs with a high degree of confidence. That's why the finished human genome reference sequence has more than 10× coverage, and it is thought that each of the sequenced fragments is accurate to at least one error in a million bp (10^{-6}).

The quality of the human genome sequence was a problem in the genome war between Celera Genomics and the International Human Genome Project (IHGP). The goal of the publicly funded project was to generate a highly accurate sequence that could be used to identify genetic diseases and other features that required a fully finished genome sequence. Celera, on the other hand, was satisfied with a "quick-and-dirty" approach using the latest Perkin

Elmer sequencing machines. Its *Drosophila* sequence, for example, was so bad that it took years to clean up. James Shreeve nicely captures the dilemma in his book *The Genome War* when he writes,

> Unfortunately, the tenacious allegiance to quality had blinded the leaders of the program [IHGP] to a fundamental weakness: for most research purposes, such a stringent standard of accuracy was overkill. In a way, the resolute, mustached Francis Collins resembled Alec Guinness's British colonel in *The Bridge on the River Kwai*, who rouses his fellow prisoners of war to build the best possible bridge they can, having left out of his mental calculus that the bridge is for the use of their Japanese captors. Collins's oversight was to forget that most potential users of the genome – academic biologists looking for genes and pharmaceutical companies eager for information that could speed up drug development – did not need a fancy bridge that would "stand the test of time." They just wanted to get across the river. And now suddenly Venter had appeared, carrying a bunch of Perkin Elmer pontoons.[34]

As it turned out, Collins and the IHGP were correct, and Venter was wrong. You can't tell a gene from a pseudogene with a sloppy sequence. An accurate sequence turned out to be essential for almost everyone, and that's why, unlike the bridge in the book and the movie, the best bridge is still standing long after the pontoons have been washed away.

The technology has improved considerably. It's now common to quickly collect billions of overlapping sequences to generate 30× coverage. The accuracy of each sequenced fragment using these rapid methods is lower than the ideal of 99.99 percent, but the extra coverage compensates to some extent. Recently a small group of labs sequenced 10,545 human genomes at this level of coverage.[35] The individual assemblies only covered 84 percent of the genomes because centromeres, telomeres, and highly repetitive regions were omitted, but the authors performed several quality-control experiments to prove that the overall accuracy of the sequenced portion was about 99.9 percent.

Many studies are not this accurate; for example, if you pay to have your genome sequenced by a for-profit company, there may be a much higher number of mistakes and you should be skeptical of predictions about your predisposition to genetic disease. Several clinical studies have used sequences with only 7× to 10× coverage, and this is not sufficient for highly accurate results. If you want to know whether you are going to die or pass on a lethal allele to your children, then you want an accurate sequence.

CHAPTER 8

Noncoding Genes and Junk RNA

[T]he sequencing community has split into two irreconcilable camps. "Determinists" ... are convinced that very low level transcripts represent a huge world of "functional" RNAs, just because they exist. Their opponents put forward serious reasons to question this view. Probably, many new biochemically active coding and ncRNAs will be discovered in the future, but it is more probable that the vast majority of low abundant transcripts are simply junk. Therefore, the first group takes a lower threshold to catch the maximum of [very low abundance] RNAs, whereas skeptics choose a higher threshold to cut off "nonfunctional" RNAs.

Eugene Sverdlov (2017)

I mostly discussed protein-coding genes in the previous chapters, but now it's time to consider noncoding genes. Recall that there are about 20,000 protein-coding genes in the human genome, but we are less certain about the number of noncoding genes. I suggested that there are probably about 5000 of these genes for a total of 25,000 genes.

We have already seen how many people were shocked when they first learned that we had so few genes. They thought we should have many more genes because our species is so much more "advanced" than other species. In order to restore their view of human exceptionalism, these scientists feel the need to come up with other explanations of genetic complexity. One of them was alternative splicing – the suggestion that each human gene can make many different proteins by splicing the primary transcript in different ways. I argued that alternative splicing is real but that there aren't very many proven examples, so alternative splicing won't solve the Deflated Ego Problem.

Those with deflated egos must look elsewhere for an explanation. The most important of these other explanations is the view that human complexity can be explained by postulating the existence of thousands of newly discovered genes that specify noncoding RNAs (ncRNAs). The idea here is that humans really do have many more genes than "lower" animals, but these genes don't code for new proteins; instead, they specify a huge number of noncoding RNAs with a biological function and that makes us much more complex than a worm or a fruit fly.

This view is very popular among those who have trouble accepting the idea that our genome could be full of junk DNA. It's a view that has been promoted in the popular press; for example, Nicholas Wade wrote the following in a *New York Times* article, "A Decade Later, Genetic Map Yeilds Few New Clues," on June 12, 2010:

> Research on the genome has transformed biology, producing a steady string of surprises. First was the discovery that the number of human genes is astonishingly small compared with those of lower animals like the laboratory roundworm and fruit fly. The barely visible roundworm needs 20,000 genes that make proteins, the working parts of cells, whereas humans, apparently so much higher on the evolutionary scale, seem to have only 21,000 protein-coding genes.
>
> The slowly emerging explanation is that humans and other animals have much the same set of protein-coding genes, but the human set is regulated in a much more complicated way, through elaborate use of DNA's companion molecule, RNA.*

* I argue in this chapter that just about everything in this *New York Times* article is either wrong or misleading.

This brings us to a discussion of the number of genes that specify functional noncoding RNAs. The question for this chapter is whether the number of noncoding genes accounts for a significant fraction of the genome – enough to challenge the idea of junk DNA. We'll discover that most of our genome is transcribed, even though only 45 percent is devoted to known genes. This raises a question about whether the additional transcripts are functional RNAs derived from unknown genes or whether they are spurious transcripts, otherwise known as junk RNA. That's the question posed by Eugene Svederlov at the beginning of this chapter – he correctly notes that scientists are split into two irreconcilable camps. My goal is to demonstrate that the scientists in one of those camps are wrong!

Different kinds of noncoding RNAs

There are so many different kinds of noncoding RNA that it's hard to keep up. The main ones are the ribosomal RNAs (rRNAs) and transfer RNAs (tRNAs) that I covered in chapter 1. Recall that there are several hundred tRNA genes and at least 300 rRNA genes. In addition there are a slew of small RNAs that perform a huge variety of different functions. Some of them are unique – they only do one job. Examples in this category include the RNA component of RNAse P, for which Sidney Altman received the Nobel Prize in 1989, and 7SL RNA, the main component of the signal recognition particle required for protein secretion, which also gave rise to *Alu* SINES. There are many others.

In addition to these unique RNAs, there are several *classes* of small RNAs in which each member of the class has a similar function. Here's a list of the most important ones.[1]

- **small nuclear RNAs (snRNAs):** These are the RNAs that help form the spliceosome. Most eukaryotic species have five of them (U1 RNA, U2 RNA, U4 RNA, U5 RNA, U6 RNA). They are the main components of a typical spliceosome. Many species have extra snRNAs that can be recruited to specific spliceosomes. The extra snRNAs have names like U4atac, U7, and others. In some species there may be as many as 20 different spliceosomal RNAs and 20 different genes.[2]

- **small nucleolar RNAs (snoRNAs):** As the name implies, these small RNAs are mostly found in the nucleolus – the site of ribosomal RNA synthesis in eukaryotic cells. The RNAs act as guide sequences directing a protein–RNA complex to a specific sequence that will become modified. For example, some bases in ribosomal RNA are methylated, and a particular snoRNA guides the methylase enzymes to the correct site. There are more than 100 members of the snoRNA family.
- **microRNAs (miRNAs):** This is a heterogeneous class of small RNAs about 22 nucleotides long produced by processing a much larger precursor RNA. The functional region is an antisense RNA because it's complementary to the sense sequence of mRNA. miRNAs regulate the translation of mRNA by binding to one end of the mRNA as part of an *R*NA-*i*nduced *s*ilencing *c*omplex (RISC, pronounced "risk"). Nobody knows for sure how many of these miRNAs are made in human cells, but there's probably at least 100 miRNA genes and possibly as many as 1000. Many putative miRNA genes have not been shown to have a function – they may be pseudogenes or misidentified sequences. miRNA genes and RISC complexes are largely restricted to animals and plants, but there are similar small antisense RNAs in bacteria with a different mechanism of regulation.
- **short interfering RNAs (siRNAs):** The siRNAs are similar to miRNAs, and they also form part of a RISC structure that interferes with translation. Whereas miRNAs block the translation of mRNAs, siRNAs stimulate the rapid degradation of mRNA and are important in blocking the expression of foreign viral RNAs in an infected cell. There are many potential genes, but few have been confirmed as functional. The siRNA class of small RNAs is only found in eukaryotes.

- **PIWI-interacting RNAs (piRNAs):** PIWI stands for *P*-element *I*nduced *WI*mpy testis. It refers to a class of proteins found originally in the fruit fly, *Drosophila melanogaster*. PIWI proteins form a complex with piRNAs creating a RISC complex that directs the destruction of other RNAs. The functional part of piRNA is a bit longer than that of miRNA and siRNA, and it forms a double-stranded region that can be chewed up by nucleases. In addition to regulating the expression of some specific mRNAs (by destroying them), they are also involved in destroying the RNAs produced by transposons (selfish genes), preventing them from inserting into new regions of the genome. These piRNAs are usually expressed in the germ line (e.g., testis), where selfish gene jumping could be passed on to the next generation. PiRNAs are only found in animals. There are several thousand potential piRNA genes in the human genome, but it's not clear how many are functional.[3]
- **long noncoding RNAs (lncRNAs):** lncRNAs (pronounced "link RNA") are longer than the other small RNAs. Most of them are more than 200 nucleotides in length, and they have many of the characteristics of mRNAs except they don't encode proteins. Different lncRNAs are known to perform a variety of functions inside the cell, but there's no common theme. It is the most heterogeneous category and is also the most controversial category. We don't know how many lncRNA genes there are in the human genome.

Many transcripts are placed in various categories because they resemble other RNAs in that category, but this choice does not mean those transcripts actually have a biological function. That's part of the problem in this field. It's a terminology issue like the one we encountered when we discussed ORFan genes and alternative splicing; for example, just because an RNA sequence *resembles* other miRNAs does not mean it *is* a miRNA. It's a *putative* or *potential* miRNA until it has been shown to have a function. It could be junk RNA.

Understanding transcription

One of the main themes in this book is that biological systems are noisy and sloppy. What this means is that all processes, including transcription, are inherently error-prone. Mistakes can be made at all three steps of transcription: initiation, elongation, and termination, giving rise to spurious transcripts or junk RNA.

I discussed this issue when I talked about alternative splicing and explained that most splice variants were probably due to errors in splicing. The same reasoning applies to errors in transcription initiation. Recall that initiation begins when RNA polymerase binds to DNA and scans it for a promoter sequence. When it lands on a promoter it unwinds a bit of the double-stranded DNA to create a transcription bubble, and then it starts copying one of the strands into RNA by moving down the gene, making a longer and longer bit of RNA. This is all that's needed if there's a really strong promoter, but such promoters are rare, especially in eukaryotic cells. The typical strong core promoter consists of a short stretch of Ts and As called a TATA ("tah-tah") box, but less than 5 percent of genes possess such a strong promoter.[4]

In most cases, a gene is regulated in some manner meaning that it is not transcribed all the time; instead, it is only transcribed when certain transcription factors are present. The gene is activated when additional factors bind to DNA near a weak promoter, creating an improved binding site for RNA polymerase – often by forming direct contacts with the enzyme – and such genes have a weak core promoter because they must be silent most of the time. (In addition to activation, transcription can also be repressed by a repressor that binds near a promoter to block initiation. This seems to be less common than activation, so I'll ignore it for now.)

There are many examples of transcription activators, and their DNA-binding characteristics have been well studied. They usually recognize short sequences such as TGTGA, the site of binding of transcription factor CRP in *E. coli*. Another well-studied example is the binding site for transcription factor cII (= TTGC) involved in regulating gene expression in bacteriophage lambda.

Many activators have somewhat more complex binding sites consisting of two separate sequences separated by a few base pairs. For example, the binding site for the estrogen receptor is

5′ ... AGGTCANNNTGACCT ... 3′
3′ ... TCCAGTNNNACTGGA ... 5′,

where "N" is any base pair. If you look closely, you'll see that the binding site is symmetrical or *palindromic* because the sequence AGGTCA on the left of the top strand is the same as the sequence from 5′ to 3′ on the right of the bottom strand. The binding site is palindromic because the estrogen receptor is actually two identical proteins attached together to form a face-to-face *dimer*. Each protein binds to one-half of the binding site, so you get twice as much binding for the same price!

The estrogen receptor protein is a classic example of a regulated transcription factor. The factor doesn't bind to DNA in the absence of the estrogen hormone, but when estrogen is present, it binds to estrogen receptor protein and converts it to a specific DNA-binding protein. The activated protein enters the nucleus and binds near the promoters of several genes, creating a favorable landing site for RNA polymerase. The result is that exposure to estrogen hormone turns on the expression of all genes that have the estrogen receptor binding site.

You might think that the transcription factor binding site must be near the promoter region of the DNA since the activator has to contact RNA polymerase, but this isn't always the case. It turns out that binding sites can be hundreds of base pairs away from the promoter region because the DNA can form large loops (Figure 8.1). This is how the activator protein can contact RNA polymerase at the promoter even if its binding site is far away.

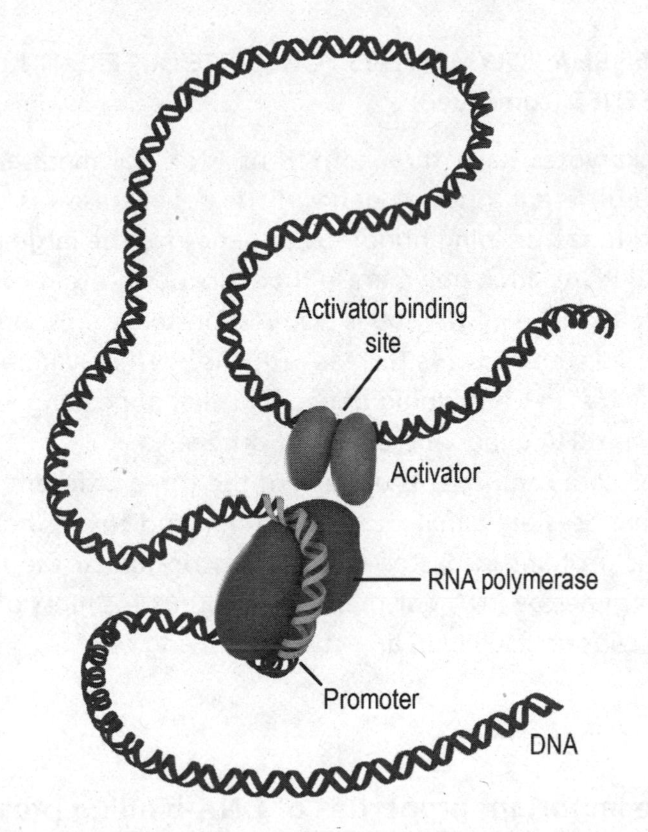

FIGURE 8.1 Transcription factor activation. The transcription factor activator binds to an activator binding site that's some distance from the promoter. The intervening DNA forms a large loop that brings the bound activator into contact with the RNA polymerase molecule sitting at the promoter. This activates RNA polymerase to begin synthesizing RNA, and the gene is expressed.

THREE RNA POLYMERASES FOR THREE DIFFERENT KINDS OF GENES

Most bacteria have just one kind of RNA polymerase that transcribes all genes. This core polymerase can associate with various transcription factors that help it to recognize different promoters.

(*continued*)

THREE RNA POLYMERASES FOR THREE DIFFERENT KINDS OF GENES (continued)

Eukaryotes have three different RNA polymerases for three different kinds of genes. RNA polymerase I is dedicated to transcribing hundreds of genes for the large ribosomal RNAs. RNA polymerase II transcribes protein-coding genes to make mRNA and is also responsible for producing some noncoding RNAs such as lncRNAs. RNA polymerase III specializes in transcribing genes for small noncoding RNAs, such as tRNA genes and a host of others.

The core components of each of the three different RNA polymerases are similar to each other and to the homologous proteins in bacteria. The transcription of the large rRNA genes by RNA polymerase I accounts for most of the transcription activity in an actively growing cell.[5]

On the important properties of DNA-binding proteins

When I was a graduate student, our lab was studying DNA-binding proteins. My thesis topic was proteins involved in DNA replication, but one of my fellow students, Keith Yamamoto, was trying to understand how the estrogen receptor worked. He purified the receptor protein from cells by taking advantage of its ability to bind to DNA, which allowed him to study the effect of estrogen on its ability to bind to specific promoter regions.

This was a time (early 1970s) when several DNA-binding proteins were being studied in great detail. In particular, it was known that *lac* repressor – a transcription factor that regulates expression of lactose genes in *E. coli* – binds with high affinity to a specific DNA sequence near the promoter of the *lac* genes, but it also binds to any random piece of DNA, albeit with a much lower affinity.[6]

Since there's a lot more random DNA than specific binding sites, it follows that most of the *lac* repressor will be bound nonspecifically at any given moment. The same is true of RNA polymerase and a host of other DNA-binding proteins. These properties of DNA-binding proteins are thoroughly covered in the best undergraduate textbooks.

Keith Yamamoto and our PhD supervisor, Bruce Alberts, published a series of papers showing that the estrogen receptor worked in a similar manner in eukaryotic cells. The estrogen–receptor complex binds to a specific DNA sequence near the promoter of those genes, but it also binds nonspecifically to the vast amount of extra DNA in the genome. Unfortunately, at the time it was impossible to detect specific binding because of the huge background of nonspecific binding, but Keith did some computer simulations (in 1973!) to show that with reasonable estimates of the two kinds of binding, 10,000 molecules of estrogen receptor protein would be needed in the nucleus in order to ensure that the specific binding sites were occupied.[7]

The important point here is that the properties of DNA binding proteins are such that even if they specifically regulate certain genes, they will always bind random DNA sequences in the nucleus. Part of this nonfunctional binding is just very weak binding to any random sequence of DNA, but the most significant part of nonfunctional binding is to sequences identical to the normal, functional binding site. Such sequences will occur by chance in large genomes because typical transcription factor binding sites are quite short. Take one-half of the estrogen receptor site as an example. The strongest binding site is AGGTCA, and these sequences will occur in a random DNA sample with a frequency of $1/4^6$ because there are six base pairs in the target site and each base can be one of four possibilities (A, G, T, C). The binding sites will occur, on average, every 2048 bp ($4 \times 4 \times 4 \times 4 \times 4 \times 4 = 2048$). There will be more than 1 million of these random sequences in a typical mammalian genome, but only a handful will be involved in biologically meaningful regulation of gene expression.[8] This is why our cells need thousands of copies of each transcription factor.

(Recall that the functional estrogen receptor binding site has two adjacent sequences consisting of AGGTCA and its palindrome TGACCT, and the transcription factor binds as a face-to-face dimer. This is a special case that limits, but does not eliminate, spurious nonfunctional binding. There will still be significant binding to a half-site.)

Many transcription factors bind to a single short sequence; for example, the transcription factor CEBPA regulates gene expression in liver cells by binding preferentially to the sequence TTGCN-NAA. There should be more than 1 million of these sequences in the human genome, but fewer than 20 or so have a biological function in regulating the transcription of liver genes. About 16,000 binding sites are easily detected in liver cells.[9] (In chapter 11 we will discuss why only a small subset of these potential binding sites are actually detected.)

Sophisticated models of transcription based on known promoters and termination sites predict that 66 percent of random DNA sequences will be transcribed due to nonspecific binding.[10] Experiments done in bacteria show that 10 percent of short random fragments of DNA can function as promoters demonstrating that spurious binding of transcription factors can induce transcription.[11]

This is all just basic biochemistry that should be familiar to most undergraduates. At any point in time, most of the transcription factors (and RNA polymerase) in a cell will be bound nonspecifically to DNA. This is true in all species, but it's especially true in species with large genomes since the nonspecific target DNA is in huge excess. The random binding of a transcription factor won't result in transcription in most cases, but in some cases a nearby RNA polymerase – also bound to a random sequence – will be triggered to begin transcription, accounting for a low level of random transcription noise throughout the genome.

Since different transcription factors are expressed in different cell types – liver, brain, muscle, and so on – we expect this sort of spurious transcription to be tissue-specific. In other words, spurious transcription that has nothing to do with biological function will have two characteristics: (1) the transcripts will be rare, and (2) they will be tissue-specific. That's exactly what we see in the transcription data.

Random transcription initiation is the rule

Although the tendency ... is to assume that any transcript represents a gene, classical genetics demands some evidence of associated function. Crucially, what is not yet established (but is implied to be relatively abundant by these studies) is the extent of biological "noise" in the transcriptome of any given cell. In other words, what fraction of transcripts which can be isolated have any meaningful function? What fraction might be mere by-products of spurious transcription, spuriously fired off ... ?

S.A. Aparicio (2000)

Not much has changed since that was written in 2000 – the year before the human genome sequence was published – when the author was reviewing conflicting reports on the number of genes. Thoughtful scientists are still warning others to keep in mind the fundamental properties of DNA-binding proteins when thinking about transcription; for example, Kevin Struhl issued this caution in 2007:

> Eukaryotes transcribe much of their genomes, but little is known about the fidelity of transcriptional initiation by RNA polymerase II in vivo. I suggest that ~90% of Pol II initiation events in yeast represent transcriptional noise, and that the specificity of initiation is comparable to that of DNA-binding proteins and other biological processes. This emphasizes the need to develop criteria that distinguish transcriptional noise from transcription with a biological function.[12]

Based on what we know about transcription, we expect that cells will contain a lot of junk RNA due to transcription noise, and that's exactly what we observe. (Another source of transcription noise comes from the 200,000 LTR promoters scattered throughout the genome and from other transposons and viral DNA sequences as described in chapter 3.)

Random transcription termination is also common

So far I've concentrated on showing that spurious transcription *initiation* should be expected in the human genome; now let's turn our attention to transcription *termination.*

In bacteria there has been strong selection for efficient transcription termination at the end of each gene because genes are densely packed and RNA polymerase will run on into the next gene if it doesn't stop when it should.

In contrast, transcription termination in eukaryotes is a very sloppy process. The consequences of run-on transcription are much less severe in eukaryotes because genes are farther apart, and RNA polymerase can easily run over a transcription termination site and keep going for a long time before it runs into a new gene. There's usually no strong selective pressure to evolve efficient transcription termination for most genes.[13]

Thus, it should not come as a surprise to learn that DNA downstream of a gene is transcribed by accident from time to time. In fact, a significant percentage of junk RNA transcription comes from regions that are downstream of active genes.

Sometimes RNA polymerase goes off in the wrong direction

Spurious transcription also occurs at the other end of the gene near the initiation site (promoter). Sometimes the assembled transcription complex fires off in the wrong direction and transcribes DNA that's upstream of the gene. The combination of spurious, accidental low-level transcription at each end of a gene accounts for the observation that a large percentage of junk RNA transcripts occur in the regions around known genes.[14]

Antisense transcription

I want to say a few words about antisense transcription if only for the sake of completeness. Recall that one strand of the double helix is copied by the transcription complex. We call this the template strand (Figure 1.8), and the resulting RNA transcript is complementary to the template strand. If it's an mRNA, then it contains codons that will be translated into protein. This is sense RNA.

But the same gene could be transcribed in the opposite direction beginning at the end of the gene and proceeding to the beginning. If the gene is transcribed in the opposite direction, then the resulting RNA will be *antisense RNA*. The antisense RNA will be complementary to the normal transcript, and if it's present in high enough concentrations, it could hybridize to form double-stranded RNA, which could prevent the normal functioning of the main transcript; for example, the double-stranded region could block translation (e.g., miRNA). Double-stranded RNA formed by antisense RNA is usually degraded by ribonuclease enzymes because double-stranded RNAs are often viruses that threaten the cell. Thus, in addition to blocking translation, another possible function of antisense RNA is to control the levels of the normal transcript by increasing turnover (e.g., siRNA).

The third possible function affects transcription directly. By transcribing the gene in the opposite direction, the cell can limit the amount of normal transcription because the colliding transcription complexes interfere with one another.

There are known examples of all three mechanisms dating back to the 1970s, so there is no question about the fact that antisense RNA can have a biological function. However, just because there are possible functions of some antisense RNAs does not mean that all antisense RNAs have a function. They could still be spurious transcripts or junk RNA; in fact, if accidental transcription is common, then we expect a significant amount of antisense RNA since random promoters should occur with equal probability on both strands of DNA.

There's another kind of transcript that could be called antisense RNA. As I mentioned earlier, there are transcripts from a normal gene promoter region that fire off in the opposite direction. This isn't unexpected since promoters are inherently bidirectional, but in most cases the arrangement of RNA polymerase at the promoter site is such that it takes off in the direction of the gene and produces a normal transcript. The result of misfiring is a low level of spurious transcription from normal promoters but in the opposite direction.[15]

In some cases, short transcripts from the promoter regions or from nearby enhancers (regulatory elements) may be necessary in order to keep the gene in an active chromatin configuration. These small RNAs aren't necessarily in the antisense direction, but that doesn't matter because these enhancer RNAs, or eRNAs, don't have a biological function on their own. Instead, it's the act of transcription that's required and the sequence of the RNA is irrelevant. The idea is that continuous transcription from these regions keeps the genes in an active open chromatin configuration. So far, there are only a few well-studied examples of this phenomenon, and the effects seem to be quite small.[16]

How much of our genome is transcribed?

The goal of the ENCODE Consortium is to find all the functional elements in the human genome. The consortium scientists published a preliminary study in 2007 that focused on just 1 percent of the genome, and one of the main conclusions of this pilot study was highlighted in the abstract of that paper: "First, our studies provide convincing evidence that the genome is pervasively transcribed, such that the majority of its bases can be found in primary transcripts, including non-protein-coding transcripts."[17]

This was not the first use of the term *pervasive transcription*, but it marks the time when most scientists became aware of the phenomenon. ENCODE looked at a wide variety of tissues and discovered that when you add up all the RNAs in all those tissues, it turns out that they are complementary to 93 percent of the genome. Since genes are DNA sequences that are transcribed, it follows that pervasive transcription suggests that genes occupy most of our genome, leaving very little room for junk DNA. Is this a valid conclusion?

No, it is not valid because it ignores a key part of our definition, namely, that the product of a true gene must be *functional*. If most of the transcripts don't have a biological function, then they aren't produced from real genes. Function is the key to understanding the controversy – you can't just assume without evidence that every detectable RNA has a biological function.

Does pervasive transcription indicate that the genome is full of unknown functional genes, or is much of this transcription just due to biochemical noise? The ENCODE researchers were remarkably coy about these possibilities in their initial publication. They mention the possibility that many of their transcripts might "reflect neutral turnover of reproducible transcripts with no biological role," but it's clear that they favor a functional explanation.[18]

As I emphasized above, the key question is whether all those transcripts are functional or not. It appears on the surface that ENCODE researchers were maintaining an open mind on this question, but subsequent statements reveal that the majority of them are not so open-minded. They believe that most of these transcripts are functional, and this belief conflicts with the view that most of our genome is junk. We know from several sources that ENCODE leaders were opposed to junk DNA back in 2007 (see below).

The media ignored the possibility that these transcripts could be noise; instead, they jumped all over the fact that most of the genome is transcribed and immediately assumed it was functional. Most of the articles in the popular press, including those from prestigious news organizations like the BBC, announced the death of junk DNA.[19]

Many of the ENCODE leaders joined in. John Mattick, for example, published an article in *The Scientist* where he "amicably disagreed with his colleagues" who expressed reservations about the importance of pervasive transcription. Mattick said, "Contrary to current dogma, most of the genome may be functional." He believes that most of the transcripts are required for gene regulation. Mattick also published a scientific review in the same year that the initial ENCODE paper came out, and in that paper he emphasizes the importance of functional noncoding RNAs and promotes the idea that junk DNA is a myth.[20]

Another ENCODE leader, Thomas Gingeras, wrote an essay the previous year with the provocative title "TUF love for 'Junk' DNA." (The acronym "TUF" stands for "transcripts of unknown function.") Gingeras and his coauthor, Willingham, conclude that "[t]he widespread occurrence of noncoding RNAs – unannotated eukaryotic transcripts with reduced protein coding potential – suggests that they are functionally important."[21]

Statements like this one and Mattick's reinforce the popular belief that the mere existence of a transcript implies function.

Many other ENCODE leaders were quoted in various newspaper articles and press releases as supporting the idea that most of our genome is functional because of pervasive transcription; for example, here's what one of the ENCODE leaders, Ewan Birney, said in 2007 to a reporter (John Lauerman) from Bloomberg.com: "People involved in genomics knew this stuff was not just hanging around for the hell of it. The junk is not junk. It is active, it does a lot of things."[22]

Birney is implying that genomics experts were skeptical of junk DNA even before the data was analyzed. That's interesting because I'm saying the exact opposite – I claim that many genomics experts were convinced that most of our genome is junk and that's why they were skeptical of the story that the ENCODE leaders were promoting. You will have to decide for yourself which scientists are more knowledgeable about the composition of our genome – is it Ewan Birney and the ENCODE scientists or those scientists whose views I described in previous chapters? Your decision is not going to be easy because articles that presume to debunk junk are pervasive and you have to look very hard to find any mention of the alternative explanation for abundant transcription, namely, transcriptional noise and no biological function.

It's true that, at the time (2007), very few scientists openly expressed doubt about the possible function of all these RNAs in the scientific literature, but there was considerable opposition on blogs and at scientific meetings, especially in the bars and pubs. There's no possibility that the ENCODE researchers were unaware of this criticism and unaware of the substantive arguments in favor of junk DNA.[23]

The pre-ENCODE history of pervasive transcription

Evidence for abundant transcription was first published in the 1960s. It was based on experiments in which isolated RNA was hybridized to DNA, and the amount of DNA found in DNA:RNA hybrids was calculated as a fraction of the genome size.[24] The results showed that most of the genome is transcribed.

This was a major paradox back then because introns had not been discovered, and it was assumed that genes occupied only a small percentage of the genome. Scientists soon realized that most of the newly synthesized RNA was in the nucleus whereas mature mRNA in the cytoplasm covered only a small percentage of the genome and this was subsequently explained by RNA processing and the removal of intron sequences. (At the time, this fraction was called heterogeneous nuclear RNA or hnRNA.) However, there was still a lot of transcription that wasn't restricted to introns in protein-coding genes.

Much of this earlier work has been forgotten, and that's why the rediscovery of pervasive transcription was a surprise to many scientists. Given that 45 percent of the genome has to be transcribed because it consists of known genes, the question is, What about the rest of the genome that's transcribed? In 2009 science writer Karen Hopkin explained the controversy like this when describing the ENCODE result:

> [I]t could be that there's just a ton of low-level transcription going on throughout the genome. Or, as someone in the "just noise" camp might put it: "The transcriptional machinery exists in the cell, DNA is in the cell, so these things happen," explains [Paul] Flicek, who does not actually subscribe to the philosophy.
>
> But some variants clearly do have biological functions, as evidenced by the activities of microRNAs. "So there are two models," [Laurence] Hurst says. "One, the world is messy and we're forever making transcripts we don't want. Or two, the genome is like the most exquisitely designed Swiss watch and we don't yet understand its workings. We don't know the answer – which is what makes genomics so interesting."[25]

Laurence Hurst is correct: we don't really know the answer for certain, although after fifty years of investigation, it looks very much like a messy genome is the answer. Nevertheless, you will certainly get the impression that scientists *do* know the answer to the question if you read the popular science literature, and the answer comports with the Swiss watch analogy. According to this view, the genome is not messy but "exquisitely designed." That's what the general public is being told, and that's what many scientists believe.

I am trying to convince you that this is probably the wrong answer. I think the genome is sloppy and most of the transcripts are nonfunctional noise. In other words, they are junk RNA. More important, I hope, at the very least, to convince you that the issue has certainly not been settled in favor of exquisite design. There is no strong consensus among biologists that one of the answers is correct despite what you might have read in the popular press, so even if you are not convinced that most of our genome is junk, you should at least know that this is still a very real possibility.

How do we know about pervasive transcription?

I haven't talked much about the development of sequencing technology, so this is a good time to mention it. The first two methods of sequencing DNA were developed in the 1970s in the labs of Walter "Wally" Gilbert at Harvard University in Boston (Massachusetts, USA), and Frederick "Fred" Sanger at the Medical Research Council Laboratory of Molecular Biology in Cambridge (UK). They were awarded the Nobel Prize in Chemistry in 1980.

The Sanger method proved to be much better and subsequent improvements led to the development of automated sequencing machines that could sequence single fragments of DNA rapidly and efficiently. This was the technology that produced the human genome sequence in 2001.

The next generation of sequencing technology was perfected in the first decade of the twenty-first century. It involves something called "pyrosequencing," which was a new way of detecting nucleotides using flashes of light. Using this method, thousands of fragments could be analyzed simultaneously with very sensitive photodetectors. This massively parallel sequencing – up to 1 million sequences at a time – opened up new avenues of research in genomics.

This was the technology that produced the complete genome sequence of Jim Watson in 2008.[26] The complete sequence of Craig Venter's genome had been published the previous year using first-generation sequencing technology.[27]

The new "next-gen" sequencing technologies allowed for massive experiments to sequence all the RNA molecules in dozens and dozens of tissues. It works like this:

1. Isolate all the RNA molecules from a tissue.
2. Convert them to DNA fragments using reverse transcriptase.

3. Sequence all the DNAs using several tricks that we needn't go into.
4. Match all the sequences to the reference human genome sequence in order to identify the DNA sequences that were transcribed.

Mapping such large data sets has been greatly aided by advances in computer technology. The complete technique of identifying transcripts is called *RNA-Seq*, and it was widely used in modern research labs in the first decade of this century.

"Next-generation" sequencing is not the latest technology. There are even better technologies available today, and a furious debate over whether they should be called third-generation sequencing.[28] The point is that with a well-characterized human reference genome, these new technologies allow for rapid sequencing of massive amounts of RNA that can be mapped back to the reference genome.

How many lncRNAs?

One of the most controversial topics these days is the number of biologically relevant lncRNAs. This is the same issue as the one concerning the general debate over pervasive transcription, but I want to specifically address lncRNAs since that's the most contentious class of RNAs. Do all lncRNAs have a biological function, or are some of them just noise?

The consensus view in the scientific literature is that most lncRNAs have a function even though the evidence doesn't support such a claim. There are many lncRNA databases. Some of them have more than 50,000 distinct RNAs, but there's not much overlap between different databases. If you combine all the databases there are more than 200,000 possible lncRNAs and the vast majority (>90 percent) are present at very low levels in the cells where they are detected. They are not conserved, and most of them have no known function, but nevertheless, they are all "lncRNAs" according to those databases.

In an attempt to make sense of the data, the RIKEN Center in Yokahama (Japan) compiled a list of 27,919 high-confidence lncRNAs and concluded that only 19,175 of these are *potentially* functional. Other databases claim that there are 50,000 "high-confidence" lncRNAs.[29] One database, EVLncRNA, suggests that there are about 2500 "experimentally validated functional lncRNAs" in the human genome using criteria designed to cover a very broad definition of function. This is not a large number given that the database contains more than 500,000 transcripts. The curators of this database say the following in the opening paragraph of a recent paper: "The discovery of one order of magnitude more transcripts coded for RNAs (i.e. non-coding RNAs) than proteins provided a paradigm shift in our understanding of genome regulation. Long non-coding RNAs (lncRNAs), in particular, have emerged as key players in essentially every biological process and associated with many human diseases including cancer, cardiovascular and neurodegenerative diseases."[30]

That's a good example of the common perspective in the scientific literature, but if there really is one order of magnitude more transcripts than proteins, then that means at least 200,000 noncoding RNAs. How many of these have a biological function? If only 2500 are functional, is that a paradigm shift?

You might think there's a serious effort underway to answer the question, but, as is so often the case, the general assumption is that everything has a function whether it has been demonstrated or not. This assumption attempts to put the onus on skeptical scientists to prove that a given lncRNA has no function, and that's the impossible task of proving a negative.

There are serious scientists who are attempting to address the question of function. For example, a group at Harvard looked at 25 well-characterized lncRNAs in zerbrafish and concluded that none of them had a biological function in spite of the fact that the scientific literature claimed that they were all functional.[31] Several other high-quality studies reach the same conclusion about potential lncRNAs in other species, but those studies are swamped by hundreds of papers promoting the idea that all lncRNAs are functional.

Does the term *lncRNA* refer specifically to an RNA with a function, or can we have nonfunctional lncRNAs? The correct null hypothesis is that these long noncoding RNAs are examples of noisy transcription, and you need evidence of function before you can conclude that they are real. For now they are just unknown transcripts. Thus, calling every long transcript a lncRNA is a bad idea when most of them are just junk RNAs. I once suggested that we could avoid the terminology by referring to new *lo*ng *l*ow *a*bundance *t*ranscripts as LOLATs and that they can only be called lncRNAs when a function has been demonstrated. That suggestion didn't catch on, and neither did the term TUF, for *t*ranscript of *u*nknown *f*unction. The latest attempt is VLA RNAs for "very low abundance" RNAs.[32] No matter what you call them, we need to have a term for transcripts that we assume are just noise until proven otherwise.

Nobody knows for sure how many of these transcripts will eventually turn out to have a biological function, but the important point is that the number has been pruned from more than 200,000 possibilities to less than 20,000. That means that the genome annotators have already concluded that 90 percent of the transcripts labeled as lncRNAs are probably just noise. So far, the number with a proven function is less than 500 in humans, and in my opinion, this number is unlikely to increase beyond 1000.

One of the ENCODE researchers, Thomas Gingeras, coauthored a major review in one of the leading scientific journals where he failed to even discuss the possibility that most of these transcripts were spurious, nonfunctional, mistakes in transcription; instead, he assumed that they had an important function and even suggested that pervasive transcription data will overthrow what he believed to be the "current dogma" on gene expression.[33]

Gingeras's assumption is another example of pervasive error in the field of genomics. It may turn out to be true that many of those low-abundance transcripts have a biological function – that's a point that's still open to debate. The scientific error occurs when you ignore alternative explanations that may conflict with your bias. That's not how science is supposed to work; you're supposed to acknowledge both sides of the argument.

One of the points I'm trying to make in this book is that science is often misrepresented in the popular press, but this isn't entirely the fault of science writers because scientists, themselves, often fail to meet the standards of science by thinking critically about their own data and what it means. I don't mean to imply that all scientists behave this way. Some scientists describe the controversy in the introduction to their scientific papers even though they may favor one side. At the end of their paper they will explain how their data support one side or the other and why they reject the evidence of their opponents. Unfortunately, these scientists are the exception, not the rule.

This dispute over the role of pervasive transcription occasionally surfaces directly in the scientific literature in the form of a debate. For example, a few years ago my colleagues Ben Blencowe and Tim Hughs at the University of Toronto (Ontario, Canada), questioned the standard interpretation of pervasive transcription in a paper they published in a widely read scientific journal.[34] They pointed out that the ability to detect very low-abundance RNAs leads to misleading statements about pervasive transcription – statements that imply biological function when, in fact, the low-abundance transcripts could be just noise. They specifically refer to the fact that many of the low-abundance RNAs are close to the ends of known genes and could be due to mistakes in transcription termination, allowing RNA polymerase to read through the normal termination signal and transcribe junk DNA.

The pro-RNA, anti-junk scientists responded with a letter to the journal in which they defend the idea that most of the transcripts are real and most have a function.[35] My colleagues responded with a brief lesson in scientific reasoning: "Of course some rare transcripts … are functional, and low-level transcription may also provide a pool of material for evolutionary tinkering. But given that known mechanisms – in particular imperfections in termination ... – can explain the presence of low-level random (and many non-random) transcripts, we believe the burden of proof is to show that such transcripts are indeed functional, rather than to disprove their putative functionality."[36]

I agree with my colleagues that the burden of proof is on functionalists. They need to show that pervasive transcription is biologically relevant given that there's a perfectly reasonable alternative explanation, namely, transcriptional noise. You can't just assume that every transcript has a function and use that to discredit the notion of junk DNA. The null hypothesis is noise.

The controversy continues, and the issue will not be resolved quickly. Most scientists in this field continue to act as though all RNAs have a function, and they continue to ignore data that indicate otherwise. They continue to selectively report data that implicate function – a classic example of the fallacy of confirmation bias.

I'm pleased to note that there's some indication that the tide may be turning since a small trickle of papers questioning this consensus has begun to appear in some of the leading journals. One of the most important of these skeptical papers is a major review published in the *Annual Review of Genomics and Human Genetics*. The authors of this "provocative review" are Chris Ponting at the University of Edinburgh (Scotland, UK) and Wilfried Haerty at the Earlham Institute in Norwich (UK). They looked carefully at the literature on lncRNAs and concluded that "[t]he majority of evidence, we argue, indicates that most lncRNA transcript models reflect transcriptional noise or provide minor regulatory roles, leaving relatively few human lncRNAs that contribute centrally to human development, physiology, or behavior."[37]

One of the most interesting parts of this review paper is the list of common "logical fallacies present in the literature on noncoding RNAs." The authors list 12 different fallacies, some of which

I've already described in this book. The exact nature of these logical fallacies isn't important – what's important is the recognition that they are so widespread in the peer-reviewed scientific literature.

REVISITING THE CENTRAL DOGMA?

The real meaning of the central dogma of molecular biology is "once information gets into protein it can't flow backward to nucleic acid" (chapter 3). The originator of the central dogma, Francis Crick, was well aware of genes that didn't encode protein. They don't figure into the central dogma.

Many scientists have a very different view of the central dogma. They were taught, incorrectly, that the real meaning of the central dogma is that DNA makes RNA makes protein (DNA → RNA → protein) and the only function of DNA is to encode protein in protein-coding genes. They were somehow led to believe that there was only one kind of gene, namely, protein-coding genes.

Apparently John Parrington was taught this incorrect view because he writes the following in his book on *The Deeper Genome*: "Surely there could hardly be a greater divergence from Crick's central dogma than this demonstration [by ENCODE] that RNAs were produced in far greater numbers across the genome than could be expected if they were simply intermediaries between DNA and protein. Indeed, some ENCODE researchers argued that the basic unit of inheritance should now be considered as the transcript."[38]

This sort of misunderstanding makes arguing the case for junk DNA all the more difficult because there are just too many misconceptions that have to be overcome all at once. In this case, it's a false understanding of the meaning of the central dogma and a false view that new discoveries call for

(continued)

REVISITING THE CENTRAL DOGMA? (continued)

revising our understanding of genes (see chapter 6). Then there's the false interpretation of pervasive transcription, namely, that all the transcripts are functional because of Swiss watches.

By the time I've finished correcting all these misconceptions, opponents of junk DNA can become very confused because they can't adjust quickly to so many challenges to their preconceptions. They need to change their entire worldview just in order to understand the arguments for junk DNA.

John Mattick proves his hypothesis?

John Mattick has been one of the most vocal opponents of junk DNA. He claims that our genome is full of regulatory sites and regulatory RNAs. To him, pervasive transcription is evidence that thousands of small RNAs are required for exquisite regulation of protein-coding genes.

His view is succinctly described in a 2012 press release from his university on the occasion of his winning the prestigious Chen Award from the Human Genome Organization. It's worth quoting:

> Mattick divided the scientific establishment in the mid-1990s with his then-radical theories about the vast stretches of DNA that do not code for proteins, but which are copied into RNA.
>
> "Genetics has been misunderstood for the past 50 years because of a fundamental assumption that turned out to be wrong – that most genetic information is transacted by proteins," he explained. "The so-called central dogma of molecular biology was that 'DNA makes RNA makes protein' and therefore that genes specified proteins through the intermediate of RNA….
>
> "It turns out that only a tiny fraction – around 1.5% – of the human genome encodes proteins. Put simply, the rest of it was condemned

as junk – because people knew, or thought they knew, that genes only made proteins, and so these vast tracts of non-coding DNA that didn't make proteins made no sense.

"My breakthrough came when I realised that not only did these non-protein-coding sections produce RNA, but if those RNAs were functional, it would mean that the system was much more sophisticated than we expected.

"The obvious and very exciting possibility was that there is another layer of information being expressed by the genome – that the non-coding RNAs form a massive and previously unrecognized regulatory network that controls human development. This is turning out to be correct, and is probably also the basis of brain plasticity and learning, although it is still early days.

"We now realise that the genome is extraordinarily complex, and the deeper we drill down, the more surprises we find. Indeed, what was dismissed as junk because it was not understood almost certainly holds the secret to understanding human development and cognition. It is also likely to hold the secret to understanding many complex diseases."[39]

I argue that this view is entirely false. It misrepresents the central dogma and the standard view of those scientists who support junk DNA, and it misrepresents the best explanation of pervasive transcription. The sad thing about this press release is that it reflects a growing trend in which university press releases are becoming major sources of scientific misinformation. I suspect this is because today's press releases are more about promoting the university and its faculty than about communicating accurate scientific information. All too often, the press release contains statements about the work being promoted that bears no relation to what's actually in the peer-reviewed literature. This is serious because universities are still widely recognized as legitimate authorities, so their press releases are not seen to be as self-serving as those from major private corporations.

Despite well-known and widespread objections to John Mattick's views, the Human Genome Organization awarded him a major prize. Here's why, according to the press release: "The Award Reviewing Committee commented that Professor Mattick's 'work on long non-coding RNA has dramatically changed our concept of 95% of our genome,' and that he has been a 'true visionary in his field; he has demonstrated an extraordinary degree of perseverance and ingenuity in gradually proving his hypothesis over the course of 18 years.'"[40]

I don't think you'll find very many knowledgeable scientists who agree with the Human Genome Organization even if they are opposed to junk DNA. Mattick has not *proved* his hypothesis. In fact, in my view, his hypothesis is almost certainly wrong.

While it's true that there are many scientists who agree with Mattick, to imply that every scientist has accepted his hypothesis is extremely misleading. There's a particular tactic used by Mattick and his collaborators in order to convince people that they are true visionaries. They set the stage by describing a (false) paradigm that supposedly dominated thinking up until recently. This makes new ideas look like a "paradigm shift" when it is overturned by these visionaries. In this case, the false paradigm is that everyone thought that protein-coding genes were the only kind of gene.

A person who goes by the name "Diogenes" is one of the frequent commenters on my *Sandwalk* blog, and he coined a very useful term for this trick of setting up a strawman to make your ideas look revolutionary. He calls it a "paradigm shaft." Let's look at another example to see how the paradigm shaft works. Here's a quotation from a paper from the Mattick lab: "Discoveries over the last decade portend a paradigm shift in molecular biology. Evidence suggests that RNA is not only functional as a messenger between DNA and protein but also in the regulation of genome organization and gene expression, which is increasingly elaborated in complex organisms."[41]

There is no paradigm shift here. Experts have known for decades that mRNA is not the only kind of RNA, and you can't have a paradigm shift if there's no paradigm to be shifted. You can, however, have a paradigm *shaft* – that's when people create a false view of the field and claim that new discoveries have overthrown the standard paradigm.

The null hypothesis

There must be a way to resolve this controversy over the possible role of pervasive transcription. There are tens of thousands of transcripts, and we would really like to know for sure if they have a biological function. It makes no sense to assume without proof that all these transcripts have a function since we know that spurious transcription is a very real phenomenon that could account for many of these transcripts. Indeed, it makes much more sense to assume they do *not* have a function until one has been demonstrated. This is what Laurence Hurst means when he wrote, "[W]e expect … accidental transcription factor-DNA binding to go on at some rate, so assuming that transcription equals function is not good enough. The null hypothesis after all is that most transcription is spurious and alternative transcripts are a consequence of error-prone splicing."[42]

This point has been made repeatedly by a number of scientists over the past few years, notably by Ponting and Haerty in their 2022 review of the lncRNA literature (see above). They quote an earlier article: "a sensible [null] hypothesis is that most of the currently long (typically >200 nt) noncoding RNAs are not functional, i.e. most impart no fitness advantage, however slight."[43]

My departmental colleagues Alex Palazzo and Eliza Lee at the University of Toronto also made this point about the null hypothesis a few years ago when they wrote a review on noncoding RNA (ncRNA) in which they said, "In the absence of sufficient evidence, a given ncRNA should be *provisionally labelled as nonfunctional*. Subsequently, if the ncRNA displays features/activities beyond what would expect for the null hypothesis, then we can reclassify the ncRNA in question as being functional."[44]

I agree that the null hypothesis should be *no function* and that the burden of proof lies on those who propose that these RNAs are functional, but how should we determine function? Palazzo and Lee discussed several guidelines that should help us decide, and I've listed my own versions of these below.

- ***Abundance:*** Most of the RNAs detected in these assays are present at less than one copy per cell. This result is consistent with random spurious transcription – the correct default hypothesis. It's extremely difficult to see how large numbers of such low-abundance RNAs could have a significant biological function. Consequently, the first step in establishing function is to show that the RNA molecules are present at a reasonable concentration, like all other functional gene products.
- ***How's it working out so far?*** We've reached the point where we can evaluate the track record of those who are looking for function. After several decades of work on tens of thousands of RNAs, scientists have only discovered about 500 new functional transcripts. This result does not look promising for those who are trying to refute the null hypothesis. Tens of thousands of putative ncRNAs have been rejected by genome annotators, establishing that spurious transcripts are far more abundant than those with established function.
- ***Conservation:*** Sequence conservation is an important indication of function (see chapter 4). There are exceptions, as Palazzo and Lee discuss in their paper, but if neither the sequence of the transcript nor its existence is conserved in other species, then it probably doesn't have a real biological function. In a recent review, only 1000 lncRNAs out of 60,000 were conserved in mammals.[45] This is consistent with the null hypothesis (spurious transcripts) and inconsistent with the view that most of these transcripts are functional. Of course, the real test of function is to delete the

suspected noncoding gene and see if it affects the fitness of the individual. Many such knockouts have been performed, and the data show that less than 8 percent of the very best candidates are functional by this criterion.[46]

- ***Specificity:*** It is often assumed that specificity is evidence of function, but this is an incorrect assumption. If most transcripts are accidental transcripts due to inappropriate binding of transcription factors, then they will only appear in cells where those transcription factors are present. Since different tissues and cell types synthesize different transcription factors, it follows that spurious transcripts will be cell-specific just like the transcripts from the real genes that are being regulated by the transcription factors.
- ***Localization:*** Many, but not all, functional transcripts are exported to the cytoplasm where they do their thing. Spurious transcripts, however, are usually confined to the nucleus, where they are rapidly degraded. In fact, mammalian cells have evolved a variety of mechanisms to rapidly eliminate most spurious transcripts.[47] Most of the transcripts from pervasive transcription are short-lived RNAs, and they are confined to the nucleus. That's consistent with junk RNA – the null hypothesis.

Let me emphasize, once again, that in order to make the case for function, you have to prove that an RNA is functional. You can't just assume that every RNA has a function and place the burden of proof on the other side, especially when there's so much evidence of spurious transcription. This emphasis on junk as the default explanation has been around for a long time, but it's routinely ignored by those who argue for function.[48]

In 2004, a group of scientists got together to make this point with respect to pervasive transcription in mice as detected by cDNAs. They pointed out that all the so-called evidence for function is consistent with junk, and they drew attention to the fact that the RNAs are not conserved: "Given that all of the best techniques for detecting RNA genes depend on sequence conservation, the absence of this cannot be summarily dismissed even if isolated examples of RNA genes being weakly conserved can be found. Extraordinary claims require extraordinary proof – this is particularly true when much of the data support an alternative interpretation that they are simply non-functional cDNAs."[49]

Unfortunately, most of the evidence for noise, such as low concentrations and a lack of sequence conservation, has been summarily dismissed by many scientists working in this field and the idea that the default assumption should be junk RNA (= noise) – is actively opposed.

THE RANDOM GENOME PROJECT

There's an experiment that scientists need to do. They need to insert a large hunk of random sequence DNA into one of the human chromosomes and see whether you can detect transcripts from this nonfunctional part of the genome. Such an experiment has been proposed by Sean Eddy, who claims that the random genome project is a missing negative control in studies of the human genome. Here's how he describes the project: "To clarify what noise means, I propose the Random Genome Project. Suppose we put a few million bases of entirely random synthetic DNA into a human cell, and do an ENCODE project on it. Will it be reproducibly transcribed into mRNA-like transcripts, reproducibly bound by DNA-binding proteins, and reproducibly wrapped around histones marked by specific chromatin modifications? I think yes."[50]

I agree with Sean. I think there will be numerous transcripts from the random synthetic DNA and the general impression from the analyses will be that the sequence is pervasively transcribed. I think a great many knowledgeable scientists would agree that this is the anticipated result. Such a result would not prove that most of the transcripts detected by ENCODE are spurious junk RNA, but it would certainly cast doubt on ENCODE's claim that most have to be functional merely because they exist.

Right now the Random Genome Project is just a thought experiment, but even as a thought experiment it has some value as noted by Sean Eddy: "Even as a thought experiment, the Random Genome Project states a null hypothesis that has been largely absent from these discussions in genomics. It emphasizes that it is reasonable to expect reproducible biochemical activities ... in random unselected DNA."[51]

On the origin of new genes

Some scientists don't like the idea of junk DNA but they're willing to concede that most of the minor transcripts are just noise. However, they note that aberrant transcription could still be advantageous because it leads to the evolution of new genes. The hypothesis is that lots of spurious transcription of junk DNA could have no direct benefit to an individual but could be a long-term advantage to a species. That's because some of these randomly transcribed sequences could evolve into functional protein-coding genes if mutations create a coding region and the RNA is translated. Alternatively, a noncoding RNA could accidentally acquire a noncoding function and become an miRNA or some other regulatory molecule.

This view is promoted by Jürgen Brosius: "In the past few years it has become clear that nonfunctional genomic DNA is a vast reservoir for co-option (exaptation) of novel genes, or parts of genes, generating significant diversity over evolutionary time. At various stages of 'decay', segments of this superfluous genomic mass have the potential to be exapted. Even though the odds are against it, exaptation is a pervasive evolutionary force that is at least as important for evolution as the adaptive changes of existing genes or regulatory elements by nucleotide exchanges."[52]

Jürgen Brosius is not the only scientist who promotes this view, but he happens to be someone who has thought about it quite a bit. He published a paper with Stephen Jay Gould in 1992 in which they proposed a bunch of names for junk DNA sequences that had the potential of becoming new genes. They suggested that DNA sequences with the potential to contribute to future use should be called "potonuons" or "potogenes."[53] That idea never caught on – it's one of the sillier ideas that Gould has supported – but it's based on the concept of exaptation that was covered in chapter 3.

Recall that exaptation refers to features that are currently maintained by selection because they have an important function but originally evolved because they had another function entirely. At some point in the past these features took on a new role.[54] An example used by Gould and Vrba was the use of feathers in flying. It seems likely that feathers first evolved as a protection from cold in ancient dinosaurs and later on became exapted for flying in birds.

Other examples of exaptation at the molecular level include the duplicated genes we talked about in the previous chapter, where one gene takes on an entirely new function. The proteins that make up the crystal structure in the lens of vertebrate eyes is another type of well-known example. Some of the crystallins in your eye are related to proteins called heat shock proteins that originally evolved to refold proteins after they were denatured by heat. Other lens crystallins are related to common metabolic enzymes, but the genes for these enzymes evolved for different reasons entirely. They were not selected to become lens crystallins; instead, they just happened to have the property of forming good transparent crystals under physiological conditions.

The idea that there might be sequences in the human genome that could become entirely new genes is similar to the idea of exaptation except that the original role was junk DNA, and the fact that these sequences are accidentally transcribed at some level provides an opportunity to evolve new functions. Perhaps maintaining a lot of junk DNA and spurious transcription is advantageous in order to allow for future evolution.

There are two main flaws in this argument. The first is that the formation of *de novo* genes is very rare, as we discussed in the previous chapter. It looks like the evolution of such genes in the human lineage occurs at a rate of about two new genes every million years, so it doesn't seem like junk DNA and spurious transcription are very efficient mechanisms for evolving new genes.[55]

The second major flaw is the same one that comes up time and time again in these discussions. Evolution can't see into the future, so there's no way for natural selection to maintain a huge amount of junk DNA – with or without spurious transcription – on the remote chance that some tiny bit of it will become functional a million years from now. These teleological arguments are inconsistent with our current understanding of evolutionary theory (see the discussion about introns in chapter 3). As the philosopher Michael Ruse once said, "Evolutionists want teleology but they cannot have it."[56] The potential for the future evolution of new genes is not a satisfactory explanation of pervasive transcription despite the fact that the idea keeps appearing in the scientific literature.[57]

The valid part of exaptation is not teleological. It does not claim that heat shock genes, for example, evolved *because* they were destined to become lens crystallins in the future. The correct view of exaptation is that it's a fortuitous event that was not predictable and the sequences that acquired a new function come from sequences that just happened to be available. In that sense, junk DNA does provide a reservoir for future evolution, but that doesn't mean that excess junk DNA is selected and maintained for future evolution.

To paraphrase the views expressed by both Richard Dawkins and Stephen Jay Gould, in one of the few times that they agree, we should not be fooled by hindsight into looking for causes when what we are seeing is historical contingency.[58]

CONSTRUCTIVE NEUTRAL EVOLUTION

A functional sequence is one that's currently preserved by purifying selection. Such sequences may have arisen by natural selection, in which case their current function is also related to their historical origin. However, they may have arisen by constructive neutral evolution, in which case their origin is not due to direct selection for their current function.

Constructive neutral evolution is a term coined by Arlin Stoltzfus in 1999 to describe the non-adaptive evolution of complexity. Similar ideas have been promoted by Michael Gray and Ford Doolittle at Dalhousie University in Halifax (Nova Scotia, Canada).[59]

Consider a simple example in which two proteins acquire mutations that allow them to stick together. The complex of two proteins does not affect the function of either one, so this interaction may become fixed in the population by chance (random genetic drift). Now imagine that one of the proteins acquires another mutation that's neutral in the context of the complex but detrimental if the proteins separate. (The mutation may destabilize the isolated protein, but this destabilization is offset by binding to the other protein.) We now have a situation in which the interaction of the two proteins has become essential, but this increase in complexity was driven by neutral mutations, not by selection for complexity.[60]

Many structures in the cell seem to be needlessly complex, especially in multicellular eukaryotes. The spliceosome is one obvious example that's often discussed in the constructive neutral evolution literature. The complexity could be accidental.

The origin of new protein-coding genes from junk RNA transcripts is another example of constructive neutral evolution. The presence of a spurious transcript in a cell is a fortuitous event that's effectively neutral. Similarly, the accidental creation of an open reading frame is another neutral event. The protein that's produced may increase fitness so that a new protein-coding gene has been created – one that will be maintained by purifying selection.[61] There was no direct selection for the new gene, and that's why it's an example of constructive neutral evolution. As Sean B. Carroll says in the title of his latest book, it's "A Series of Fortunate Events."[62]

What the scientific papers don't tell you

The modern scientific literature is full of papers on noncoding RNAs, and the general theme is that there's a vast array of important functional RNAs waiting to be discovered. The papers usually don't tell you about the abundance of the RNAs they've detected. They don't tell you that the concentrations are often less than one molecule per cell, and they don't tell you that the putative genes are not conserved. They certainly don't tell you that the presence of most of these RNAs is adequately explained as transcriptional noise.

In addition to hiding this information, these authors often employ another sleight of hand. They frequently begin their paper with a description of the huge number of transcripts detected by ENCODE and then go on to present evidence that one particular transcript has a function. The implication is clear – if one of these transcripts has just been found to have a function, then probably all of them have a function, and we just need to keep working to uncover their unknown functions.

That argument is illogical (see *post hoc fallacy* in chapter 3).[63] Nobody is saying we know the function of every bit of the genome – what we are saying is that there's good evidence that most of it is junk and most transcripts are probably junk as well. That doesn't rule out the possibility that a few functional transcripts might be discovered from time to time. The exception doesn't prove the generality. It's like looking at a pile of broken old jalopies in the junkyard and concluding that they are all junk, but then you happen to discover a brand-new Fiat hidden under a pile of rusty Fords, and you immediately assume that all the rest of the pile must also be working automobiles. That doesn't make sense.

Many scientists emphasize what we don't know and downplay what we do know. This is why so many recent papers in this field talk about the mysterious "dark matter" of the genome. But we actually know a great deal about our genome, including abundant evidence that most of it is junk.

Elizabeth Pennisi is a science writer who is fond of summarizing and promoting these distorted views of the genome in her *Science* articles. (*Science* is one of the top two science journals in the world.) Let's see what this looks like with respect to pervasive transcription – the topic of this chapter. Here's what she wrote in "Shining a Light on the Genome's 'Dark Matter.'"

> Genetic dark matter also loomed large when scientists surveyed exactly which DNA was being transcribed, or decoded, into RNA. Scientists thought that most RNA in a cell was messenger RNA generated by protein-coding genes, RNA in ribosomes, or a sprinkling of other RNA elsewhere. But surveys by Thomas Gingeras, now at Cold Spring Harbor Laboratory in New York, and Michael Snyder, now at Stanford University in Palo Alto, California, found a lot more RNA than expected, as did an analysis of mouse RNA by Yoshihide

> Hayashizaki of the RIKEN Omics Science Center in Japan and colleagues. Other researchers were skeptical, but confirmation soon came from Ewan Birney of the European Bioinformatics Institute and the Encyclopedia of DNA Elements project, which aims to determine the function of every base in the genome. The 2007 pilot results were eye-opening: Chromosomes harbored many previously unsuspected sites where various proteins bound – possible hotbeds of gene regulation or epigenetic effects. Strikingly, about 80% of the cell's DNA showed signs of being transcribed into RNA. What the RNA was doing was unclear.[64]

Do you see all the problems with this account? The first problem is her distortion of history. Pervasive transcription is not a new discovery that calls into question all previous thinking. Yes, it's true that the extent of pervasive transcription is greater than what we previously thought, but that's not a big deal.

Another problem is when scientists turn this result into a mystery called "dark matter." In the real world, it turns out that many knowledgeable researchers were "skeptical" about all these RNAs because they had a better explanation. They believed that much of the transcription was due to intron sequences and that much of the rest was spurious transcription due to the nonfunctional binding of transcription factors and RNA polymerase. This was consistent with a genome full of junk DNA – a view strongly supported by the data.

When Pennisi says that what the RNA was doing was "unclear," she is ignoring a large group of scientists who think they have the answer. The answer is that they aren't doing anything. There is no mysterious "dark matter" that needs explaining.

This is what the scientific papers and science journalists aren't telling you. What's wrong with telling the general public that pervasive transcription might just be noise due to sloppy transcription? Isn't that idea just as exciting as calling it "dark matter" and pretending that it's a mystery waiting to be solved?

THE FALSE LOGIC OF THE ARGUMENT FOR NONCODING RNAs

If you look back on the Nicholas Wade quotation at the beginning of this chapter, you'll see a succinct description of an attempt to solve the Deflated Ego Problem. According to Wade, and many scientists, the low number of protein-coding genes in humans can be explained by postulating that we have huge numbers of genes for noncoding RNAs and that explains how humans can be "much higher on the evolutionary scale" than fruit flies even though we have the same number of protein-coding genes.

As is the case with alternative splicing, the logic of this argument requires that "lower" species don't have all these extra things that are supposed to make us special. But pervasive transcription is ubiquitous. All species produce spurious transcripts, especially those with large genomes such as plants but even those with a small genome such as yeast, so you can't argue that humans are special just because we have more RNAs.

The only way around this faulty logic is to add an extra caveat to the argument, namely, that those extra RNAs have a function in humans but not in the "lower" species. This requires that you admit that other species may have lots of spurious nonfunctional transcripts but not humans. That's a tacit admission that junk DNA is common in other species, and it's a rather extreme argument for human exceptionalism. It requires that thousands of new noncoding genes have evolved in the lineage leading to humans but the spurious transcripts in worms and fruit flies have remained nonfunctional.

I don't think this makes a lot of sense. Noncoding RNA genes are not going to solve the Deflated Ego Problem. The best way to solve that problem is to simply admit that humans aren't that special.

Biochemistry is messy

Biochemical reactions are not perfect. Like all chemical reactions, there's an intrinsic error rate associated with biochemical reactions. This error rate is very low in the case of DNA replication, where incorrect nucleotides are only inserted once every million times, but even this low error rate is too high for effective evolution, so additional proofreading and repair reactions are required to correct most of the replication errors. As I explained in chapter 4, this process gives rise to an acceptable mutation rate of one in 10 billion for DNA replication. Other enzymes are more error-prone. For example, we've seen that the error rate in RNA splicing is as high as one in 10,000 or even one in 1000. This gives rise to many spurious, incorrectly spliced RNAs.

In this chapter I've described errors in initiating transcription and explained that RNA polymerases can often make transcripts using random DNA sequences. These errors are rare, but collectively they mean that almost every part of the genome will be transcribed at some time or other.

The evidence shows us that biochemistry is messy and that it's wrong to think of the cell as a finely tuned Swiss watch where every piece has an important function that has to work perfectly in harmony with every other part. This concept is difficult to accept because there's been so much emphasis on design in nature and because we were taught as undergraduates that all biochemical reactions work perfectly.

There's no denying that evolution has produced sophisticated molecular structures, but that does not mean they are perfect; instead, the reactions of transcription and splicing are only as good as they need to be to get the job done. François Jacob said it best when he described evolution as a tinkerer: "Natural selection has no analogy with any aspect of human behavior. However, if one wanted to play with a comparison, one would have to say that natural selection does not work as an engineer works. It works like a tinkerer – a tinkerer who does not know exactly what he is going to produce but uses whatever he finds around him whether it be pieces of string, fragments of wood, or old cardboard; in short it works like a tinkerer who uses everything at his disposal to produce some kind of workable object."[65]

Tinkerers are not artisans. Tinkerers make mistakes as anyone who has seen my home improvement projects can attest. Evolution is a tinkerer, not a sophisticated watchmaker (or an intelligent designer).

Change your worldview

Late adopters tend to resist novel ideas that they were not, in general, enculturated into when they developed their professional belief set. We might think of this as the "undergraduate effect": what one learns first tends to be more deeply entrenched in one's overall belief development, much as an earlier developmental process in biology affects the downstream phenotype in ways that are hard, if not impossible, to reverse. Many scientists assert as statements of faith things they learned in freshman year and that they have not since needed to revise in the light of empirical evidence.

John S. Wilkins (2013)

I remember the first time I read *The Selfish Gene* by Richard Dawkins and then started to read Stephen Jay Gould's essays in *Natural History*. Both men were advocating a new way of looking at evolution. Dawkins wanted us to see that evolution was primarily a change in the frequency of alleles in a population – the so-called gene-centric view of evolution. He also wanted us to recognize the overwhelming importance of natural selection and the power of adaptation – a point of view known as adaptationism.

Gould, however, was a powerful advocate of a pluralistic view of evolution where natural selection was just one of many mechanisms of allele frequency change and maybe not even the most prominent one. (Random genetic drift is now known to be the most common mechanism.) Gould also took a hierarchal position on the unit of evolution where not only genes but also organisms and populations (species) could be important. In addition, he emphasized the role of chance and accident (contingency) in determining the history of life, especially in his book *Wonderful Life*.

Dawkins and Gould understood that they weren't promoting any new facts or any new information that wasn't already widely available. Instead, they wanted us to change the way we thought about biology. Dawkins makes this very clear in the opening passage of *The Extended Phenotype* when he writes,

> This is a work of unabashed advocacy. I want to argue in favor of a particular way of looking at animals and plants, and a particular way of wondering why they do the things that they do. What I am advocating is not a new theory, not a hypothesis which can be verified or falsified, not a model which can be judged by its predictions. If it were any of these things, I agree with Wilson (1975) that the 'advocacy method' would be inappropriate and reprehensible. But it is not any of those things. What I am advocating is a point of view, a way of looking at familiar facts and ideas, and a way of asking new questions about them.[66]

Dawkins is a wonderful writer. You may agree with his point of view or disagree – I fall into the latter category – but there's no denying that his prose is spellbinding. I'm applying Dawkins's advocacy approach to a different subject. I'm trying to get you to look at certain facts about biochemistry from a different perspective. I want to change your point of view so that you see the messiness and sloppiness of molecular processes instead of imagining them to be the workings of a fine-tuned Swiss watch. This view is diametrically opposed to the view that Dawkins espouses in his many books, especially *The Blind Watchmaker*.

Gould also wants to change your worldview, and he doesn't disguise his goals. He writes in an entirely different style than Dawkins –

a style that turns off a great many people because it sounds elitist. In fact, it's just Gould's way of conveying complex and controversial ideas using the same kinds of allegories and metaphors that Dawkins uses.

Dawkins claims that his way of looking at things is neither better nor worse than any other point of view, but Gould takes a far different stance. He argues that some worldviews (e.g., adaptationism) are wrong and dangerous and need to be changed. He argues that there are better worldviews (e.g., his own worldview!), and I agree with him.

Gould explains this argument in an essay titled "In the Mind of the Beholder." It first appeared in the magazine *Natural History* in 1994, but you can read it in the anthology *Dinosaur in a Haystack*: "Nothing is more dangerous than a dogmatic worldview – nothing more constraining, more blinding to innovation, more destructive of openness to novelty. On the other hand, a fruitful worldview is the greatest shortcut to insight, and the finest prod for making connections."[67]

I think the dogmatic, constraining, worldview is that every fact about our genome has to be explained as an adaptation. For example, if there are multiple splice variants from the same gene, then there must be alternative splicing leading to many different proteins produced from a single gene, and if the genome is pervasively transcribed then, according to this worldview, the genome must be chock-full of noncoding genes.

The fruitful and liberating worldview, in my opinion, is that biochemistry is messy and mistakes are common. Rare transcripts can simply be accidents of transcription and processing. If this isn't the way you were taught to think about biochemistry, then I urge you to think again and, hopefully, change your worldview. Doing so may be difficult because of the undergraduate effect described by my friend John Wilkins in the quotation at the beginning of this section. If you were taught in school that Dawkins's view of marvelous design by natural selection is correct, then discarding that long-entrenched viewpoint is going to be hard.

What I've described so far was the consensus view of the human genome in 2012, keeping in mind that when I use "consensus," I mean the majority view of the experts in molecular evolution who were knowledgeable about the subject of genome evolution. I thought that the controversy over the origins of pervasive transcription had been solved. The balance of evidence suggested that the ENCODE researchers were wrong to assign functions to the abundant transcripts they had detected in their 2007 paper. I continued to teach my students that messy genomes and spurious transcription account for pervasive transcription. This is what I was teaching in my molecular evolution course for students in their final year of university, and it's what I was teaching in my second-year course on critical thinking.

A prominent intelligent design creationist named Jonathan Wells had just published a book called *The Myth of Junk DNA* in 2011. Intelligent design creationists have "predicted" that all this junk will turn out to be functional because a sloppy genome full of junk isn't consistent with an intelligent designer. Fighting with creationists is one of my hobbies – I've been at it for almost 30 years on the internet, and I had recently posted a series of articles on my blog refuting all their arguments against junk DNA.[68]

I was feeling pretty good. It looked like real science and critical thinking were winning, and people were beginning to understand that most of our genome is junk.

I was very wrong.

CHAPTER 9

The ENCODE Publicity Campaign

I would be quite proud to have served on the committee that designed the E. coli genome. There is, however, no way that I would admit to serving on a committee that designed the human genome. Not even a university committee could botch something that badly.

David Penny[1]

September 5, 2012, started out as a normal day. It was a Wednesday, and I was preparing for the start of my class on critical thinking in science. The first sessions were on the evolution/creation controversy during which I hoped to show my students that creationists don't understand science. As usual, I checked the blogs and my RSS feeds to see what was going on.

It was not going to be a normal day. The first thing I saw that day was a post by science journalist Ed Yong on his blog *Not Exactly Rocket Science*. He was announcing the publication of a massive number of papers by the ENCODE Consortium.

Ed Yong, along with many other science writers, had received advance notice of the papers that were about to appear in *Nature*, and they had been given the opportunity to interview the ENCODE Consortium leaders. This is quite normal – the publication of any information on what's about to appear in a science journal is under an embargo until the papers are published, and journalists must promise to respect that embargo in order to gain preliminary access to the papers. That's why there was a deluge of press reports appearing everywhere on September 5, the day the papers were released in the September 6, 2012, issue of *Nature*. (The papers were rejected by *Science* because the authors made unreasonable demands concerning peer review.[2]) The science writers had had a couple of days to do their research and write their summaries. This procedure ensures that the science journals get to put their spin on the research that's coming out before critics have a chance to see the papers. It's free publicity that will hopefully bring in more ad revenue.

Almost every single science writer reported the "amazing" news that ENCODE had discovered a function for 80 percent of the human genome. The headline news was that most of the genome was not junk according to ENCODE. Even Ed Yong, normally a skeptical writer, was sucked into the publicity campaign launched by Ewan Birney, the ENCODE Consortium leader, in collaboration with *Nature* and the complicity of *Science*. Here's what Ed Yong announced on that day:

> According to ENCODE's analysis, 80 percent of the genome has a "biochemical function". More on exactly what this means later, but the key point is: It's not "junk". Scientists have long recognised that some non-coding DNA probably has a function, and many solid examples have come to light. But, many maintained that much of these sequences were, indeed, junk. ENCODE says otherwise. "Almost every nucleotide is associated with a function of some sort or another, and we now know where they are, what binds to them, what their associations are, and more," says Tom Gingeras, one of the study's many senior scientists.
>
> And what's in the remaining 20%? Possibly not junk either, according to Ewan Birney, the project's Lead Analysis Coordinaor and self-described "cat-herder-in-chief." ... It's likely that 80 percent will go to 100 percent," says Birney. "We don't really have any large chunks of redundant DNA. This metaphor of junk isn't that useful."[3]

To his credit, Ed Yong qualified his reporting over the next few days by pointing out that he was only repeating what the ENCODE leaders had said. He even added an update where he quotes Ryan Gregory and me saying that there's much more to this than meets the eye and it's premature to dismiss junk DNA. In contrast, other science writers have still not admitted that they were taken in by the publicity campaign.

In my first blog post on Wednesday, September 5, 2012, I noted that the ENCODE claim was certainly wrong, and I linked to my article on *What's in Your Genome?* I ended the post with "This is going to make my life very complicated."[4]

That was an understatement. The published papers were quite technical and difficult to understand, even for someone like me who is familiar with the type of experiments they did. Recognizing this, the ENCODE leaders published a summary of the work in the lead paper titled "An Integrated Encyclopedia of DNA Elements in the Human Genome." Here's the abstract of that paper – I've highlighted the "amazing" part for emphasis: "The human genome encodes the blueprint of life, but the function of the vast majority of its nearly three billion bases is unknown. The Encyclopedia of DNA Elements (ENCODE) project has systematically mapped regions of transcription, transcription factor association, chromatin structure and histone modification. *These data enabled us to assign biochemical functions for 80% of the genome, in particular outside of the well-studied protein-coding regions.*"[5]

The take-home message is that 80 percent of the genome has a function. Quotations from many ENCODE Consortium leaders make it very clear that by saying "function," they meant *real function*, not junk. Two of those leaders (Tom Gingeras and Ewan Birney) were quoted in Ed Yong's article. Another, John Stamatoyannopoulos, says the same thing in a *Science* article written by Elizabeth Pennisi with the provocative title "ENCODE Project Writes Eulogy for Junk DNA."

> This week, 30 research papers, including six in Nature and additional papers published by Science, sound the death knell for the idea that

> our DNA is mostly littered with useless bases. A decade long project, the Encyclopedia of DNA Elements (ENCODE), has found that 80% of the human genome serves some purpose, biochemically speaking. "I don't think anyone would have anticipated even close to the amount of sequence that ENCODE has uncovered that looks like it has functional importance," says John A. Stamatoyannopoulos, an ENCODE researcher at the University of Washington, Seattle.
>
> Beyond defining proteins, the DNA bases highlighted by ENCODE specify landing spots for proteins that influence gene activity, strands of RNA with myriad roles, or simply places where chemical modifications serve to silence stretches of our chromosomes. These results are going "to change the way a lot of [genomics] concepts are written about and presented in textbooks," Stamatoyannopoulos predicts.[6]

This statement is important because many of those same ENCODE leaders said later on that they were misquoted and that they never meant to say that 80 percent of the genome had a real function. They were talking about "biochemical function," and that's not real function – at least that's what they want you to believe now that the dust has settled on this fiasco. It's also important to note that many science writers, like Elizabeth Pennisi, were more than willing to "sound the death knell" for junk DNA and promote the views of the ENCODE leaders because that's what they wanted to believe. As we shall see, reports of the death of junk DNA have been greatly exaggerated.

ENCODE results

The ENCODE researchers looked at 147 different types of human cells. Almost all of them were cell lines grown in culture in the laboratory. They created a total of 1640 data sets consisting of results from various types of experiments on various types of cells. Each lab group usually concentrated on a particular kind of experiment that was repeated over and over until the entire genome of a cultured cell line was covered. Then they repeated the analysis with a different cell line. Some bioinformatics lab groups were assigned the task of collecting all this data from different labs and assembling it into a manageable database.

ENCODE looked at transcribed regions of the genome by identifying RNAs using techniques such as RNA-Seq and others. They also looked at all the sites in the genome where transcription factors were bound.

Actively transcribed genes are found in "open" chromatin where the DNA is more accessible to binding transcription factors and RNA polymerase. There are several assays for open chromatin domains, and they all detect slightly different regions of the genome. The ENCODE data reveal that much of the genome is in an open domain in at least one cell type. In addition, researchers looked at regions of the genome that were modified by attaching methyl groups ($-CH_3$) to certain bases (methylation) because changes in methylation patterns correlate with changes in gene expression. They identified 20,687 protein-coding genes, with an average of 6.3 alternatively spliced transcripts per gene. Because of alternative splicing, the average gene makes about four different proteins, according the ENCODE Consortium. I didn't believe them for the reasons outlined in chapter 6.

The consortium identified 8801 small RNA genes and 9640 genes for long noncoding RNAs (lncRNAs), but this is misleading. What they actually identified is thousands of regions of DNA that are transcribed without offering any additional evidence that these are actually functional genes.

Recall that about 45 percent of the genome is transcribed at one time or another because known genes often have large introns. The ENCODE Consortium researchers detected most of these transcripts, including the ones corresponding to well-known RNAs like tRNA, snRNA, snoRNA, and miRNA. They also detected many additional DNA regions that were transcribed to produce potential lncRNAs. In total, 62 percent of the genome was transcribed in their assays.

This is pervasive transcription, and they claim that this is evidence of function just as they proposed in 2007.

This claim conflicts with the criticisms described in the previous chapter, but the newspapers and the associated publicity campaign ignored all those earlier criticisms of the 2007 results.

There are hundreds of transcription factors encoded in the human genome, but the ENCODE Consortium researchers looked at "only" 120 different transcription factor binding sites. They assayed 72 different cell types and found 636,336 binding regions covering 231 million base pairs, or 8.1 percent of the genome. ENCODE researchers believe that all these binding sites are indications of biological function. They are wrong.

The researchers identified almost 5 million sites of open chromatin domains, and as expected, these sites overlap the transcription factor binding sites. The sites cover a bit more than 40 percent of the genome. This finding isn't a big surprise since known genes account for this much of the genome and those regions will be in an open chromatin domain when the genes are expressed. But regions of junk DNA will also be detected from time to time in open chromatin domains, so this assay doesn't mean that all regions in open domains are functional.

Here's how the ENCODE researchers summarize their results: "Accounting for all these elements, a surprisingly large amount of the human genome, 80.4%, is covered by at least one ENCODE-identified element. ... Given that the ENCODE project did not assay all cell types, or all transcription factors, and in particular has sampled few specialized developmentally restricted cell lineages, the proportions must be underestimates of the functional bases."[7]

They don't mention possible *over*estimates of functional bases. In fact, the authors make just one single attempt at critically evaluating their data. They note that only 3 to 8 percent of the human genome is conserved; however, if 80 percent of the genome is functional but only 8 percent is conserved, then 72 percent is not conserved. What could all that non-conserved DNA be doing?

One possibility, according to ENCODE, is that the 72 percent of non-conserved DNA is "biochemically active with little overall impact on the organism." That's a nice way of spinning the result while avoiding the term *junk*. The ENCODE researchers did not take this possibility seriously even though it's the correct conclusion from their data.

Another possibility is that 72 percent of the genome has evolved new functions since we diverged from chimps about 6 million years ago. The ENCODE researchers like this explanation because it solves the paradox and allows them to reconcile two contradictory claims about function. But this solution is not credible since it is not possible for 2 billion base pairs of DNA to evolve new biological functions in only 6 million years.

Over the next few days, dozens of scientists attacked the ENCODE conclusions on blogs, Twitter, and Facebook and in the popular press. A few months later the harsh criticisms began to appear in the scientific literature.

The ENCODE publicity campaign

There's no doubt that *Nature* editors meant for the ENCODE results to be interpreted as evidence for function and against junk DNA. Brendan Maher, writing for *Nature*, made it clear that the goal of ENCODE is to find functional DNA and he went on to describe the vast amount of functional DNA discovered by ENCODE.[8]

There's another editorial in the September 6, 2012, issue written by Magdalena Skipper,[9] who, at the time, was a senior editor at *Nature*. (In 2018 she became the first woman to become editor in chief of *Nature*.) Skipper joined ENCODE leader Ewan Birney in a video produced by Illumina – one of the companies that makes DNA sequencers.[10] She reported in the video that the most striking result of ENCODE is that it assigns function to 80 percent of the genome. The reason this is so striking, she says, is because not so long ago we still thought that most of the genome was junk.

Skipper also introduced us to five expert researchers who were given advance access to the data and were invited to participate in the publicity campaign by sharing their views on the ENCODE results. None of these scientists raised any concerns about ENCODE's interpretation of the data.[11] The first one to comment was Joseph Ecker, and he says that the finding of function in 80 percent of the genome is remarkable because it dispatches the widely held view that most of the genome is junk. The other experts went along with this view at the time, but I'm sure they are now somewhat embarrassed about being taken in by the hype. The point is that the ENCODE Consortium leaders knew exactly what they were doing when they said that 80 percent of the genome is functional. As I said earlier, this is important because later on we will see the ENCODE leaders claiming that they never meant to pronounce the death of junk DNA in spite of the fact that in September 2012, they actively participated in the hype and none of them ever complained about being misrepresented.

Another video was produced by the National Human Genome Institute, part of the National Institutes of Health (NIH) in Bethesda (Maryland, USA).[12] This is a prestigious scientific organization with a lot of credibility. In this video, Ewan Birney, the ENCODE leader, says, "This metaphor about junk DNA has become, I think, very entrenched. It's been entrenched publicly, entrenched scientifically. And ENCODE totally challenges that. We just don't have big, blank, boring, bits of the genome; all the genome is alive at some level." Another ENCODE leader in this video is Mike Pazin, who states clearly that most of the genome is functional: "Very little of our genomes are junk. 80% of the genome is engaged in at least one biochemical activity."

Nature and Illumina teamed up to make a cartoon video with Tim Minchin as the narrator.[13] He explains that the whole genome is just buzzing with activity, even the parts that we used to think of as junk. The intended take-home message is obvious – the ENCODE result is a breakthrough that changes the way we think about the human genome.

The European Molecular Biology Laboratory (EMBL) also got into the act with another publicity video promoting ENCODE.[14] The blurb for this one says, "Ewan Birney of EMBL-EBI, Tim Hubbard of the Wellcome Trust Sanger Institute and Roderic Guigo of CRG talk about ENCODE, an international project which revealed that much of what has been called 'junk DNA' in the human genome is actually a massive control panel with millions of switches regulating the activity."

Tim Hubbard says, "Before we had the human genome sequence we didn't really know what most of the letters of DNA were actually doing. Now we know something around 80% of those letters, as a result of the data that ENCODE's collected. That's kinda removed the definition of junk DNA in some ways. A lot of what we previously we didn't know what it was, and we thought it might be junk, now clearly is involved in some sort of function."

Recall that Elizabeth Pennisi wrote the eulogy for junk DNA in the September 7, 2012, issue of *Science*.[15] That same issue has an article on "Ewan Birney: Genomics' Big Talker," also written by Elizabeth Pennisi. The standard view in the popular press was that Ewan Birney, the ENCODE Consortium spokesperson, was an insightful and careful scientist who had revolutionized the field of genomics by discovering that most of our genome was functional. "Big talker" was meant to be a complement back in those first few heady days of excitement. Today, it means something very different.

None of this is normal – most papers don't get promoted by such a deliberative and expensive publicity campaign. All this hype proves that the leaders of the ENCODE Consortium wanted to advertise their results as proof that most of our genome is functional. They orchestrated a publicity campaign to convince the general public that their work was revolutionary, and in this they were largely successful, but, unfortunately, they were also wrong. The human genome is still mostly junk DNA.

Now we must deal with the consequences of ENCODE's behavior.

Criticisms of ENCODE

Biochemists, geneticists, evolutionary biologists, and molecular biologists from all over the world reacted almost immediately to the ENCODE publicity campaign. They complained that ENCODE had misled everyone by choosing to redefine function in order to make their work look sexier.

The ENCODE Consortium claimed that a DNA sequence has a function as long as it bound a transcription factor, was in a region of modified chromatin, or was transcribed – that's how they got to 80 percent. But none of those features are reliable indicators of true biological function as many of us were quick to point out. ENCODE was often looking at spurious binding and spurious transcripts and assuming that they indicated function.

The editors of *Nature* soon realized they had a serious problem on their hands. They tried to rationalize their position by posting rebuttals on their website in the days immediately following the big splash. Brendan Maher, the feature editor for *Nature*, took the lead in an article published the very next day, saying, "First up was a scientific critique that the authors had engaged in hyperbole. In the main ENCODE summary paper, published in Nature, the authors prominently claim that the ENCODE project had thus far assigned 'biochemical functions for 80% of the genome.' I had long and thorough discussions with Ewan Birney about this figure and what it actually meant, and it was clear that he was conflicted about reporting it in the paper's abstract."[16]

That's an astonishing admission from *Nature*! A *Nature* editor is responding to blogs and Twitter the day after the papers were published and admitting that the most important conclusion is questionable and that the leader of the ENCODE Consortium was "conflicted." He doesn't tell us if the scientists who reviewed the paper were also "conflicted," which raises a question about how rigorously the paper was peer-reviewed before publication.

Brendan Maher then goes on to describe some comments made by Ewan Birney on his blog, where he says that 1 percent of the genome encodes proteins and 8 percent of the genome binds transcription factors. This gives a total of 9 percent. According to Maher, Birney and his colleagues estimate that there's another 11 percent of the genome with regulatory activity that ENCODE didn't detect because its sampling was incomplete.[17] He concludes, "That gets them to 20%. So, perhaps the main conclusion should have been that 20% of the genome in some situation can directly influence gene expression and phenotype of at least one human cell type. It's a far cry from 80%, but a substantial increase from 1%."[18] *

Yes indeed; it's true that "possibly 20% functional" is a far cry from 80 percent functional, and it certainly doesn't refute the junk DNA argument! Remember that Brendan Maher is a senior editor at *Nature* writing only 24 hours after the papers appeared, but by then it was too late because the worms were already out of the can. The damage had been done, and journalists all over the world were proclaiming the death of junk DNA. Nobody bothered to read the *Nature* website for the semi-retraction.

How does Brendan Maher defend *Nature* in light of such criticism? Here's how:

> ENCODE was conceived of and practised as a resource-building exercise. In general, such projects have a huge potential impact on the scientific community, but they don't get much attention in the media. The journal editors and the authors at ENCODE collaborated over many months to make the biggest splash possible and capture the attention of not only the research community but also of the public at large. Similar efforts went into the coordinated publication of the first drafts of the human genome, another research-building project, more than a decade ago. Although complaints and quibbles will probably linger for some time, the real test is whether scientists will use the data and prove ENCODE's worth.[19]

---

* I remind you, once again, that no knowledgeable scientist ever claimed that only 1 percent of our genome is functional – not in 1970, not in 2001, and especially not in 2012.

Nature and the ENCODE Consortium collaborated to "make the biggest splash possible" and, as a result, misled the public about their findings. The general public and science writers now believe that most of the human genome is functional. Don't ENCODE and *Nature* have a responsibility to correct their mistake?

Members of the ENCODE Consortium should have known that their claims of "biochemical function" were nothing more than evidence for spurious binding and spurious transcription, but there's very little evidence in the published papers that they seriously considered these possibilities. The irony is that this is a case of what Yogi Berra calls "déjà vu all over again." Many of us went through the same fight back in July 2007 when the ENCODE Consortium published (in *Nature*) a preliminary study of only 1 percent of the human genome.

I don't understand why the ENCODE Consortium didn't address these issues in 2012. Scientists are supposed to deal with criticism, not ignore it. John Timmer probably didn't understand either because the day after the 2012 papers come out he posted the following on the *Ars Technica* website:

> [T]he third sentence of the lead ENCODE paper contains an eye-catching figure that ended up being reported widely: "These data enabled us to assign biochemical functions for 80 percent of the genome." Unfortunately, the significance of that statement hinged on a much less widely reported item: the definition of "biochemical function" used by the authors.
>
> This was more than a matter of semantics. Many press reports that resulted painted an entirely fictitious history of biology's past, along with a misleading picture of its present. As a result, the public that relied on those press reports now has a completely mistaken view of our current state of knowledge (this happens to be the exact opposite of what journalism is intended to accomplish). But you can't entirely blame the press in this case. They were egged on by the journals and university press offices that promoted the work – and, in some cases, the scientists themselves.[20]

How did John Timmer get it right when so many other science journalists (and *Nature* publicists) were getting it wrong? It's because he has an unfair advantage – he's a biochemist/molecular biologist, and he knows what he's talking about!

Mike White is another one of those science writers who happens to be a scientist and an expert on the subject. He published an article in the *Huffington Post* just a few days after the big splash. Here's what Mike White said:

> If you read anything that emerged from the ENCODE media blitz, you were probably told some version of the "junk DNA is debunked" story. It goes like this: When scientists realized that classical, protein-encoding genes make up less than 2% of the human genome, they simply assumed, in a fit of hubris, that the rest of our DNA was useless junk. (You might have also heard this from your high school or college teacher. Your teacher was wrong.) Along came the ENCODE consortium, which found that, far from being useless, junk DNA is packed with functionality. And so everything scientists thought they knew about the genome was wrong, wrong, wrong.
>
> The Washington Post headline read, "'Junk DNA' concept debunked by new analysis of human genome." The New York Times wrote that "The human genome is packed with at least four million gene switches that reside in bits of DNA that once were dismissed as 'junk' but that turn out to play critical roles in controlling how cells, organs and other tissues behave." Influenced by misleading press releases and statements by scientists, story after story suggested that debunking junk DNA was the main result of the ENCODE studies. These stories failed us all in three major ways: they distorted the science done before ENCODE, they obscured the real significance of the ENCODE project, and most crucially, they mislead the public on how science really works.
>
> What you should really know about the concept of junk DNA is that, first, it was not based on what scientists didn't know, but rather on what they did know about the genome; and second, that concept has held up quite well, even in light of the ENCODE results.[21]

Note that John Timmer and Mike White are making the same point that I have been emphasizing in the previous chapters. They point out that many scientists do not understand the history of this subject and the substantial amount of evidence in support of junk DNA.

The reason why I was so annoyed on September 5 was because the ENCODE hype conflicted with the history that I had been teaching in my courses and because I had been making the case for junk DNA in my debates with creationists. I had been using the junk DNA debate as an example of a scientific controversy to teach students the importance of critical thinking. I told my students that when you look at all the evidence the only logical and scientific conclusion is that most of our genome is junk. But here was a group of prominent scientists – who were presumably experts on the subject of the human genome – saying that 80 percent of it had a function and that junk DNA was dead. The popular press went crazy because nobody wanted to believe we have a genome full of junk. The creationist bloggers were ecstatic because they thought that "Darwinism" and science have been disproved, once again. It made a mockery of my claim of critical thinking. That's why I said, "This is going to make my life very complicated."

It took a few months, but papers attacking the ENCODE Consortium claims began to appear in the scientific literature. Here's a list of the most important ones.

- Eddy, S.R. (2012). The C-value paradox, junk DNA and ENCODE. *Current Biology, 22,* R898. doi:10.1016/j.cub.2012.10.002
- Niu, D.-K., and Jiang, L. (2013). Can ENCODE tell us how much junk DNA we carry in our genome? *Biochemical and Biophysical Research Communications, 430,* 1340–3. doi:10.1016/j.bbrc.2012.12.074
- Doolittle, W.F. (2013). Is junk DNA bunk? A critique of ENCODE. *Proceedings of the National Academy of Sciences, 110,* 5294–300. doi:10.1073/pnas.1221376110

- Graur, D., Zheng, Y., Price, N., Azevedo, R.B., Zufall, R.A., and Elhaik, E. (2013). On the immortality of television sets: "Function" in the human genome according to the evolution-free gospel of ENCODE. *Genome Biology and Evolution, 5*, 578–90. doi:10.1093/gbe/evt028
- Eddy, S.R. (2013). The ENCODE project: Missteps overshadowing a success. *Current Biology, 23*, R259–61. doi:10.1016/j.cub.2013.03.023
- Hurst, L.D. (2013). Open questions: A logic (or lack thereof) of genome organization. *BMC Biology, 11*, 58. doi:10.1186/1741-7007-11-58
- Morange, M. (2014). Genome as a multipurpose structure built by evolution. *Perspectives in Biology and Medicine, 57*, 162–71. doi:10.1353/pbm.2014.000
- Palazzo, A.F., and Gregory, T.R. (2014). The case for junk DNA. *PloS Genetics, 10*, e1004351. doi:10.1371/journal.pgen.1004351

This reaction is truly extraordinary. I've never seen such a strong and forceful response to the claims made by prominent scientists in a leading journal. All these papers (and more) are objecting to the idea that most of our genome has a function as claimed by ENCODE. They defend the view that most of our genome is junk.

The criticisms are sometimes harsh. ENCODE researchers are accused of exaggerating their results both in their scientific papers and in the publicity campaign that followed. The scientific criticisms focus on three points:

1. ENCODE ignored all the earlier scientific evidence and data showing that most DNA is junk.
2. ENCODE ignored all the scientific evidence indicating that much of their "biochemical activity" is spurious and not an indication of biological function.
3. ENCODE and *Nature* collaborated to deliberately hype their results, thus misrepresenting to the general public the actual conclusions of the experiments.

I think Ford Doolittle does an excellent job of describing the main objections to the ENCODE publicity campaign in an interview he gave in 2015.[22] He rightly criticizes the ENCODE scientists for their behavior:

> [I]n my mind, there are two devastating things you can say about the ENCODE people.
>
> One is that they completely ignored all that history about junk DNA and selfish DNA. There was a huge body of evidence that excess DNA might serve some structural role in the chromosomes, but not informational. They also ignored what philosophers of biology have spent a lot of time asking: what do you mean by "function?" And you can mean one of two things: we might mean either what natural selection favored, which is what I think most biologists mean, or we might mean what it does. Some people might say, "Well the function of this gene is in the development of cancer," but they don't really mean that natural selection put it there so that it would cause cancer. These are not-so-subtle differences.
>
> I think many molecular biologists and genomicists, in particular, think that each and every nucleotide is there for a reason, that we are perfect organisms. It is almost as if we were still theists thinking God doesn't make junk; we just now think natural selection doesn't make junk. I think there is a deep issue about the extent to which we are noisy creatures and the extent to which we are finely honed machines. I think the latter view informs much of genomics, and I think it is false.[23]

What he is saying is that, in addition to ignoring the history, the ENCODE scientists were probably misled because they interpreted their results in terms of a finely tuned Swiss watch and not a noisy, sloppy genome. That's an explanation of their behavior but not an excuse.

The scientific criticism of ENCODE was not widely reported in the popular press and not widely read by scientists since these papers are in less important journals and were not promoted by a massive publicity campaign. As a result, science writers and the general public are not generally aware of the criticism, and neither are most scientists.

Science journal doubles down

In December 2012, three months after the ENCODE papers were published in *Nature*, the editors of *Science* chose ENCODE as one of the top 10 breakthroughs of 2012. Their summary demonstrates that the editors and writers were prepared to downplay all the criticisms that had appeared over the past few months. Here's what they wrote:

> ENCODE researchers took an intensive look not just at genes but at all of the DNA in between. Their results drive home that much of the genome that at one time was dismissed as "junk DNA" actually seems to play an essential role, often by helping to turn genes on or off. They pinpointed hundreds of thousands of landing spots for proteins that influence gene activity, many thousands of stretches of DNA that code for different types of RNA, and lots of places where chemical modifications serve to silence stretches of our chromosomes, concluding that 80% of the genome was biochemically active.[24]

The figure accompanying this description states that 80 percent of the genome is functional. To be fair, *Science* was honest enough to note the controversy surrounding this "breakthrough," although the editors forgot to mention that they had contributed to the misleading hype and they forgot that most of our genome is still junk.

> When these papers were published in September, the media went wild. ENCODE was hailed in The New York Times as a "stunning resource" and "a major medical and scientific breakthrough" with enormous and immediate implications for human health. The Guardian called it

"the most significant shift in scientists' understanding of the way our DNA operates since the sequencing of the human genome."

But several scientists in the blogosphere called the coverage overhyped and blamed the journals and ENCODE leaders for overplaying the significance of the results. For example, ENCODE reported that 76% of DNA is transcribed to RNA, most of which does not go on to help make proteins. Various RNAs home in on different cell compartments, as if they have fixed addresses where they operate, suggesting that they play a role in the cell. Critics argue, however, that it was already known that a lot of RNA was made, and that many of these RNAs may be spurious genome products that serve no purpose. Likewise, one ENCODE researcher found 3.9 million regions across 349 types of cells where proteins called transcription factors bind to the genome – but again, it's unclear how much of that binding is functional.

Nonetheless, ENCODE stands out as an important achievement that should ease the way for more insights into the genome.

This is ridiculous. How can the ENCODE results be a major breakthrough if their main conclusions are wrong? How can it be an "important achievement" if the ENCODE researchers can't tell the difference between functional DNA and junk DNA and their conclusions are being actively challenged by the scientific community?

ENCODE backpedals

Faced with all this criticism, the leaders of the ENCODE Consortium were forced to respond, but they didn't respond in the popular press where they might have had as much impact as their initial publicity campaign. They didn't respond in the leading science journals, *Nature* and *Science*, where they might have reached a majority of scientists who were taken in by the 2012 papers. They didn't make a new video starring Tim Minchin (or take down the old one). Instead, they responded at the end of April 2014 with a wishy-washy paper in the *Proceedings of the National Academy of Sciences (USA)*.[25]

They began by recognizing that there was a "discrepancy" between biochemical activity as an indication of function and evolutionary conservation as evidence of function. This was their way of conceding that their definition of function was flawed. They now admit that their original claim of 80 percent function has certain limitations that have been highlighted by other scientists. Later on in the paper they discussed, for the first time, the possibility that biochemical activity may not be an indicator of function because it can occur "stochastically." This is an obtuse way of admitting that much of their biochemical activity may be noise or artifact.

Sixteen months after they launched their publicity campaign, the ENCODE researchers were backpedaling by saying that evolutionary lines of evidence should be considered more seriously – something that critics had stated from the beginning. Evolutionary lines of evidence emphasize sequence conservation. However, according to ENCODE, there are problems with using evolutionary conservation as an indicator of function because it's hard to detect conserved regions, especially if they are short. In addition, as they claimed in the original paper, newly evolved, human-specific functional sequences will not be conserved, by definition. Thus, according to ENCODE, the lack of conservation isn't a problem. Very few experts in molecular evolution accept this rationale (see chapter 4).

The ENCODE leaders admitted that the case for junk DNA is much stronger than they knew, or acknowledged, back in 2012. They finally described the correct history and the evidence for junk DNA from the C-Value Paradox, genetic load, an understanding of modern population genetics, and the fact that half of our genome consists of bits and pieces of defective transposons. In other words, they admitted that there was substantial evidence for junk DNA that they had ignored two years earlier.

Having paid lip service to the views of their opponents, the ENCODE leaders try to defend their original claim by emphasizing how difficult it is to prove function. The case for function takes up three times as much space in the paper as the case for junk, but it doesn't tell us anything that wasn't in the 2012 original papers.

It's the very next section of the paper where the furious backpedaling began in earnest. They mentioned, for the first time, that much of their pervasive transcription – which they said covers 75 percent of the genome – consisted of very rare transcripts and that 70 percent of this coverage is from transcripts present at less than one copy per cell. What does this mean? According to the new view from the ENCODE leaders, it means that not all transcripts are functional because they have come to the realization that transcription (and splicing) is error-prone. Then there's this remarkable statement – a statement that should have been in the original 2012 paper: "In general, sequences encoding RNAs transcribed by noisy transcriptional machinery are expected to be less constrained, which is consistent with data shown here for very low abundance RNA."[26]

This assertion is an admission that most of what they called pervasive transcription is noise – just like their critics were saying back in September 2012 and in 2007. Furthermore, most of those transcripts are not conserved – just as you would expect if it came from junk DNA. Presumably they knew back then that most of their transcripts were from non-conserved regions of the genome and were present at very low levels in the cells, but I guess they just forgot to mention it because it wasn't important.

The ENCODE leaders also admitted that many of their chromatin marks and transcription factor binding sites are just noise and not indicative of function.

Given all the problems with interpreting their data, what conclusion do they now reach concerning biological function in the human genome? Surely, it's reasonable that they should reach some conclusion about function given that ENCODE spent several hundred million dollars in order to find out which parts of the genome are functional, right?

Wrong! The new ENCODE goal, according to Kellis et al. in 2014 is not to find out how much of the genome is functional. Instead, it's the following: "[T]he Encyclopedia of DNA Elements Project [ENCODE] was launched to contribute maps of RNA transcripts, transcriptional regulator binding sites, and chromatin states in many cell types."[27]

Really? The new revisionist goal of ENCODE was just to complete a bunch of experiments mapping all the spurious transcripts, all the spurious transcription factor binding sites, and all the insignificant, nonfunctional chromatin states in the genome? Given this new (retroactive) goal, the project is an unqualified success! Here's how Kellis et al. now describe their new goal, "The major contribution of ENCODE to date [2014] has been highly-reproducible maps of DNA segments with biological signatures associated with diverse molecular functions. We believe that this public resource is far more important than any interim estimate of the fraction of the human genome that is functional."[28]

So, all that talk about function and the death of junk DNA back in 2012 was a mistake because the real value of the ENCODE data is simply to provide maps of all the biochemical noise scattered throughout the genome! Contrast this goal with what was said in the 2012 summary paper "The Encyclopedia of DNA Elements (ENCODE) aims to delineate all the functional elements encoded in the human genome."

This semi-retraction has not been as widely publicized as the original claims that they now disown. I'd love to see *Nature* make a new video in which Tim Minchin recites an updated version of "Storm,"* emphasizing the importance of critical thinking about the human genome.

Dan Graur has written a quick summary of the ENCODE backpedal paper of 2014: "My interim summary: In 2012, ENCODE lied and deceived so well that a few ignorant dimwits from *Science* to the *New York Times* fell for it big time. Now that their stupid

* "Storm" is a 9½-minute jazz-backed beat poem about critical thinking.

claims came back to bite them in the proverbial junk, ENCODE proclaims that ENCODE never gave a shit about the fraction of the human genome that was functional. ENCODE in its new incarnation is presented as merely a huge exercise in stamp collecting. Actually, a huge tentative exercise in stamp collecting. Surprise, surprise ... some of their 'biochemical signals' are now claimed to be 'low.'"[29]

For the record, I agree with Dan Graur, and I think the strong language is justified and necessary. If this sounds too harsh, consider what the lead author, Manolis Kellis, says in a *Nature* editorial that appeared a short time later: "Kellis says that ENCODE isn't backing away from anything. The 80% claim, he says, was misunderstood and misrepresented. Roughly that proportion of the genome might be biochemically active, he explains, but some of that activity is undoubtedly meaningless leaving unanswered the question of how much of it is really functional."[30]

Hmmm ... I don't know about you, dear reader, but that seems disingenuous to me. Judge for yourself whether the 80 percent claim was misrepresented or whether the ENCODE Consortium is now trying to escape responsibility by blaming it all on the media.

Personally, I don't believe for a second that the 80 percent claim of function was misrepresented in the media. I think the leaders of the ENCODE Consortium really believed that most of the human genome was functional. I think they really believed that they had refuted the evidence that our genome is full of junk DNA. I think most of them still believe that.

After the dust had settled, a group in France published a summary of the history of the fiasco that was intended to correct the false impression created by the media in 2012. The abstract says it all.

> In September 2012, a batch of more than 30 articles presenting the results of the ENCODE (Encyclopaedia of DNA Elements) project was released. Many of these articles appeared in Nature and Science, the two most prestigious interdisciplinary scientific journals. Since that

time, hundreds of other articles dedicated to the further analyses of the Encode data have been published. The time of hundreds of scientists and hundreds of millions of dollars were not invested in vain since this project had led to an apparent paradigm shift: contrary to the classical view, 80% of the human genome is not junk DNA, but is functional. This hypothesis has been criticized by evolutionary biologists, sometimes eagerly, and detailed refutations have been published in specialized journals with impact factors far below those that published the main contribution of the Encode project to our understanding of genome architecture. In 2014, the Encode consortium released a new batch of articles that neither suggested that 80% of the genome is functional nor commented on the disappearance of their 2012 scientific breakthrough. Unfortunately, by that time many biologists had accepted the idea that 80% of the genome is functional, or at least, that this idea is a valid alternative to the long held evolutionary genetic view that it is not.[31]

ENCODE III in 2020

The ENCODE Consortium continues to be well funded to the tune of millions of dollars. They published their latest results in August 2020 without a massive publicity campaign. The latest results are that the human genome contains 20,225 protein-coding genes and 37,595 noncoding genes.[32] (Both of these numbers are likely to be overestimates, especially the number of noncoding genes.)

The latest results do not establish how much of the genome is functional and how much is junk. They simply extend the ENCODE data to more tissues and more transcription factors with a new goal stated as "The purpose of ENCODE is to provide valuable, accessible resources to the community."[33]

There are now 2,157,387 open chromatin regions and 1,224,154 transcription factor binding sites, but ENCODE avoids telling us what this means. According to the revisionist goals of the ENCODE Consortium researchers, the effort up until now has been just a huge exercise in identifying meaningless features of the human genome (stamp collecting). They say, however, that the next phase (ENCODE IV) promises to be much more informative because new techniques will help identify functional elements more accurately and that's going to "greatly enhance our understanding of human biology and disease." I'm pretty sure that ENCODE IV is going to require millions of dollars of new grants, but I'm skeptical that it will advance our understanding of function because I think we already know the answer – 90 percent of our genome is junk.

Over the past 20 years the big ENCODE labs have trained hundreds of graduate students and postdocs. Many of them have taken up research positions at major universities all around the world, so we now have a new generation of research scientists who are experts in the latest technologies but have come from labs that, in my opinion, have misrepresented their results. The ultimate effect of this on the science of genomics remains to be seen, but I don't think it will be good.

What went wrong?

Part of the problem with the 2012 ENCODE fiasco was the pressure to make their work seem more important than it really was, but that's not the whole story. It's clear that the ENCODE Consortium leaders actually believed they were uncovering real biological function in most of the genome. The problem is not that they had opinions and biases because we all have opinions and biases. The real problem is that they forgot how to be scientists. Here's how Stephen Jay Gould explains the special obligations of scientists: "we scientists are no different from anyone else. We are passionate human beings, enmeshed in a web of personal and social circumstances. Our field does recognize canons of procedure designed to give nature the long shot of asserting herself in the face of such biases, but unless scientists understand their hopes and engage in vigorous self-scrutiny, they will not be able to sort unacknowledged preference from nature's weak and imperfect message."[34]

Gould was writing about another situation in which scientists behaved badly, but this is good advice for all scientists in all situations. We scientists need to be extra careful about examining our conclusions to make sure they don't just reflect our biases and prejudices. "Vigorous self-scrutiny" is one way to describe this rule of science.

This failure of self-scrutiny has been pointed out by many ENCODE critics, including Laurence Hurst in one of the early criticisms that I quoted in the last chapter. The ENCODE scientists forgot that there's a "null hypothesis" that needs to be taken into account when you undertake self-scrutiny. In this case, the null hypothesis is to assume no function, meaning you have to provide convincing evidence of function before you can legitimately conclude that a given sequence has a real function. In the case of pervasive transcription or alternative splicing, the null hypothesis would be that these are the result of spurious accidents and not an indication of function.

The National Academies of Sciences (USA) published a report on scientific integrity in June 2015, and it's worth quoting from that report: "Like all human endeavors, science is imperfect. However, as Robert Merton noted more than a century ago 'the activities of scientists are subject to rigorous policing, to a degree perhaps unparalleled in any other field of activity.' As a result, as Popper argued, 'science is one of the very few human activities – perhaps the only one – in which errors are systematically criticized and fairly often, in time, corrected.' Instances in which scientists detect and address flaws in work constitute evidence of success, not failure, because they demonstrate the underlying protective mechanisms of science at work."[35]

What this means is that all scientists should be willing to accept criticism because this is how science works. Not only should they be *willing* to accept criticism, but they also should *expect* it, especially when they say something controversial.

If all the ENCODE critics are right, then we have witnessed a snow job that's unprecedented in biology. It has set back the field of genomics for more than a decade.

So, who do we blame for this fiasco? I blame the ENCODE Consortium leaders, especially the principal investigators who wrote the summary paper in *Nature,* but there are 20 other principal investigators who share the blame. Many of them spoke out in support of the extravagant claims of functionality.

The ENCODE leaders knew they were saying something controversial so they must have known what to expect. And they should have taken steps to deflect criticism by being open and honest about what their data said and didn't say.

Another part of the National Academies report is important: "All researchers need to realize that the best scientific practice is produced when, like Darwin, they persistently search for flaws in their arguments. … the need for vigilant self-critique may be especially great in research with direct application to human disease."[36]

The ENCODE Consortium leaders did not exercise "vigilant self-critique." For that reason they are the ones who bear the most responsibility for the fiasco.

The leading science journals share the blame because they did not exercise due diligence in fact-checking before publication and their peer-review process didn't work. Yes, it's true that they were deceived by some big-name scientists, but that's not an excuse because leading science journals like *Nature* and *Science* are supposed to be aware of this possibility. In this case, not only were the journal editors duped, but they also actively participated in publicizing the misinformation.

CHAPTER 10

Turning Genes On and Off

Trying to conceptualize the forces that act on genome evolution is not just a matter of semantics. We can envision the human genome as a perfectly honed machine, or we can think of it as a wild landscape littered and layered with successions of decomposing molecular replicators like dead weeds decaying into a fertile soil.

Sean Eddy (2013)

ENCODE scientists promoted the idea that much of our genome is devoted to regulating the expression of our genes, but I've left coverage of regulation to the end of this book because the discussion has to include a critical analysis of the ENCODE results from 2007, 2012, and 2020. Before delving into that discussion, it's best to review the history of gene regulation.

By the end of the 1960s, most molecular biologists would have agreed with Jacques Monod's declaration from 1961: "anything found to be true of *E. coli* must also be true of elephants."[1] They were expecting that bacteria and eukaryotes would share a great many genes, and they were anticipating that the regulation of gene expression in eukaryotes would be similar to mechanisms being discovered in bacteria and bacteriophage.

Initially, the idea that elephants were similar to *E. coli* was challenged by the discovery that mammals had large genomes. All this extra DNA seemed to require an explanation – perhaps it was required for regulation? But, as we have seen in earlier chapters, the puzzle was solved with the development of the neutral theory and the realization that large genomes need not be an adaptation, leading to the conflict between two differing views of biochemistry – the Swiss watch view and the messy view – as highlighted by Sean Eddy in his (2013) criticism of ENCODE (see the chapter-opening epigraph).

Like many of the protagonists in this clash of worldviews, Sean Eddy is a veteran of debates with Intelligent Design Creationists – a group that sees Swiss watches everywhere they look.[2] The arguments for sloppiness were honed in those debates, but while it was easy to refute the arguments of intelligent design proponents, it turned out to be far less easy to convince the average scientist to change their worldview. This is why many scientists still think that large genomes have a function and one of those functions is regulating gene expression.

Ohno made the case in 1972 that the regulation of eukaryotic genes cannot be much more complicated than the regulation of classic prokaryotic genes, and by 1980 there were enough examples of mammalian gene regulation to indicate that expression was controlled by a small number of regulatory elements in the genome. There is no need to postulate that gene regulation in elephants is vastly more sophisticated than in *E. coli*.

Nevertheless, the conflict between differing views of regulation continues to this day in spite of the evidence; for example, Francis Collins, the former head of the International Human Genome Project and recently retired director of the American National Institutes of Health, believes that the latest genome data points to complex regulation and refutes junk DNA, saying in 2015:

> I would say, in terms of junk DNA, we don't use that term any more 'cause I think it was pretty much a case of hubris to imagine that we could dispense with any part of the genome as if we knew enough to say it wasn't functional. There will be parts of the genome that are

just, you know, random collections of repeats, like Alu's, but most of the genome that we used to think was there for spacer turns out to be doing *stuff* and most of that stuff is about regulation and that's where the epigenome gets involved, and is teaching us a lot.[3]

Collins believes that the concept of junk DNA was originally based on a lack of understanding, but now he thinks we know enough to say that the extra DNA is required for regulation. Collins is wrong because the case for junk DNA was not based on ignorance but on solid evidence that most of the genome cannot be functional, and as we shall see, I think he is also wrong about regulation because the evidence does not support his claim that a substantial fraction of our genome is devoted to regulation.

What is regulation?

Different genes can be transcribed at different rates because of the strengths of their promoters. Genes that need to be expressed at high rates will have strong promoters, and those whose products are not required in abundance will have weak promoters. The strength of promoters may be something that's selected over the course of many millions of years, but it's not an example of regulation according to the definition I'm using here. When I use the term "regulation" I mean that gene expression can be modified (regulated).

Some genes, especially in bacteria, may be constitutively expressed, meaning that they are transcribed all the time at the same basic rate. There may even be a substantial number of eukaryotic genes that fall into this category, such as the genes for basic metabolic enzymes (housekeeping genes), but I suspect that eukaryotic examples of nonregulated genes are rare.

Most eukaryotic genes seem to be expressed under some conditions but not under others. They might be transcribed in response to external triggers such as the presence or absence of certain metabolites (e.g., glucose) or hormones, such as estrogen. Other genes are developmentally regulated – they are activated in certain tissues at various times during development.

Stochastic gene expression

Before continuing, I want to say a few words about stochastic gene expression because it fits in with the main theme of this book. Consider the classic example of gene activation in *E. coli*: the transcription of the *lac* operon genes in response to the presence of lactose. It has been known for many years that some bacteria cells transcribe the *lac* genes at high frequency whereas other cells in the same environment barely transcribe the genes at all. This variation in expression rates from cell to cell is known as stochastic gene expression because it reflects the chance events that have to occur in order to transcribe a gene. These events include the probability that an activator will find the right promoter, the probability that RNA polymerase will bind, and the probability that a transcription initiation event will occur after binding.[4]

This brings up a point that I glossed over in chapter 8 when I described an important property of DNA-binding proteins. The emphasis there was on the fact that transcription factors will bind rapidly to any old sequence that resembles their standard binding site. This explains why most transcription factors will be bound to nonfunctional sites when you take a snapshot view of the human genome.

The point I skipped was the *kinetics* of binding or the rate at which a DNA-binding protein finds a sequence and how long it stays there. These rates have been measured in the test tube for a variety of different DNA-binding proteins, and it turns out that they can find their binding site within a few seconds or minutes by attaching to DNA and sliding along it until they find the right site. But they don't stay bound to that site forever – they fall off within a few seconds or minutes and the search must start all over again. This contributes to the randomness of gene expression in each individual cell because in some cells the transcription factors will, by chance, find their sites more quickly than in other cells.[5]

Stochastic gene expression is a common feature of transcription in all species, and it's likely to be even more variable in species with large genomes full of junk. It means that adjacent cells in the same tissue might be producing very different amounts of RNA from a given gene just by chance. This is not regulation.

What do we know about regulatory sequences?

The main question I'm addressing in this chapter is how much of our genome is required for the regulation of gene expression. There are basically two ways in which a large proportion of eukaryotic genomes could be devoted to regulation. The first way assumes that much of the genome is dedicated to producing abundant regulatory RNAs that target individual genes. This was not an outlandish idea because we know that the genome is pervasively transcribed, but many scientists are certain that this idea is wrong because the majority of those RNAs are accidental transcripts present at very low concentrations (chapter 8).

The second way of devoting a large proportion of the genome to regulation is to require that each gene is controlled by a huge number of transcription factors covering as much DNA sequence as the gene itself. In other words, the number of functional DNA binding sites is so large that it accounts for a significant percentage of the genome. Is this true?

There are two different ways of approaching this problem. The most common way is the approach taken by thousands of biochemists and molecular biologists who are studying the expression of their favorite gene. We'll call this the reductionist approach. It's the approach I took when I was studying the expression of *my* favorite gene, the major heat shock gene (HSP70). This is a gene that is normally silent, but it's turned on in response to stress, such as a sudden rise in temperature (heat shock). Transcription of this gene is induced by specific activator proteins that bind near the promoter in response to an environmental stimulus.

We did the standard promoter bashing to see how much of the DNA around the promoter was required in order to get proper regulation. Our results were similar to those of many other researchers studying the same gene or any other eukaryotic gene: about 1000 bp of DNA were all that was necessary to bind all of the transcription factors that control expression.

We could transfer the mouse HSP70 gene, along with this upstream region, to other chromosome locations in the mouse genome and the gene worked just fine. Not only that, but we could also attach the upstream region to a completely different gene and show that transcription of that foreign genes was now induced by high temperature (heat shock).[6] That's pretty good evidence that all the regulatory sequences are close to the promoter. Similar results have been described in the scientific literature for many other genes.

In the case of our HSP70 gene, we were able to identify a number of DNA-binding sites for a variety of transcription factors using this approach. There were only a small number of these sites, and they were all within the 1000-bp region that we had previously identified as important.

The second approach is to look for sequence conservation on the understanding that regulatory DNA will be conserved between species. The idea here is that you can estimate the amount of DNA devoted to regulation by seeing how much is subject to negative selection in order to preserve transcription factor binding sites.

Hundreds of individual genes have been studied by these approaches, and the findings all support the view that most of the conserved regulatory sites are near the promoter. There have been some studies that identify regulatory regions at distant sites, but these are rare and, as we shall see, they fall into a special category.

In "simple" species, like yeast, the regulatory sequences only take up about 50 to 100 bp of the DNA that's found between genes. One study shows that the conserved regulatory regions in fruit flies are found within 1400 bp upstream of the promoter, and for mammals this expands to about 2000 bp. On average.[7] If this is a true representation of the DNA required for regulation, then it's a maximum of 2000 × 25,000 ≐ 50 million bp, or 1.6 percent of the genome. Clearly this doesn't account for all the excess DNA in our genome, and it certainly doesn't refute the idea that most of our genome is junk DNA. We can conclude that only a small fraction of our genome is required for regulation.

THE MAKING OF A QUEEN BY REGULATING GENE EXPRESSION

It's obvious that morphological differences between species are inherited, and this is correctly interpreted to mean that different species have different alleles and different regulatory sequences. It's also obvious that morphological differences between individuals *within* a species have a genetic component – that's why you resemble your parents and why people in Japan are more similar to each other than to people in Nigeria or Iceland.

However, we cannot ignore the effect of chance and environment because they can also influence gene expression. Height is one obvious example because it's well known that adult height is influenced by nutrition during childhood.

A more extreme example is seen in honeybees (*Apis mellifera*). Newly hatched female larvae will usually be fed a diet of pollen and sugars and they develop into sterile worker bees, but some larvae are fed royal jelly and they develop into queen bees. Several of the components in royal jelly contribute to switching gene expression in the lava so that queens are produced. The exact mechanism hasn't been worked out. (Earlier claims that the effect is entirely due to a protein called royalactin in royal jelly have been disputed.)[8]

Regulation and evolution

What makes one vertebrate different from another is a change in the time of expression and in the relative amounts of gene products rather than the small differences observed in the structure of these products. It's a matter of regulation rather than of structure.

François Jacob (1977)

Let's think about the differences between humans and other animals. The old way of thinking was that these differences must be due to the evolution of new genes in each lineage. As I explained earlier, that's why some scientists were surprised to learn that humans had about the same number of genes as many other animals that seemed to be far less complex. (This is the Deflated Ego Problem.) I've already covered two attempts to solve this "problem": (1) alternative splicing generates multiple proteins from a single gene, and (2) much of the genome is devoted to thousands of new noncoding genes that are responsible for human complexity. As we've seen, neither of these solutions stands up to close scrutiny.

The last attempt is to postulate that humans are more complex because they have a much more sophisticated network of regulatory sequences that fine-tune gene expression. This idea is based on fundamental principles of evolutionary-developmental biology (evo-devo) that were described more than fifty years ago and whose molecular mechanisms were elucidated, mostly in *Drosophila*, in the 1980s.[9]

The basic concept was outlined in an important book by Stephen Jay Gould back in 1977. The book was called *Ontogeny and Phylogeny*, and it described all the work that had been done on the problem up to that point in time. Gould discussed the differences between chimpanzees and humans:

> What, then, is at the root of our profound separation? King and Wilson argue convincingly that the decisive differences must involve the

> *evolution of regulation;* small changes in the timing of development can have manifold effects on a final product: "Small difference in the timing of activation or in the level of activity of a single gene could in principle influence considerably the systems controlling embryonic development. The organismal differences between chimpanzees and humans would then result chiefly from genetic changes in a few regulatory systems, while amino acid substitutions in general would rarely be a key factor in major adaptive shifts."[10]

Gould referred to this idea as *heterochrony* – changes in the timing of gene expression leading to changes in phenotype. That word hasn't caught on, but this idea is still at the heart of evo-devo and has been well covered in popular books such as Sean B. Carroll's *Endless Forms Most Beautiful.*

In most cases the differences in regulation are due to relatively small changes in the genome, but if you believe that humans are a lot more complex than other animals, then you don't think that's sufficient to account for such complexity. However, in my opinion, there's no evidence to support the idea that humans are much more complex than other animals and no evidence that we need a lot more regulatory DNA to explain this postulated complexity. Our current evo-devo model works just fine.

By the way, now that I've brought up Gould's book, I can't help but mention another of his important ideas. He summarizes the evidence that adult humans look very much like immature chimpanzees: we have less hair, heads that are large compared to our bodies with a shape characteristic of juvenile chimps, late eruption of teeth, a prolonged period of infant dependency, and so on. Many evolutionary biologists attribute this to "neoteny" in the human lineage, where the term is broadly used to describe the retention of juvenile characteristics in adults coupled with the premature development of sexual characteristics. (*Paedomorphosis* may be a slightly more accurate term, but I don't want to get into this particular semantic brouhaha.)

One of Gould's most popular essays was "A Biological Homage to Mickey Mouse."[11] He pointed out that Mickey Mouse has "evolved" considerably from the 1930s to the present day by becoming progressively more juvenile in appearance; for example, his head has become larger relative to his body. The essay is about neoteny and the importance of small genetic changes affecting large differences in appearance. It's possible to explain much of the difference between adult chimps and adult humans by altering just a few regulatory steps that delay the development of most adult features leading to juvenile-looking apes that are sexually mature.

The point I'm making is that you can get all the differences between species such as humans and chimpanzees with only a few small changes in gene regulatory sequences – most of which will simply alter the timing of gene expression during development. You do not need to postulate huge changes in the amount of the genome devoted to regulation. Do not be fooled into thinking there's some mysterious missing information required to specify each new species.

CAN COMPLEX REGULATION EVOLVE BY ACCIDENT?

Our basic understanding of gene regulation comes from studies of bacteria and bacteriophage genes in the 1960s and 1970s. Those studies showed that transcription is controlled by a small number of transcription factors that bind to short DNA sequences just upstream of the transcription start site. The factors could be activators, which stimulate transcription, or repressors, which inhibit transcription. Combinations of these factors can produce sophisticated developmental and temporal regulation of prokaryotic genes.

Initial studies of eukaryotic gene regulation revealed that similar processes were at work in controlling gene expression in eukaryotes. In some organisms, like yeast, the

(*continued*)

CAN COMPLEX REGULATION EVOLVE BY ACCIDENT? (continued)

basic features of gene regulation were very much like those in bacteria, but the process is more complicated in large multicellular organisms with big genomes. In these species, there are many more transcription factors controlling gene expression because there are more types of cells and each tissue expresses a distinct subset of genes. In theory, this could be accomplished by simple mechanisms like those in yeast and bacteria, but it turns out that regulation in plants and animals is a lot messier than you would think. There appear to be far more factors than necessary, including many that are part of the transcription machinery itself.[12]

This plethora of factors seemed to indicate a much more sophisticated form of gene regulation in these species and this view aligns with the Swiss watch model. But that isn't the only possibility. There are many scientists who argue that complicated regulation in plants and animals can arise by accident. Consider the example of a slightly harmful mutation in a transcription factor binding site that reduces its affinity and leads to lowered expression of an important gene. Enhanced expression could then be restored by a mutation that creates a new binding site for another transcription factor or by a mutation that leads to the binding of a new protein to the transcription complex making it more efficient. The overall complexity of the process has increased by restoring full expression, but there was no selection for complexity. Instead, it just happened because of an accidental mutation that was slightly deleterious (nearly neutral).[13] This model of the evolution of complicated regulation is related to constructive neutral evolution (see the "Constructive Neutral Evolution" box in chapter 8).

The point is that complexity isn't always indicative of sophisticated fine-tuning because it also can arise by accident. The appropriate metaphorical image, in my view, is a Rube Goldberg/Heath Robinson machine as in the Mousetrap game for children. These contraptions are extraordinarily complex, but they are not sophisticated or elegant. There are much simpler ways of achieving the objective.

Regulating gene expression by rearranging the genome

A very strange fruit fly was discovered in the late 1940s – it had legs growing out of its head instead of antennae. The mutation was called *Antennapedia*, and it was associated with a chromosome inversion where a segment of a chromosome has been broken in two places and the piece in between has been flipped or inverted. This apparently led to deregulation of a gene that caused legs to grow in place of antennae.[14]

This hypothesis was confirmed 40 years later when DNA sequencing revealed that one of the breakpoints was in the regulatory region of the *Antp* gene. This gene is normally under the control of a promoter that is expressed in the thorax, and the product of the gene leads to the development of legs in the appropriate places. The inversion flips the DNA and lands the gene next to a promoter that's active in the head where antenna normally develop. The *Antp* gene in the mutant falls under the control of this new regulatory site, and it is expressed in the head, leading to the growth of legs in the wrong place.[15]

There are many other examples of such misregulation by gene rearrangement. Many cancers, for example, are due to inappropriate expression of a gene in tissues where they are not normally expressed, and some of the most famous examples are due to chromosome rearrangements like the one in the *Antennapedia* mutation.

Burkitt lymphoma is a classic example; this aggressive B-cell cancer is caused by inappropriate expression of a gene called *Myc*, and one of the common causes is due to a chromosome rearrangement that attaches the *Myc* gene to a new regulatory region causing expression in B cells.

These mutants help us to understand how genes are regulated by short DNA regions just upstream of the promoter, but this switch in regulation isn't confined to mutants. There are several examples of deliberate chromosome rearrangements coupled to regulation. It's not a common phenomenon, but it's worth mentioning in order to give you a complete picture of gene regulation.

The classic examples come from bacteria, and the molecular mechanisms have been known since the 1970s. Many of them involve a phenomenon called *phase variation* that's associated with a switch from one form of the bacterium to another.[16] The most popular one in the textbooks is the flagellar switch in *Salmonella typhimurium*.

Bacteria flagella are flexible whip-like structures that are driven by a small molecular motor. They act like propellers to move the bacterium through liquid media. The proteins of the flagellum are targets for the immune response in humans and other mammals; for example, upon infection by *Salmonella*, the organism mounts an immune response to kill the bacteria.

Salmonella expresses two different types of flagellar protein, and the switch between the two types is thought to help the bacteria escape the immune response. This back-and-forth switch is what gave rise to the term *phase variation*. The molecular details of the switch mechanism are somewhat complicated, but a simplified version goes like this. The two genes controlling the flagellar proteins are called *fliC* and *fljB*, and there's a 996-bp region of DNA containing a promoter located between the two genes. In one orientation the promoter drives the expression of the *fliC* gene, but when the DNA is flipped (inverted) the *fliC* gene is silent and the *fljB* gene is expressed. Thus, expression of the two genes is regulated by rearranging the genome, in this case by a specific set of recombination proteins that mediate inversion of the DNA containing the promoter.

Regulation by genomic rearrangement has also been observed in eukaryotes. The best example is the production of the antibody protein immunoglobulin G (IgG). There are millions of different IgG proteins that can recognize a huge repertoire of antigenic sites on potential invaders such as *E. coli*, *Salmonella*, and viruses. The proteins are produced by B cells and, as a general rule, each B cell makes only one kind of IgG protein.

The region encoding the heavy chain protein of IgG covers a huge stretch of DNA on human chromosome 14. It consists of the main part of the gene known as the constant region I and several upstream segments known as variable regions (V), diversity regions (D), and joining regions (J). There are usually hundreds of V segments and four copies each of the D and J segments. As B cells mature, they randomly join one V segment to one D segment and one J segment, and this VDJ piece is fused to the constant region I (C), creating a heavy chain gene composed of different combinations of VDJC pieces. The process is known by the very descriptive term *V(D)J recombination*.[17]

The result is a very distinctive antibody with a distinctive binding site for antigens formed by the VDJ arrangement. Each B cell will produce a different antibody so that the organism can mount an immune response to a wide variety of foreign antigens. This is another example of regulating gene expression by rearranging the genome.

Open and closed domains

Recall that our DNA is tightly associated with packaging proteins called histones. The histone proteins form a disc-like structure called a core nucleosome and each of them binds about 200 bp of DNA to form the complete nucleosome. The resulting protein–DNA complex is called chromatin (chapter 1).

A loose association of nucleosomes gives rise to an "open" domain because the DNA is accessible to transcription factors and RNA polymerase. Genes that are being actively transcribed are found in open domains.

Higher-order structures can form when individual nucleosomes bind to each other to form more compact forms of chromatin, often referred to as heterochromatin. This form is called a "closed" domain because the DNA is not accessible to binding by transcription factors. Gene expression is repressed when the gene is embedded in a closed domain.

The shift from one form of chromatin to another is correlated with two types of chemical modification. The first is the modification of DNA by attaching a methyl group ($-CH_3$) to one of the bases, namely, cytosine (C). Methylated DNA is associated with closed chromatin domains and demethylation (removing the methyl group) is associated with gene activation. The second modification affects the histones; they are modified to prevent their association into higher-order structures, thus forcing chromatin into an open domain. There are many types of histone modifications, but one of the most common ones is acetylation due to attachment of an acetyl group ($-CH_2-CH_3$) to one of the histone proteins.

One of the tests for open domains involves digesting chromatin with the enzyme DNase I (DNase "one") – an enzyme that cuts DNA. DNA located in an open domain will be hypersensitive to DNase I, while DNA in a closed domain will be protected. You can identify the parts of the genome that lie in open domains by the fact that they exhibit DNase I hypersensitivity.

Only a fraction of the genome is exposed in an open chromatin domain at any one time, and this fraction will include all the genes that are being expressed in that tissue. However, it will also include random sites that are accidentally in open domains for various reasons that have nothing to do with gene expression. A spurious open domain can form because the open and closed forms exist in dynamic equilibrium so that even if a given region is tightly bound to histones it may transiently "breathe" from time to time.

During that brief spontaneous opening, the DNA might be recognized by specific DNA-binding proteins that shift the equilibrium to the open form even if there's no accompanying expression of a nearby gene. This is why most of the transcription factor binding sites mapped by ENCODE are not important. It also means that just mapping an open domain region does not mean that it contains an active gene.[18]

As I pointed out in chapter 8, many scientists seem unaware of the fact that transcription factors can bind to sequences occurring by chance in the genome and that open domains do not necessarily indicate regulation of gene expression.[19] The null hypothesis should be that open domains and bound transcription factors are nonfunctional as a result of spurious events.

X-CHROMOSOME INACTIVATION

You will often see a small dark mass in the cell nucleus if you look at mammalian cells under a microscope. It's called a Barr body after its discoverer, Murray Barr.[20] Later on, scientists realized that these Barr bodies only appear in cells from females. It represents one of the two X chromosomes that has been partially inactivated and condensed into a heterochromatic state. The phenomenon of X-chromosome inactivation was discovered by Mary Lyon, and it's often called "Lyonization." This is an extreme example of a closed domain.

There are about 700 genes on the X chromosome. Males have a single copy of those genes since they have only one X chromosome (and one Y chromosome), and females have two copies of each gene because they have two X chromosomes. The expression of some X-chromosome genes can be very active in males because there's only one copy of each gene, but in females this can result in too much product

(*continued*)

X-CHROMOSOME INACTIVATION (continued)

being synthesized because there are two copies of each gene. Consequently, the genes on one of the chromosomes in females are turned off to keep expression to the same level seen in males.

Either the maternal or the paternal version of the X chromosome can be randomly inactivated. The crucial step occurs early on in embryogenesis, giving rise to patches of cells in the adult where either the paternal or maternal X chromosome has been inactivated. Normally this doesn't matter, but there are cases where the two X chromosomes contain different alleles of a gene. In humans, this is seen as different protein variants being made in different parts of the body depending on which X chromosome was inactivated.

The most spectacular example of this phenomenon is the different patches of hair color on the backs of calico cats. One X chromosome has the allele for black hairs, and the other has the allele for orange hairs. Calico cats are always female.

About 500 of the 700 genes are inactivated in the heterochromatic version of the X chromosome.[21] Inactivation is controlled by a gene called *Xist* and a few others. *Xist* stands for "X-inactive specific transcript," and as the name implies, the product of the gene is a regulatory RNA that's only expressed on the inactive X chromosome. It's one of the few examples of a lncRNA with a well-established function (see chapter 8). The RNA transcript coats the X chromosome, leading to its condensation into heterochromatin and the shutdown of most genes. This is a good example of a regulatory noncoding RNA.

X-chromosome inactivation illustrates the correlation between gene expression and chromatin state. Genes on the inactive X chromosome are repressed because, in part, they are in a closed chromatin domain.

The recruitment model of gene expression

The thing that nature figured out – it's kind of amazing, actually – is that once you have all the reading machinery, it's just a question of recruiting it to the right place. And to do that we have evolved these very simple little factors that get together and attract the RNA polymerase to the gene.

Mark Ptashne[22]

Gene expression can be regulated at many levels: transcription initiation, transcription elongation, transcription termination, splicing, and RNA stability, to name just a few that apply to RNA. I am concentrating on transcription initiation because that's the one that's most directly relevant to a discussion of what's in your genome.

One of the controversies in this field concerns the relationship between the activation or repression of transcription initiation and the structure of the surrounding chromatin. There's no debate over the correlation between transcription initiation and an open chromatin domain, and there's no debate over the presence of modified nucleosomes and demethylated DNA at those sites. What's debatable is cause and effect; the question is whether the shift from closed to open domains is separable from the binding of transcription factors, or whether the binding of a transcription factor causes the change in chromatin leading to an open domain.

Those who argue in favor of the primacy of chromatin will often refer to the *histone code,* which serves as a metaphor for the idea that changes in chromatin structure are what ultimately controls whether a gene is transcribed or not. Proponents of this idea also tend to emphasize epigenetics as an important player in gene regulation and evolution. (see "What the heck is epigenetics?").[23]

Mark Ptashne is not a big fan of the histone code. He is a strong, and often acerbic, advocate of the idea that transcription is initiated by the binding of activators to specific sites near the promoter and these activators subsequently bind to and recruit RNA polymerase to the promoter. This binding event triggers histone modifications and demethylation of DNA to create an open chromatin domain. In other words, the open domain is a consequence of transcription factor binding and not a cause. This is an extremely important point, so let me restate it another way. Ptashne's model proposes that transcription factors can recognize and bind to DNA as long as that DNA is not embedded in a heterochromatic region (closed domain). But such heterochromatic regions will "breathe" from time to time, exposing the DNA to binding.[24] The binding of a transcription factor triggers histone modifications that lock the chromatin into an open domain.

This model of regulation is known as the *recruitment model,* but it's not simply a model about a single activator recruiting RNA polymerase. It's much more than that.[25] If we think of transcription binding sites as switches that are turned on by binding, then the model includes the presence of multiple switches for every gene. In this case the binding of transcription factors is often cooperative, meaning that these factors bind to each other and assist in recruiting multiple factors to the promoter region. The classic examples were worked out fifty years ago when the regulation of *E. coli* and bacteriophage lambda genes was being elucidated – that's the subject of Mark Ptashne's famous book *A Genetic Switch.*

The same process occurs in eukaryotes, but usually there are multiple activators and multiple switches, and the switches can be located farther away from the promoter than in bacteria. (In both cases, loops of DNA can form when a distant switch is activated, but the loops can be bigger in eukaryotes.) It's important to note that when I say "multiple switches," I'm not referring to dozens of switches; I'm referring to 10 or fewer – this is more than enough to regulate all the genes in our genome. The amount of DNA devoted to this kind of regulation is only a small percentage of the genome.

ENCODE promotes regulation

Now what we're finding is that the whole genome is alive with switches. You can't move for switches; there are switches everywhere. There's not a single place in the genome that doesn't have something that you might think could be controlling something else.

Ewan Birney (2012)[26]

Ewan Birney and his ENCODE collaborators claim to have identified more than 600,000 regulatory sites accounting for 8 percent of the genome.[27] That implies a large number of sites for each gene, but the problem with the ENCODE conclusion is that the researchers didn't distinguish between random binding sites for regulatory proteins and the much smaller subset that actually functions in regulation. The important question for this chapter is not how many binding sites there are but how many *functional* regulatory sites are present in the genome.

In chapter 6, I presented evidence that protein-coding exons account for 1.0 percent of the genome. The ENCODE researchers used a different value of 1.2 percent of the genome, and they claim that the total amount of the genome devoted to regulation is "significantly higher" than this amount. In fact, they speculate that it may be 8 percent or even 20 percent of the genome.[28]

Let's assume the lowest value of 8 percent – that's 256 Mb of DNA. Assuming there are 25,000 genes, this means an average of about 10,000 bp of regulatory sequence per gene. If we take an upper estimate of 10 bp per regulatory site, then there will be roughly 1000 regulatory sites for each gene if the ENCODE speculation is correct.

I don't think this is what they mean since they seem to be using a rather broad definition of regulatory site – one that covers much more than 10 bp. Thus, they seem to be suggesting something like 50 or so sites per gene. Still, the bottom line is that ENCODE researchers claim, based on their data, that 8 to 20 percent of the genome is directly involved in regulation, and that's the claim that was widely reported as if it were a fact. For example, Brendan Maher, a writer for *Nature*, summarized the ENCODE results in the same issue in which they were reported (September 6, 2012), and he said there were 70,000 promoter regions and 400,000 enhancer regions. (An enhancer is a transcription factor binding site, usually one that is some distance from the promoter.) This number of enhancers works out to about 16 per gene, but according to Maher, the analysis is only 10 percent complete, meaning there are actually 160 sites per gene.

Keep in mind that ENCODE researchers are trying to solve the Deflated Ego Problem. Here's how Maher summarizes their position:

> The real fun starts when the various data sets are layered together. Experiments looking at histone modifications, for example, reveal patterns that correspond with the borders of the DNaseI-sensitive sites. Then researchers can add data showing exactly which transcript factors bind where, and when. The vast desert regions have now been populated with hundreds of thousands of features that contribute to gene regulation. And every cell type uses different combinations and permutations of these features to generate its unique biology. This richness helps to explain how relatively few protein-coding genes can provide the biological components complexity necessary to grow and run a human being. ENCODE "is much more than the sum of its parts," says Manolis Kellis, a computational genomicist at the Massachusetts Institute of Technology in Cambridge, who led some of the data analysis efforts.[29]

The ENCODE scientists want you to believe that they have solved the problem of human complexity and that the answer is that humans can get away with the same number of genes as other animals because our genes are much more highly regulated. This is also the conclusion promoted by the editors of *Nature* in their unprecedented publicity campaign, but this conclusion is not supported by the data they published. It's the same issue I talked about in chapter 8 when I discussed pervasive transcription: ENCODE researchers didn't seriously consider evidence that their transcripts could be junk RNA, and in the case of regulation, they didn't seriously consider that their transcription factor binding sites could be spurious and nonfunctional. In other words, they proposed an extraordinary idea (huge amounts of regulation) that conflicts with the evidence from other studies, is not proved by their data, and fails to take into account alternative explanations.

Furthermore, all the evidence for junk DNA and spurious binding needs to be refuted or challenged if you are going to take seriously the idea that a substantial proportion of our genome is required for regulation. If you don't do that, then you may be practicing what Richard Feynman calls "cargo cult science": "Details that could throw doubt on your interpretation must be given, if you know them. You must do the best you can – if you know anything at all wrong, or possibly wrong – to explain it. If you make a theory, for example, and advertise it, or put it out, then you must also put down all the facts that disagree with it, as well as those that agree with it."[30]

Does regulation explain junk? How can we test the hypothesis?

If much of the genome is bound by transcription factors, as the ENCODE data demonstrate, then how can we decide whether this is just due to sloppy binding or real regulation? One way to resolve the controversy is to demonstrate that all those binding sites are really necessary for controlling the expression of particular genes. As I pointed out earlier, doing so requires a detailed analysis of individual genes and candidate regulatory sites – a reductionist approach to the problem. That's not going to happen because it requires considerable investment in money and considerable investment in graduate students and postdocs. The most likely result of all that investment is the realization that most transcription factor binding sites don't do anything; no papers would be published, countless graduate students will not get their PhDs, and hundreds of postdocs will never get research positions. No responsible scientist is going to waste time on such a fool's errand.

Some labs are looking to genomics to resolve the issue even though this isn't the best way to answer the question. The main difficulty with this approach is that scientists can't make a direct connection between abundant transcription factor binding sites and the expression of individual genes.

Yoav Gilad's group at the University of Chicago (Illinois, USA) come up with an idea to get around this problem. They engineered cell lines where the level of individual transcription factors was reduced drastically by genetic manipulation.[31] The trick is to eliminate individual transcription factors and then assay the cell lines to see which RNAs are affected. If the level of a particular RNA decreases following the knockdown, then this indicates that the transcription factor is probably activating that gene. Conversely, if the amount of transcription increases then the transcription factor may be repressing gene expression.

They looked at 59 different transcription factors in human cells and found plenty of examples of regulation and strong correlations between affected genes and nearby transcription factor binding sites, just as you would expect. However, the number of genes that showed an effect were only a small fraction of the total number of genes. Most of the transcription factor binding sites were near genes that were unaffected by eliminating binding, leading the authors to conclude that most transcription factor binding doesn't result in measurable changes in gene expression, an idea that's consistent with nonfunctional binding, as discussed above. By contrast, the result is not consistent with the hypothesis that most binding sites are functional.

There are caveats associated with such genomic approaches; for example, it's very likely that the authors missed some functional interactions, but this is unlikely to affect the general conclusion that most DNA binding by transcription factors has no effect on regulation.

The crucial element that's missing in most genomic experiments is the negative control. I mentioned this in chapter 8 in the box on "The Random Genome Project" when I described Sean Eddy's random genome experiment. He wonders what would happen if you insert a large amount of random DNA sequence into a genome, and he predicts that much of it would be transcribed, indicating that pervasive transcription is not an indication of function.

Mike White is also a vocal critic of projects that assume function in the absence of a negative control. He actually did an experiment to see whether random DNA fragments could promote transcription, and the answer is yes they can.[32] Presumably, there are transcription factors that can bind to the random sequences and stimulate the transcription of nearby DNA, demonstrating that junk DNA can be mistaken for a functional promoter.

What this result shows is that sorting out regulatory events is not easy because there's a background of transcription factor activity that is accidental or spurious. This is the same effect that gives rise to spurious transcription, but in this chapter, we are discussing it in terms of regulatory sequences, not the production of junk RNA.

A THOUGHT EXPERIMENT

Ford Doolittle has proposed a different thought experiment to help us see the problem raised by ENCODE's focus on open domains and regulation.[33] He asks us to imagine that massive data collection experiments were conducted using the pufferfish and the lungfish. The pufferfish genome is only one-eighth the size of our genome, and the lungfish genome is 40 times larger than the human genome. All three genomes have about the same number of genes.

Imagine that all transcription factor binding sites and chromatin marks have been mapped in the two fish genomes. Doolittle imagines two possible outcomes.

(*continued*)

A THOUGHT EXPERIMENT (continued)

The first is that all three genomes (human, pufferfish, lungfish) have the same number of sites. Since the three species have the same number of genes, this implies a constant (large) number of sites for each gene as ENCODE predicts.

The second, more likely, result is that the three species will have a very different number of sites. Pufferfish will have the smallest number of sites, and lungfish will have about 40 times more than humans. If there's a correlation between the number of sites and genome size, then it suggests that many of the sites are spurious and not required for regulation. There's no reason to think that lungfish need 40 times more regulatory sequences than humans.

This is a thought experiment* that's closely related to the Onion Test. Many biochemists and molecular biologists would agree with Ford Doolittle – their money is on the second outcome because most transcription factor binding sites are not involved in regulation.

* The experiment will never be done because it would cost at least US$400 million to repeat the ENCODE results with pufferfish and lungfish.

3D chromosomes

The idea that chromatin is organized into higher-order three-dimensional structures has garnered a lot of attention in recent years. I remind you that the entire human genome is organized into large chromatin loops (see chapter 1), and there are about 100,000 of these loops – enough to accommodate one gene per loop. The loop is formed by proteins that bind to scaffold attachment regions (SARs) at the base of each loop and I estimated that these SARs account for about 0.3 percent of the genome.

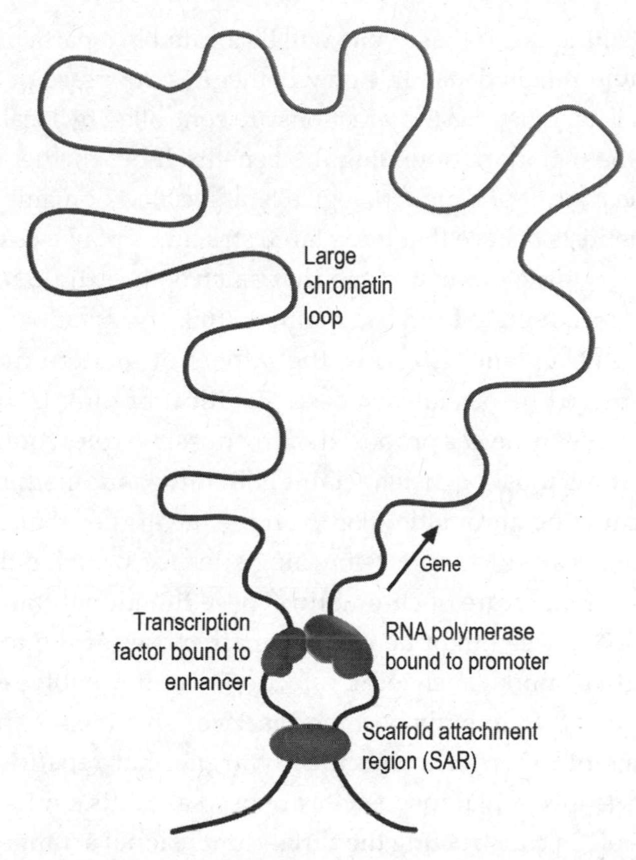

FIGURE 10.1 Long-range interactions. Some enhancers can be located at great distances from the promoter. One model of such interactions is that the promoter and the enhancer are close to each other at the base of a large chromatin loop.

There has been a lot of speculation about the relationship between these large chromatin loops and gene expression. One thing that seems very clear is that there can be long-range interactions between promoters and transcription factors that are bound to DNA near the base of the loop, as shown in Figure 10.1. This model explains some initially puzzling results showing that some enhancers (transcription factor binding sites) could be located tens of thousands of base pairs from the promoter region of the gene they regulate.

There's little doubt that a gene will be located in a particular chromatin environment depending on whether it is expressed or not. The question is whether these interactions are controlled by local interactions in the regions surrounding the gene itself or whether they are under the control of large topologically associated domains (TADs). Some scientists believe that these large structures play a crucial role in regulating gene expression and that much of the extra DNA in our genome is required to form and maintain this organization.[34]

Emile Zuckerkandl is one of the fathers of modern molecular evolution, and he has always been skeptical of junk DNA. Over the past 50 years he has proposed many possible roles for the extra DNA, but his most persistent claim is that it has something to do with organizing chromatin. For example, he argues that transposon sequences could serve as binding sites for proteins that play a role in the structure of chromatin. These functional transposon-related sequences might be only a small subset of the total, and they might be under such weak selection that they evolve at a very rapid rate and frequently become inactive. This idea is similar to the claims of others who specifically argue that repetitive DNA elements (especially transposons) are used as SARs, and they play a major role in constructing the three-dimensional arrangement of chromatin.[35] There's very little support for the idea that transposon sequences play a direct role in organizing chromatin. Degenerate transposons are much more likely to be exactly as they seem – once-active transposons that have been degraded by mutation.

Another idea is that the higher-order structure of chromatin could impose functional constraints on the nucleotide composition of various parts of the genome. Zuckerkandl proposed that some stretches of DNA are more compatible with nucleosome binding and positioning, so there's selection for bulk properties of DNA at the level of GC content but not necessarily sequence.

He argued that no region of the genome evolves at the neutral rate as expected if it were junk DNA and that's because of functional constraints on the binding of nucleosomes and formation of higher-order chromatin structures. The extra DNA might be required for proper packaging of the sequences that are evolutionary, conserved but not necessarily compatible with chromatin structure. For example, coding regions are severely constrained because they have to encode amino acids and the resulting sequence may not be able to form efficient higher-order chromatin fibers. However, if the coding sequence is broken up into small fragments separated by introns, then there could be selection on the base composition of the intron's DNA to overcome the negative effects of the coding sequence.[36]

This perspective is different from the one advanced by ENCODE since the ENCODE scientists claim that much of the extra DNA is directly involved in gene regulation by binding to transcription factors.

Another speculation by Zuckerkandl (and others) postulates a different indirect role for the extra DNA through structuring chromatin. They suggest that the presence of transposon sequences can result in a transient opening and closing of chromatin domains because transposons contain transcription factor binding sites and promoters. Thus, the first step in regulation is already accomplished, making it easier to evolve new genes. According to this hypothesis, pervasive transcription is a feature, not an accident. Hence, species that bulk up their genomes with transposon sequences become more evolvable, and the species is more likely to survive.

It all sounds very plausible until you realize that it's a teleological argument, and it doesn't pass the Onion Test. Not only that, but these speculations also fall into the category of a solution looking for a problem. A relatively simple model of gene regulation, such as the recruitment model, seems quite capable of explaining most gene regulation. There's really no need to go looking for more complicated solutions as long as you realize that spurious transcription is common in a sloppy genome. The real motivation behind these speculations is not to account for some mysterious feature of gene regulation that requires an explanation – it's to explain away junk DNA.

What the heck is epigenetics?

But, as sometimes happens in science, [epigenetics is] also a field that has been exposed and leapt upon in a press frenzy to attempt to explain all sorts of as yet unsolved mysteries of biology. The legion purveyors of ackamarackus' love a real but tricky scientific concept they can bolt their quackery onto. It happens with words like "quantum" which offers up some magical scienceyness, none more so than in "quantum healing" – an unfathomable extension of reiki, which, let's face it, is a load of old cobblers already.

Adam Rutherford (2016)

Epigenetics is the new buzzword in biology. It has something to do with the regulation of gene expression, but nobody's quite sure about the connection because the term *epigenetics* means different things to different people. The word was created in 1942 by Conrad Waddington to describe the process by which genes bring the phenotype into being, but that broad definition covers a lot of territory, including the regulation of *E. coli* and bacteriophage genes by combinations of transcription activators and repressors. Few scientists described their discipline as epigenetics during the 1950s, 1960s, and 1970s when those fundamental aspects of regulation were being worked out.

The term *epigenetics* only caught on when it became apparent that gene expression in eukaryotes was coupled to modifications of histones and methylation of DNA. Proponents of epigenetics are attracted to these modifications because they seem to be different than the ordinary types of regulation seen in bacteria and bacteriophage, suggesting that elephants might be different than *E. coli.* One particular feature of these modifications is that they could be heritable, and a great deal of the hype promoted by epigenetics enthusiasts is focused on its potential to define a new way of evolution – one that's distinctly Lamarckian.

Robin Holliday proposed that epigenetics should be restricted to changes that could be inherited by daughter cells following cell division, and there's a growing consensus that this the proper domain of epigenetics.[37] The following definition is as good as any. "An epigenetic trait is a stably heritable phenotype resulting from changes in a chromosome without alterations in the DNA sequence."[38]

There's no doubt that histone modifications and DNA methylations are important features in the shift from open to closed chromatin domains but whether they are truly heritable is another thing entirely. There's hardly any evidence that histone modifications are heritable in any meaningful sense, but methylation and demethylation are more complicated because there are known mechanisms for passing this feature on to daughter cells following cell division.[39] (See the next section.)

But epigenetic enthusiasts go much farther than simple cellular inheritance because they claim that epigenetic markers can be passed from mothers to their offspring, and they quote extraordinary claims such as evidence that Dutch mothers who suffered through a famine during World War II could pass on the effects to their children and even their grandchildren. Such claims should be treated with skepticism because there's no obvious mechanism for transferring chromatin markers from somatic cells (e.g., stomach, liver, and intestinal cells) to the egg cells that form the offspring, especially since those egg cells had already formed when the mother was born.[40]

The importance of epigenetics is absolutely dependent on whether you accept Ptashne's recruitment model of regulation. If you do, as I do, then histones and DNA modifications are epiphenomena and not crucial steps in the regulation of gene expression. In order to inherit a gene expression pattern, a cell must pass on transcription factors, which can then cause demethylation and histone modifications.[41]

Restriction/modification and the inheritance of methylated nucleotides

The idea that methylated DNA can be inherited from one cell to its descendants strikes many people as strange and mysterious, but its mechanism has been known for a very long time. The key discoveries were made in the 1970s by Werner Arber's group at the University of Geneva in Geneva (Switzerland) and by Hamilton Smith's group at Johns Hopkins University in Baltimore (Maryland, USA). They were working with restriction enzymes, and they discovered that these enzymes cleave DNA at a particular sequence of DNA.

Let's look at one example of a restriction enzyme. *Eco*R1 is made in *E. coli* cells, and when the cell is invaded by foreign DNA (e.g., a bacteriophage/virus) the enzyme cuts that DNA at multiple sites having the palindromic sequence GAATTC (Figure 10.2). This is a very effective way of restricting infection by the bacteriophage.

Here's the problem. The *E. coli* genome contains hundreds of GAATTC sequences, so why doesn't *Eco*R1 cleave its own DNA? The answer is that the *E. coli* DNA is methylated at one of the adenylate (A) residues in the GAATTC sequence as shown in the figure. The restriction enzyme doesn't recognize the methylated DNA, so it won't be cut.

Methylation is a key part of the restriction enzyme defense mechanism and the coupled system is known as restriction/modification to reflect this association. When DNA is copied during cell division, the newly synthesized strand will not be methylated, but the parental strand will still have the methyl group. Such sites are said to be hemimethylated; they are not recognized by the restriction enzyme, but they are recognized by the methylase enzyme that rapidly converts it to a full methylated site. As a consequence, both daughter cells will inherit the methylated site following DNA replication and cell division, and the methylated site will be passed on to future generations.

Werner Arbor and Hamilton Smith received the 1978 Nobel Prize in Physiology or Medicine for their work on restriction/modification. Later on, Smith hooked up with Craig Venter at The Institute for Genomic Research and led the effort to sequence the first bacterial genome. He then moved with Venter to Celera, where he was deeply involved in assembling the human genome

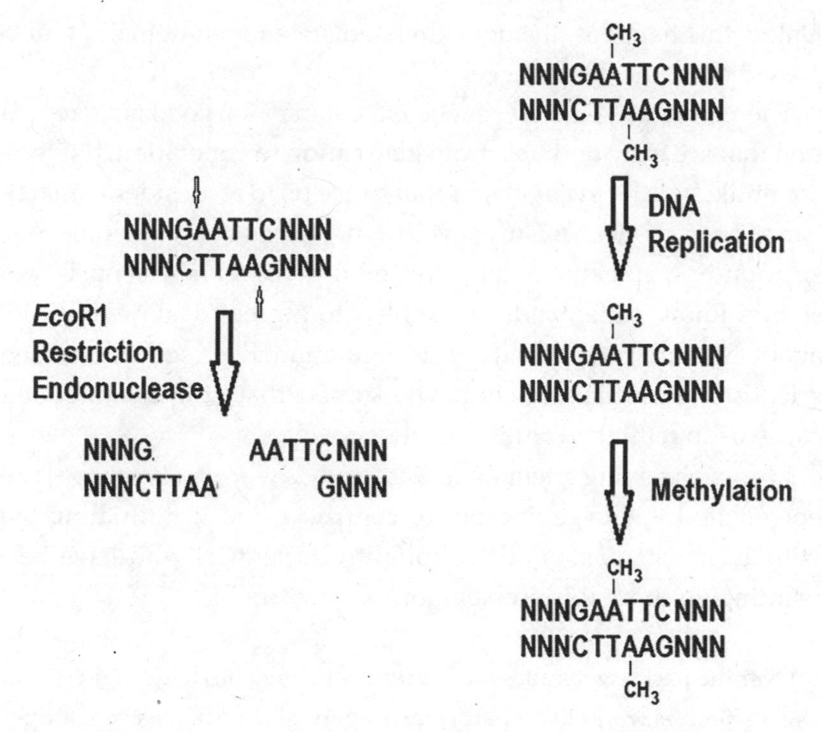

FIGURE 10.2 Restriction/modification. The restriction enzyme *Eco*R1 binds to the sequence GAATTC and cleaves it at the sites shown by the two arrows. The cell's own DNA is protected from cleavage because two of the A residues in the sequence are methylated and *Eco*R1 doesn't bind to methylated DNA. Following DNA replication, one of the strands will contain unmethylated DNA. However, these hemimethylated sites are efficiently recognized by a specific methylase, and the methylation of both strands is restored.

sequence. More recently, he has been working with Venter on synthetic organisms.

The flip side of the methylase reaction is that sites that are fully demethylated will not be methylated by an enzyme that only recognizes the hemimethylated state. Thus, when human DNA is demethylated during transcription, this state will be preserved until other methylases – those involved in gene regulation – are triggered.

Unless this happens, the demethylated state (open domain) will be passed on to the daughter cells.

There may be some epigenetic effects in eukaryotic gene regulation that are truly heritable from generation to generation, but these are unlikely to be major effects that cause us to reconsider Lamarckism as another mechanism of evolution. What's clear is that the hype surrounding epigenetics has infected a great many scientists who should know better, and this has led to a great deal of confusion about evolution and how it affects regulation. Even some prominent scientists and philosophers have claimed that epigenetics should cause us to rethink evolutionary theory.

It's worth paying attention to Kat Arney's warning below because epigenetic hype is getting out of control. I find it infuriating that nobody is listening to other thoughtful scientists who have been warning us about this problem for many years.[42]

> Over the past few decades the concept of epigenetics has caught on in the scientific world like a particularly aggressive rash. It's even nudging its way into the public consciousness thanks to breathless articles warning of the epigenetic effects of everything from stress to sunshine, exhorting us to pimp our genomes by drinking green tea or munching broccoli. References to epigenetics have leaked into newspapers, seeped into medicine, and contaminated fields such as psychology and even sociology. More alarmingly, purveyors of pseudoscience are jumping on the epigenetic bandwagon, and the word seems to be used increasingly in the same way that certain people bandy about the term "quantum" as a hand-waving non-explanation for mysterious things they don't really understand. As a scientist, I worked on epigenetics before it was cool, and I find this infuriating.[43]

CHAPTER 11

Zen and the Art of Coping with a Sloppy Genome

> *[T]he "design" of organisms is not "intelligent," but rather quite incompatible with the design that we would expect of an intelligent designer or even of a human engineer, and so full of dysfunctions, wastes, and cruelties as to unwarrant its attribution to any being endowed with superior intelligence, wisdom, and benevolence.*
>
> Francisco Ayala (2004)

In *Zen and the Art of Motorcycle Maintenance*, the protagonist goes to university to become a biochemist, but he flunks out of college, joins the army, goes to Korea, quits the army, gets a degree in philosophy, and teaches writing and rhetoric at a small college in Montana. He struggles with trying to reconcile science, logic, technology, and rational thought with another view of life that's roughly equivalent to the romantic side of humans. The attempt to reconcile these two views drives him insane. The book is said to be the best-selling book of philosophy that's ever been published! It also teaches you about motorcycle maintenance.

This book isn't about that kind of struggle or why philosophy might drive you insane. I just like the title, which was borrowed from *Zen in the Art of Archery* by a German philosopher. Dozens of others have coined variations on those titles using the word *zen*, and the common thread is about acceptance and understanding – a simplistic view of some aspects of Buddhism.

The theme of this book is that life at the molecular level is very messy and disorganized. Molecular interactions are stochastic and error-prone – a view that contrasts with what many of us were taught in school. Evolution is also a stochastic process whereby natural selection is frequently overwhelmed by random genetic drift. This leads to the evolution of organisms with features that look nothing like adaptations and nothing like well-constructed Swiss watches where each gear is fine-tuned to perform a precise function.

In previous chapters, I've described the neutral theory and its offspring, the nearly neutral theory, and explained that their discovery came about at the same time scientists were uncovering some puzzling aspects of genomes. Those scientists knew that large functional genomes were improbable because of mutation load. They knew that genome sizes varied considerably among groups of related species. They knew that mRNAs covered only a small part of the genome. And they knew that most of our genome was composed of repetitive DNA that was largely made up of defective transposons.

All of these observations could be explained if most of our genome is junk, and this explanation came to be the standard among those whose main interest was the evolution of genomes. However, that view was not widely accepted by other scientists who felt very uncomfortable about having so much junk in our genome.

The turn of the twenty-first century saw the publication of the human genome sequence, and that triggered more debate about function. By then we had a whole generation of scientists who had gone to graduate school after 1980 and were pretty much unaware of the early evidence for junk DNA and the modern view of evolutionary theory. Thus, the genomics era restarted the debate from scratch under the assumption that natural selection is the dominant mechanism of evolution.

I've described that new debate in the last few chapters and tried to explain why the evidence for a genome full of functional DNA doesn't hold up. I hope I've succeeded in convincing you that the evidence for a poorly designed genome full of junk DNA is persuasive.

One of the features of *Zen and the Art of Motorcycle Maintenance* is Chautauquas – small essays or thoughts that the author shares with us each day while traveling with his son on a motorcycle vacation. Chautauquas were touring adult education and entertainment shows that were popular in the United States in the late 1800s and early 1900s. The name survives today in the Chautauqua Institute in upstate New York – I think it's one of the most interesting and delightful places in the entire country.

I'm sympathetic to the Chautauqua concept because I've taught courses on evolution at the Chautauqua Institute and met dozens of interesting people there, including several who are part of this book. In keeping with that theme, allow me to close with a few Chautauquas of my own because, as the protagonist in *Zen* says, "when you've got a Chautauqua in your head, it's extremely hard not to inflict it on ordinary people."[1]

The limitations of genomics

> *If one surveys the so-called 'new way of doing biology' that is omic science, it has several characteristics; it is based on high-throughput methods, on making observations on as much as possible at the same time, and on its reliance on technological improvements to enhance, improve and often automate many old methods. ... I am all for these technological advances but what dismays me about omic science is its departure from the hypothesis-generating-experimental basis of scientific investigation. I have even heard claims that it will liberate us from the domination of hypothesis, that is, thinking, in biology.*
>
> Sydney Brenner (2000)

The genomics era began in the 1990s. Before then, researchers concentrated on specific genes and specific functions such as DNA replication, transcription, translation, and regulation. This led to enormous advances in biochemistry and molecular biology and our solid understanding of gene expression grew out of this work and forms the core of modern textbooks.

We gain deep understanding when many labs work on the same gene. We learn when and how the gene is expressed and what factors control its transcription and translation. Regulatory sequences are identified and defined, and in most cases, we can be confident that there are no great mysteries waiting to be uncovered.

Genomics takes a different approach. Instead of analyzing specific genes or functions, genomic researchers focus on a global analysis of the entire genome. For example, they want to know how many genes are expressed in liver cells or brain cells, and they are less interested in the details of how each individual gene is regulated. Genomics spawned proteomics, transcriptomics, and a host of other omic fields.

Some scientists regret this shift in emphasis – Sydney Brenner was one, but he's not alone. They think of genomics as simply an expensive exercise in data collection ("stamp collecting" is the derogatory description), and they don't think it leads to deep understanding or insight. This criticism was leveled at the ENCODE Consortium, and I think it is mostly correct while, at the same time, missing the point. In theory, there's nothing wrong with collecting data – doing so can identify problems with our current models and provide data for proposing new models.

Most genomics researchers will tell you that this is exactly what they have accomplished. They believe they have uncovered a huge amount of novel activity in the human genome and this discovery overthrows old models of gene expression. We learned about some of these "paradigm shifts" (or "shafts," depending on your perspective) in earlier chapters. They include abundant alternative splicing, pervasive transcription, pseudogenes with mysterious functions, and millions of regulatory sequences.

Now, it's true that working on single genes can be misleading – it's possible to not see the forest for the trees – but genomics may have the opposite problem, seeing a forest but ignoring the trees. What's happened in the genomics approach is that models of gene regulation are being promoted and accepted without ever confirming them at the level of individual genes. This flawed approach is most obvious when it comes to alternative splicing, where global claims of abundant alternative splicing are based on genomics but never confirmed by looking carefully at individual genes. This is what Sydney Brenner meant when he said that omic science "liberates" us from thinking.

The same flawed approach applies to several other topics that I've covered in earlier chapters. For example, ENCODE claims that there are dozens of regulatory sites per gene, but very few genes have been shown to have this many switches.

There is no easy solution to this problem. Genomics is big science, and it's sucking up a lot of research funding and being promoted by a lot of hype. Small labs can't compete for funding, and when their work on individual genes contradicts the genomics models, it is relegated to obscure journals where it is ignored. I've struggled with the problem of explaining genomic results repeatedly while writing this book. Take pervasive transcription as another example; the genomics labs have claimed that our genome is full of genes producing noncoding RNAs, and this claim is backed up by data showing thousands of transcripts. This result is taken to be evidence against junk DNA and in favor of a functional genome, but very few of these transcripts have been looked at to see if they really do have a function. Nobody is going to randomly pick several hundred of these transcripts and do the work required to decide whether they have a function or whether they are spurious transcripts because the most likely answer is that they are spurious. All I can do to counter the genomics claim is to point out that transcription errors are common and that the claim contradicts what we know about sloppy genomes full of junk DNA. And I've also pointed out that the null hypothesis is junk, and the onus is on those who make the claim of function to provide solid evidence to back up their claim. Very few genomics labs do this.

I suppose the best solution to the problem is to convince the genomics labs to stop making broad claims that can't be easily tested. There's some evidence that this is happening with ENCODE since the latest (2020) results just report the data and say nothing about function. More genomics labs need to do this.

Meanwhile, we are stuck with one of the unintended consequences of genomics research, namely, the fact that our databases are full of incorrect and/or unsubstantiated data. I was reminded of this recently when I tried to present data on the amount of coding sequence in the genome and the size of a typical protein-coding gene (chapter 6). You would think that all you have to do is check the RefSeq database and tabulate all the coding regions and introns to get the answer, but that doesn't work because the database contains multiple mRNA variants for each gene, and there's no way of telling which one is correct. The number of transcripts from protein-coding genes in RefSeq has almost doubled in the past 10 years so that now there's an average of 2.6 mRNA variants per gene, and it's very likely that many of these are not biologically relevant.[2]

More than two decades have passed since Sydney Brenner published his essay "Biochemistry Strikes Back" in which he warned us about the demise of biochemistry and the increasing emphasis on omics. If he were still alive, I'm sure he would be disappointed that his worst-case scenario – that genomics would supplant hypothesis-driven research – has come true. At the time, he expressed confidence that most of the flaws of omic science would vanish when scientists realize that their results must be interpreted in an evolutionary framework – a framework that includes junk. That hasn't happened.

The function wars

[M]any of the most heated arguments in biology are not about facts at all but rather about the words we use to describe what we think the facts might be. However, that the debate is in the end about the mean-

ing of words does not mean that there are not crucial differences in our understanding of the evolutionary process hidden beneath the rhetoric.

W. Ford Doolittle (2013)

The world is not inhabited exclusively by fools and when a subject arouses intense interest, as this one has, something other than semantics is usually at stake.

Stephen Jay Gould (1982)

The ENCODE publicity campaign triggered a debate about function that can be split into two separate tracks. First, there's the scientific debate about our understanding of things like transcription factor binding sites, alternative splicing, and pervasive transcription. This debate relies on data such as the frequencies of transcription and splicing errors. It is resolved by evidence such as experiments showing that a particular RNA has a biological function. In the absence of evidence, the only reasonable null hypothesis is that the feature has no function.

The second track is much more metaphysical, or should we say "metabiological"? It's concerned with the correct definition of the word *function*, and it assumes that we can help resolve the controversy by coming up with a rigorous definition that will always allow us to distinguish functional DNA from junk.

This second track has attracted the attention of philosophers, leading to debates that can be quite out of touch with facts and evidence. My colleague, Alex Palazzo, calls these debates the "function wars."

Let me start by saying that the history of debates about the meaning of words in biology has taught us one important lesson: that biology is far too messy to resolve most of these controversies. There is no rigorous, universally accepted definition of *gene*, for example, and there's no definition of *species* that covers all examples.[3] Everybody disagrees with everybody else about the meaning of *Darwinism*, and nobody likes to be called an adaptationist.

Philosophers have usually not been helpful in resolving definition debates – they can't even agree on how to define "science." Nevertheless, philosophers and some scientists have waded in where many other biologists fear to tread, and dozens of papers on the meaning of "function" have been published. They are now up to their necks in the quicksand of biological semantics, so let's see what they've accomplished.

The first battle was over two historical definitions of function called the causal role (CR) function and the selected effect (SE) function. The difference between CR functions and SE functions often means simply the sorts of evidence that you can rely on to determine function. At the risk of oversimplifying, a genomic CR function is one that's identified simply by observing that a given stretch of DNA does something. Thus, if it's transcribed or it binds a transcription factor, then it's functional – this was the definition used by ENCODE. An SE function is identified by the fact that it arose by natural selection. This is a very specific definition in the philosophical literature because it focuses on traits, not genes, and the definition only applies to those traits that were clearly the result of fixation by natural selection. This view reflects the dominant view of evolution in most of the philosophical literature – a view that emphasizes phenotypic change and natural selection. The philosophical distinction between CR and SE functions was borrowed by molecular biologists to help resolve the debate over the function of DNA sequences in our genome.[4] However, as I discussed in chapter 4, molecular biologists are not interested in restricting true biological function to just those traits that have been recently fixed by natural selection.

The most obvious way of identifying molecular function is to look for conservation: if a given sequence is conserved, then it's very likely to have a biologically significant function no matter how it originated. I've used this criterion many times in the book to distinguish between things that have a function and things that might have arisen by accident. But this is not the SE definition used in the philosophical literature. Philosophers use a definition like this:

"Traits are selected because of their positive effects on the fitness of the organisms that have them. These 'selected effects' are the Proper functions of the traits."[5]

In order to avoid confusion, molecular biologists should stop describing their definition of function as an SE function and instead use *maintenance function* to specify elements that are currently being maintained by natural selection (purifying selection).

There's general agreement that the CR definition is just about useless and a definition that uses conservation as a criterion is the one we should prefer. There would not have been a debate over the 2012 ENCODE results if those researchers had recognized this, now obvious, fact. But some issues with the conservation definition need to be cleared up.

One problem is that "conservation" is not the only way to identify a stretch of DNA that has a biologically relevant function as I explained in chapter 4 and chapter 6. In the case of *de novo* genes, for example, you can't use conservation as a determining criterion since, by definition, a new gene has no history of selection. In this case you have to use evidence of purifying selection within a species to identify whether a given stretch of DNA is functional. It's important to specify that the maintenance function is defined as DNA that is *currently* being preserved by natural selection (purifying selection) regardless of whether it has a history of selection or not.[6] The distinction may seem like quibbling, but it covers the evolution of new functional sequences, and it also resolves questions about historically conserved sequences that have become useless, for example, pseudogenes and genes that are in the process of being eliminated from a population.[7]

The philosophical literature contains many papers addressing whether *de novo* genes count as SE functions and, if not, what other kind of function they might have. Traits with biologically relevant functions that are not SE could have a "propensity function," but I'm not sure what that means, and, in any case, it just gets us into another semantic brouhaha about historical functions. Best to avoid all that by setting up a molecular definition of function (maintenance function) that just helps us decide whether a given sequence has a biologically relevant function or not and avoid the semantic quibbling.

There's a less contentious argument over the term *junk,* as in junk DNA. I'm not talking about disputes over whether junk DNA exists but rather disputes over the usefulness of the word. There are scientists who would rather drop the word *junk* because it's provocative and because it doesn't sound very scientific. Sean Eddy, for example, once wrote that we could use "nonfunctional DNA" as long as everyone understands the implications. However, he also defends the use of "junk DNA": "To me, 'junk DNA' is a colloquial term of endearment, and a reminder of the history of ideas in the field. You can forget the polarizing term so long as you remember the data it stands for: astonishing genome size variation, mutational load, a small fraction of conserved DNA, and the large fraction of eukaryotic genomes that is composed of neutrally decaying transposon relics. These data support a view that eukaryotic genomes contain a substantial fraction of DNA that serves little useful purpose for the organism, much of which has originated from the replication of transposable (selfish) elements."[8]

I agree with Sean that "junk DNA" is a convenient stand-in for the historical data, and that's exactly why it should be retained. There's no other term that conveys all of the meaning that he describes; certainly not "nonfunctional DNA." Furthermore, I use "junk" DNA precisely because it's provocative; hopefully, it will encourage people to come to grips with neutral evolution and a sloppy genome. However, you can't keep using a term like *junk* just because you like the sound of it. That's why I define it as, "any stretch of DNA that can be deleted from the genome without reducing the fitness of the individual." That may not be a precise definition that covers all possibilities (see "Levels of Selection" in chapter 8), but nothing in biology is perfect. We can resolve disputes on a case-by-case basis and change the definition if someone comes up with good evidence for substantial amounts of biologically relevant DNA that isn't under purifying selection (with an emphasis on *evidence*, as opposed to *speculation*).

As far as I'm concerned, the function wars are over. We've quibbled and nitpicked our way to the realization that no definition of molecular function is going to cover every possibility and that some version of a definition requiring purifying selection is the best working definition we have. However, I'm pretty sure that my opponents aren't going to agree with me and more papers on the meaning of function will be published, especially if they cast doubt on the concept of junk DNA.

Scientific revolutions

Most revolutions are pretty obvious – you know when you're in the middle of one. One of my wife's ancestors was present at the storming of the Bastille on July 14, 1789, and some of my own ancestors took part in the American Revolutionary War (on the British side). Some scientific revolutions are like that; for example, if you were living in England in the 1860s, you knew about Charles Darwin and his revolutionary views of evolution, and if you were alive in the 1950s, you learned that genes were made of DNA. Those scientific revolutions are the exception because, in most cases, a true scientific revolution sneaks up on you over the course of many decades of slow and steady scientific advances.

This book is about one of those revolutions, namely, the idea that evolution can produce messy systems that do not appear to be well designed. It's the idea that there's more to evolution than natural selection and adaptation, a view that was formulated in the late 1960s when the neutral theory and the importance of random genetic drift were recognized by molecular evolutionary biologists and rammed into the scientific consciousness by the most famous paper in the entire field of evolution, "The Spandrels of San Marco and the Panglossian Paradigm: A Critique of the Adaptationist

Programme."[9] One of the consequences of adopting that view is the realization that most of our genome is junk.

If most of the important revolutions in science are so subtle that you don't even know they are happening, then why do there seem to be so many scientific revolutions every year? Hardly a month goes by when the popular press isn't announcing a new revolution. University press releases promote revolutions almost every week, and my local bookstore is full of popular science books describing the latest revolutions. Even the scientific literature is full of supposed revolutionary ideas that are about to overthrow old paradigms. How many times have you heard that the textbooks need to be rewritten?[10]

The answer is simple: most of those revolutions are fake or, at best, misleading. Some of the examples are obvious, such as when a false paradigm is set up as a strawman and then disproved in what's best described as a paradigm *shaft*. Others are less obvious, especially those where the underlying science is complicated.

This book contains plenty of examples of fake revolutions, the most important of which is the idea that ENCODE refuted junk DNA. Let me remind you once again what that kind of fake revolution looks like by quoting what the Sanger Institute in the United Kingdom – one of the leading centers of genome research – announced in a press release on September 5, 2012: "The ENCODE Project today, announces that most of what was previously described as 'junk DNA' in the human genome is actually functional. The ENCODE Project has found that 80 percent of the human genome sequence is linked to biological function."[11]

My goal has been to show you that the evidence supporting this so-called revolution is very weak and the idea is almost certainly incorrect. It's not true that most junk DNA is actually functional, and it's not true that there are thousands of newly discovered genes for small RNAs. We don't need a huge amount of regulatory sequence to explain the complexity of humans. Most of our genome is still junk in spite of what you might have read on Twitter or in press releases.

Whether we are scientists or not, we all face the problem of distinguishing truth from fantasy and real news from fake news. I have belonged to several different chapters of the Center for Inquiry, an organization of skeptics promoting critical thinking. Skepticism is one of the tools needed to make intelligent choices, but it has to be balanced with a mind that's open to new ideas or it becomes constraining. The balance is tricky; as the old cliché goes, "keep your mind open to new ideas but not so open that your brains fall out."

Carl Sagan is one of the heroes of the skeptical community. He has warned us on many occasions to be wary of false claims: "What skeptical thinking boils down to is the means to construct, and to understand, a reasoned argument and – especially important – to recognize a fallacious or fraudulent argument. The question is not whether we *like* the conclusion that emerges out of a train of reasoning, but whether the conclusion *follows* from the premise or starting point and whether that premise is true."[12]

I wish it were easy to follow this advice, but it isn't. It's hard to be skeptical when one is being bombarded with "scientific" information about the latest revolutionary discoveries concerning the human genome.

I offer some nuggets of advice. Keep in mind that scientific revolutions are very rare, so you should be skeptical about all claims of revolution. Be especially cautious about claims of paradigm shifts – they are much more likely to be paradigm shafts. Remember that scientists will, on occasion, come up with a truly revolutionary idea, but just because some presumably smart person makes a good-sounding argument doesn't mean they are geniuses and doesn't mean they are correct. We need to remember something else that Carl Sagan told us: "Where skeptical observation and discussion are suppressed, the truth is hidden. The proponents of such borderline beliefs, when criticized, often point to geniuses of the past who were ridiculed. But the fact that some geniuses were laughed at does not imply that all who are laughed at are geniuses. They laughed at Columbus, they laughed at Fulton, they laughed at the Wright brothers. But they also laughed at Bozo the Clown."[13]

Who is Bozo in the debate over function? I hope I've convinced you that a lot of biochemical reactions, such as transcription and splicing are sloppy and error-prone. I hope I've convinced you that your genome really is full of junk DNA. With respect to junk DNA, I'm not talking about a sudden revolution here, although it may seem like that to some of you, I'm talking about the slow, quiet kind of revolution that sneaks up on you over a generation. Junk DNA has been around for fifty years.

For another example of the other kind of revolution, in this case a fake revolution, let me get near the end of my list of quotations by coming full circle back to Francis Collins:

> It turns out that only about 1.5 percent of the human genome is involved in coding for proteins. But that doesn't mean the rest is "junk DNA." A number of exciting new discoveries about the human genome should remind us not to become complacent in our understanding of this marvelous instruction book. For instance, it has recently become clear that there is a whole family of RNA molecules that do not code for protein. These so-called non-coding RNAs are capable of carrying out a host of important functions, including modifying the efficiency by which other RNAs are translated. In addition, our understanding of how genes are regulated is undergoing dramatic revision, as the signals embedded in the DNA molecules and the proteins that bind to them are rapidly being elucidated. The complexity of this network of regulatory information is truly mind-blowing, and has given rise to a whole new branch of biomedical research referred to as "systems biology."[14]

This is what a fake revolution looks like, but it's very hard for the average person to recognize that it's fake, especially when it comes from a famous authority like Francis Collins. But, in fact, it's a paradigm shaft. Nobody ever said that all noncoding DNA was junk – we all knew about noncoding RNAs in 1970. Today's knowledge of noncoding RNA and regulatory sequences is not new and not "mind-blowing."

No comfort for Intelligent Design Creationists

I began this book by quoting Francis Collins, an evangelical Christian,[15] who was, at the time, director of the International Human Genome Project. He said that the publication of the draft sequence of the human genome was awe-inspiring because it gave us the first glimpse of God's instruction book. I opened this final chapter by quoting another well-known religious scientist, Francisco Ayala, who tells us that the instruction book doesn't look anything like the work of an intelligent designer (see the quotation at the beginning of this chapter).

Intelligent Design Creationists have been predicting for years that most of our genome would turn out to be functional. It's easy to see why they believe this since a genome full of junk DNA doesn't look much like the work of an intelligent designer. The ENCODE publicity campaign of 2012 was a big boost for them, and that was one of the reasons why I said on that day that this was going to make my life very complicated. Here's a prominent intelligent design proponent praising ENCODE:

> The demise of the idea of junk DNA illustrates too, in a more positive way, how a competing perspective can inspire research that contributes to a new discovery. Although clearly not every scientist who performed research helping to establish the functional significance of nonprotein-coding DNA was inspired by the theory of intelligent design, at least one noteworthy scientist was. During the early part of the decade, before ENCODE made the headlines, [Richard Sternberg] published many articles challenging the idea of junk DNA based on genomics research that he was conducting at the National Institutes of Health. After publication of ENCODE in 2012, his coauthor on many of

> those articles, the prominent University of Chicago geneticists James Shapiro, wrote an article in the *Huffington Post* commending [Richard Sternberg] for his groundbreaking research and for anticipating the ENCODE results years before.[16]

I don't think this book is going to convince Stephen Meyer that he is wrong about the demise of junk DNA. I have met him and many of the other leading intelligent design proponents, and I can assure you that they are not coping with a poorly designed genome as well as they should. The only thing I want to say in this Chautauqua is that incorrect scientific information has an audience that's more than happy to amplify it to suit their own purpose. It's one of the consequences of bad science, and it makes correcting it much more difficult.

Scientific controversies

It's time to end this book now that I've got those Chautauquas out of my head. My main goal was to tell you about your genome and what's in it. Along the way, we learned that our genome contains about 25,000 genes along with lots of other useful things like regulatory sequences, centromeres, origins of replication, telomeres, and scaffold attachment regions. We also learned that it's full of very useless things like pseudogenes and fragments of defective transposons.

The useless things outnumber the useful things by a ratio of 9:1, and that's why our genome is 90 percent junk DNA. As it turns out, there are several different ways I could have presented the case for junk DNA, and after much reflection, and several false starts, I settled on a historical presentation. I did this partly because the historical evidence has been ignored in recent years and partly because it provides a solid foundation for discussing the modern controversy over junk DNA. That controversy takes up a large part of the book because I tried to present the functionalist arguments as best I can.

I hope I've been fair to opponents of junk DNA even though, in my opinion, they are horribly misguided.

Some of the controversies may have been new to you. I bet many of you didn't know there was a serious controversy about the prevalence of alternative splicing, for example. And you may not have known that what you've heard about the definition of a gene is probably wrong or that you were likely not taught the correct view of the central dogma of molecular biology. On the other hand, I suspect you were at least vaguely aware of the controversy over the amount of junk DNA in our genome even though you may have been swayed by the massive publicity campaign in favor of function.

When discussing scientific controversies, it's easy to segue into analyzing how they are covered in the scientific literature and the popular press because that's a major source of news. I conclude that the scientific literature has failed to address most of these controversies, pointing to a massive failure of the peer-review process. Science writers have also failed us since they haven't done their due diligence in ferreting out this failure (with some notable exceptions) – the most obvious example is the misguided publicity campaign launched by *Nature* in 2012.

This led me to wonder about motives. Why do so many scientists resist the idea that our genome is full of junk? One of my colleagues has been warning me for years not to go down this path. I'm told it's bad form to try to attribute motives to scientists who disagree with me because I'm bound to be wrong and likely to be insulting as well.

I'm not listening to him. I can't resist trying to explain why so many scientists seem to be so unscientific, and I think I've come up with a few answers. First, we have a whole generation of scientists who haven't been taught correctly as undergraduates. They simply don't know that transcription factors can bind to nonfunctional DNA and that RNA polymerase and splicing enzymes can make mistakes. This is the "undergraduate effect" described by John Wilkins in chapter 8. This is a kind way of explaining the behavior of the ENCODE researchers; I'd rather not consider other, less kind motives.[17]

Second, the majority of today's biochemists and molecular biologists don't have a firm grasp of modern evolutionary theory. They are stuck with the old-fashioned concept that the only important mechanism is natural selection and modern organisms occupy the top of fitness peaks. It's inconceivable to them that evolution would have produced species with junky genomes. Combining the first and second problems leads to the two competing worldviews, messy versus Swiss watch, that have come up at several places in this book.

A third problem may explain why so many scientists don't appreciate the arguments for junk DNA, and that's an ignorance of history. We live in a time where everything seems so new and exciting and baby boomers are supposedly out of touch with reality. According to this view, looking back at the work of boomers and their parents in the 1960s and 1970s is a waste of time because there can't be anything of value in that stuffy old literature. After all, they didn't even know about epigenetics and CRISPR, so how could they have understood genomes!

This ignorance of history is exacerbated by a deliberate misrepresenting of the history (where I'm using "misrepresent" as a euphemism for a stronger word that might be insulting). The best examples are when modern scientists claim that the old fogies thought there were more than 100,000 protein-coding genes and all noncoding DNA was junk. (Both claims are demonstrably false.) The misrepresentation of the central dogma falls into this category as well since it's often used to bolster the claim that old fogies didn't know about noncoding genes. I find this ignorance of history to be particularly infuriating because it implies that the scientists of the 1960s and 1970s were really stupid.

I grew up with the 'phage group, and I knew most of the scientists who thought about these ideas. Quite a few of them are Nobel laureates. Those scientists were (many still are) among the best and brightest scientists I've ever met, and I'll match them against the ENCODE leaders any day.

Finally, there's the Deflated Ego Problem. It's a separate issue because it's based on human exceptionalism, and that's a different motivation. Many scientists seem convinced, contrary to all evidence, that humans must be special in some way that's visible in our genome. This viewpoint is what prompts these scientists to look for ways of accounting for that difference, such as alternative splicing and a plethora of regulatory RNAs.[18] The main problem here is not that these scientists are ignoring data but that they are looking for the solution to a problem that doesn't exist. From their perspective, one of the explanations just has to be right, and that's very similar to the behavior of Intelligent Design Creationists.

Coping with a sloppy genome

If you are skeptical of junk DNA, then you are probably sympathetic to many of the views that I've just described. That makes it very difficult to adjust to the fact that biochemical reactions are very messy and error-prone and that our genome is full of junk. I understand. I'm asking you to change your entire worldview – a view that you've probably never questioned. That's going to be hard because as that famous philosopher, Yoda, once said, "You must unlearn what you have learned."

If you were already a believer in junk DNA, then I hope I've convinced you that this is a reasonable stance despite the fact that it's under constant attack. Maybe I've told you a few things you didn't know. I hope so.

Either way, if you adopt the sloppy view of genomes then, rest assured, there's a state of zen that you can achieve by accepting this fact and learning to live with the idea that 90 percent of your genome is junk.

Notes

Preface

1 Gould (1985), p. 16.
2 Dawkins (1982), p. 8.
3 Cook (2019).

Prologue

1 Quoted in Collins (2006), p. 3.
2 *Haemophilus influenzae* sequence: Fleischmann et al. (1995). *Drosophila melanogaster* sequence: Adams et al. (2000).
3 The International Human Genome Project paper is International Human Genome Sequencing Consortium (2001) and the Celera paper is Venter et al. (2001). "The sequence data have been made available without restriction and updated daily throughout the project" (International Human Genome Sequencing Consortium, 2001). The original plan was to publish both papers in the American journal *Science*, but the International Human Genome Project decided to publish in the British journal *Nature* when the editors of *Science* violated their public policy by allowing Celera to keep the data private. *Nature* is actually a more suitable journal for an international effort because about one-third of the final draft sequence was contributed by The Sanger Centre in Cambridge, England (UK).
4 Dunham et al. (1999).
5 International Human Genome Sequencing Consortium (2004).

1. Introducing Genomes

1 Judson's book *The Eighth Day of Creation* was first published in 1979, and it's the definitive history of the early days of molecular biology.
2 The Crick announcement is from Judson (1996), p. 149, and the reference to the famous paper is Watson and Crick (1953).
3 The numbering convention specifies that the atoms in the bases are numbered first: 1, 2, 3, and so on. The numbering starts over again with the sugar molecules, so you have to add a "prime" to distinguish these atoms from the numbered atoms in the bases (1′, 2′, 3′, etc.). Look at the 2′ position in Figures 1.1 and 1.2. Notice that there's no oxygen attached at that position. That's why the sugar is 2′-*deoxy*ribose.

4 There are two major classes of bacteria called Archaea and Eubacteria. The classes are usually referred to as domains, and they are part of the Three Domain Hypothesis, along with the third domain, Eukaryota. Some scientists who are strong advocates of the Three Domain Hypothesis object to the use of the word *prokaryote* since, according to their view, it artificially groups together two of the three domains of life (Pace, 2006). That view has not caught on. Other scientists want to restrict the use of *bacteria* to members of the Eubacteria domain so that there are now two kinds of prokaryote: archaebacteria and bacteria. In this book, I use the word *bacteria* to refer to all members of Archaea and Eubacteria, and I use it as a convenient synonym for *prokaryote*. This is consistent with common usage.

5 Bacteria account for 70 gigatons of biomass, while animals take up only 2 gigatons. Plants account for 450 gigatons (Bar-On et al., 2018).

6 Charles Darwin wrote the following in his 1837 notebook: "It is absurd to talk of one animal being higher than another. We consider those, when the cerebral structure/intellectual faculties most developed, as highest. A bee doubtless would when the instincts were." He also wrote, "Never use the words higher or lower" in the margin of his copy of *Vestiges of the Natural History of Creation* by Robert Chambers (1844).

7 Sometimes it gets a bit more complicated in bacteria because many of them also have small circular bits of DNA (*plasmids*) that carry only a few genes. There are often many copies of these plasmids in a bacterial cell, and sometimes the genes on them are important. They count as part of the genome if the plasmids are big enough and if they carry genes essential for survival – in that case, they also count as a chromosome. Consequently, the genome of a typical bacterium is usually a single DNA molecule, but sometimes there are two or three chromosomes.

8 The exact relationship between the amount of stain and the amount of DNA was not known with as much precision in the middle of the twentieth century, so the value of 3 Gb for the size of the human genome was often used as a more or less accurate estimate. The 1991 reference for the amount of DNA in each chromosome is Morton (1991). More recent values for females are 3.19 Gb and 3.14 Gb for males for a total diploid length of DNA of 208 centimeters for females and 205 centimeters for males (Piovesan et al., 2019).

9 Michael Lynch's calculation can be found in Lynch (2007). The website of the Animal Genome Size Databases is at genomesize.com. See Carl Zimmer's excellent article on Ryan Gregory in the March 15, 2015, issue of *The New York Times Magazine*, http://www.nytimes.com/2015/03/08/magazine/is-most-of-our-dna-garbage.html?_r=0.

10 See Baldi et al. (2020). The somewhat regular structure shown at the bottom of Figure 1.5 was called the 30 nm structure because of its width. We now know that these regular 30 nm structures are just one of many forms that can be formed by groups of nucleosomes.

11 See Cattoni et al. (2015) for a discussion of DNA loops.

12 Thymine (T) is the modified base made from uracil by attaching a methyl group ($-CH_3$), so RNA is certainly the more primitive nucleic acid because it contains ribose and uracil. DNA probably evolved from RNA about 3.5 billion years ago.

13 There are a few exceptions. Some virus genomes consist of single-stranded DNA, and some are composed of single- or double-stranded RNA.

2. The Evolution of Sloppy Genomes

1 It's unfortunate that Thomas uses the terms *primitive* and *lower* to describe modern species. He should have said that amphibians and fish seem to be less complex than mammals, but even this is debatable.

2 "C" stands for "constant" amount of DNA. Diploid cells are 2C, and haploid cells are 1C. The term *C-Value Paradox* was first used by Charles Thomas in 1971.

3 The classic work on the composition of genomes and the discovery that many of them were composed of repetitive DNA is Britten and Kohne (1968). The Lewin references are Lewin (1974a; 1974b), and the discussion of the C-Value Enigma is in Gregory (2001).

4 See the excellent discussion of the difference between the C-Value Paradox and the C-Value Enigma in Gregory (2005b).

5 See Sean Eddy's 2012 essay for an excellent summary of these points.

6 The *Brassica* data is from Brassica rapa Genome Sequencing Project Consortium (2011) and He et al. (2021).

7 Gregory (2007); see also Palazzo and Gregory (2014).

8 This view of evolution is often referred to as a gene-centric view, especially by those who specialize in looking at phenotypic changes in whole organisms. It's important to recognize that changes in allele frequencies (genotypic changes) underlie all of evolution even if you don't know which alleles are causing the phenotypic variation that you are examining. The term *gene-centric* is often used in a pejorative manner, and it is frequently misapplied, as we shall see later on in this book. The consensus view that I'm describing here is the population genetics view, and it is the very best way of looking at evolution at the molecular level.

9 Stated in a letter to Ernst Mayr on May 8, 1963. Quoted in Rao and Nanjundiah (2011), p. 1.

10 The most important papers are Hubby and Lewontin (1966) and Lewontin and Hubby (1966).

11 The first use of the term *"beanbag" genetics* comes from a talk given by Ernst Mayr at a symposium celebrating the 100th anniversary of the publication of *The Origin of Species*. Mayr stated that during the 1930s "evolutionary change was essentially presented as an input or output of genes, as the adding of certain genes to a beanbag and the withdrawing of others." The dispute between Mayr and J.B.S. Haldane, one of the leading proponents of a mathematical approach to population genetics, is covered in Rao and Nanjundiah (2011). Mayr's point is that there's more to evolution than just the shifting of alleles of single traits, and while he is correct, he failed to appreciate the significance of simple models to understand non-intuitive processes such as random genetic drift.

12 The original papers that documented this variation looked at differences in the sizes of proteins in fruit flies using the new technique of gel electrophoresis (Hubby and Lewontin, 1966; Lewontin and Hubby, 1966).

13 There are many evolutionary biologists who disagree about the revolutionary importance of the neutral theory and random genetic drift. They believe that these ideas were easily incorporated into the Modern Synthesis developed in the 1940s. There are also many evolutionary biologists who aren't interested in molecular evolution, and they continue to focus on natural selection as the most significant mechanism of evolution.

14 King and Jukes (1969).

15 Kimura (1991).

16 Lynch (2007), p. 370.

17 The relevant number in population genetics is the effective population size (N_e), which is always less than the actual population size. N_e takes into account the fact that not all individuals reproduce and not all alleles segregate randomly. I'm going to simplify things by assuming that $N_e = N$, but keep in mind that this is an unrealistic assumption.

18 The probability of fixation is twice the selection coefficient (2*s*) so, strictly speaking, an allele with a selection coefficient of s becomes effectively neutral when $2s = 1/2N$ or $s = 1/4N$. Another way of saying this is when $4Ns = 1$.
19 Michael Lynch has done much to promote these well-established concepts in his book on *The Origins of Genome Architecture* (2007), but they have been described in many textbooks over the past 40 years (e.g., Li and Graur, 1991).
20 Ridley (1997), p. 78.
21 The Francis Crick Papers, National Library of Medicine (USA), https://profiles.nlm.nih.gov/spotlight/sc/catalog/nlm:nlmuid-101584582X199-doc.
22 See Vollger et al. (2021) for a brief review.
23 Kronenberg et al. (2018).
24 Kirchberger et al. (2020). The actual cost of junk DNA is hotly debated, and I've oversimplified somewhat to avoid making this description too complicated. The real cost might have more to do with the extra deleterious mutations that can occur in junk DNA. This is the mutation-hazard theory (Lynch et al., 2011).
25 See Blommaert (2020) and Lynch et al. (2011) for thorough reviews of the evidence and theories of genome evolution.
26 Lynch et al. (2011).

3. Repetitive DNA and Mobile Genetic Elements

1 See Schueler et al. (2001), Bloom (2013), Hartley and O'Neil (2019), and Miga (2019) for excellent reviews of centromeres. Centromeres are very complicated structures and it's not clear how much of the satellite DNA is necessary for them to function in chromosome segregation. See chapter 5 for a brief discussion. Each alpha-satellite repeat is 171 bp long, and there are about 500,000 of them altogether if you add up all the contributions from each of the 24 different chromosomes. The Hsat 2,3 repeats are CATTC, and there are about 5 million of these repeats, mostly in the Y chromosome and autosomes 1, 9, and 16 (Altemose et al., 2022).
2 See Mudd et al. (2019), Kim et al. (2017), and Kronenberg et al. (2018) for recent reviews of chromosome evolution.
3 See Gymrek et al. (2016) for a review of STRs and Kronenberg et al. (2018) for the total number of STRs.
4 Reviewed in Coursey and McBride (2019). The classic example of a dormant viruses is bacteriophage lambda in *E. coli*. The study of gene expression in lambda was one of the main research efforts in molecular biology in the 1960s and 1970s, and most of the basic principles of transcription and regulation were discovered or elucidated by groups working on bacteriophage.
5 Hoyt et al. (2022).
6 Johnson (2019).
7 De Parseval and Heidman (2005). The principle behind the molecular clock is described in the next chapter.
8 Goodier (2016).
9 Aziz et al. (2010).
10 Genes with internal promoters are characteristic features of genes transcribe by RNA polymerase III (see "Yeast Loses Its Introns" in chapter 6). This why SINES in different species are all derived from class III genes.
11 See Carl Zimmer's (2018) book *She Has Her Mother's Laugh* for a description of his genome and Konkel et al. (2015) for some of the results from the 1000 Genomes Project.

12 De Koning et al. (2011).
13 Cordaux et al. (2006).
14 See Blommaert et al. (2019) for a recent summary of papers on the causes of gene expansion.
15 Ibarra-Laclette et al. (2013).
16 Two important papers on selfish DNA were published together in *Nature* (Orgel and Crick, 1980; Doolittle and Sapienza, 1980). Several other papers on this topic appeared in *Nature* in the following months (Dover, 1980; Cavalier-Smith, 1980; Orgel et al., 1980; Dover and Doolittle, 1980; Jain, 1980). The idea that selfish DNA is selected at the level of genes instead of organisms fits with Gould's view of evolution at many levels (hierarchical theory). See Gould (2002), pp. 693–4.
17 Dawkins (1976, 1989a). See Dover (1980) for a thorough discussion of this point. Dawkins briefly referred to selfish DNA in his book, and that has caused considerable confusion about the distinction between selfish DNA and selfish genes. The authors of the original selfish DNA papers deny that they were influenced by that single sentence in Dawkins's book *The Selfish Gene* (Doolittle, 2022).
18 Reviewed in Moschetti et al. (2020).
19 The exaptation paper is Gould and Vrba (1982), and they used transposons as one of their examples of possible exaptation. The idea that repetitive DNA could be important in regulating gene expression dates back to Britten and Davidson (1969, 1971), but recent data show that there are only a small number of confirmed examples spread across many species (Britten, 1997). A survey of hundreds of species uncovered only 94 cases where a fragment of the transposase gene was co-opted over a period spanning 300 million years of evolution (Cosby et al., 2021). Smit (1999) was one of the first to describe the exaptation of transposon-related sequences.
20 The term *post hoc* is a shortened version of *post hoc ergo proper hoc*, which means "after this, because of this." This fallacy, and many others, are described in Chris DiCarlo's excellent books *How to Be a Really Good Pain in the Ass* (2011) and *Six Steps to Better Thinking* (2017).
21 Shapiro and Sternberg (2005); Shapiro (2011); Shapiro (2022). See my review of his 2011 book (Moran, 2012a).
22 Gould (1983) in "Quick Lives and Quirky Changes," p. 63.
23 Gould (2002), p. 1275. I'm specifically referring to the situation in which evolvability is selected because that's the argument used to argue against the idea that transposons are junk. However, there are other situations in which the evolvability might arise without being directly selected (Dawkins, 1988; Pigliucci, 2008; Doolittle, 2022).
24 Brenner (1998).
25 Calabrese et al. (2012); Lang et al. (2012); Simone et al. (2011).
26 Orgel and Crick (1980).
27 Ohta (1983); Charlesworth et al. (1994).
28 Kapusta et al. (2017).

4. Why Don't Mutations Kill Us?

1 Bhattacharyya et al. (1990).
2 J.B.S. Haldane, along with Sewall Wright and Ronald Fisher, are the founders of population genetics.
3 Haldane (1949); Muller (1950).

4 Muller (1966); Ohno (1972a); Ohno (1972b); Kimura (1968); Ohta and Kimura (1971); Crow and Kimura (1970); King and Jukes (1969).
5 Nachman and Crowell (2000).
6 We ignore large deletions and insertions since each one is a result of a single mutation event and there are only a small number of them compared the vastly greater number of other mutations.
7 The generation times are from Langergraber et al. (2012). The value of 30 years per generation may seem high, but remember that we are looking at the average number of years to produce an offspring that will go on to reproduce. This is very different than the age at which a first child is born.

The calculated mutation rate of 134 mutations per generation is just a rough estimate that illustrates the agreement between the estimates from the biochemical method and the phylogenetic method. It does not take into account population size and other parameters that are known to affect the calculation. Other estimates may use a different value for the percentage difference between the human and chimpanzee genomes (Suntsova and Buzdin, 2020), or they may assume a higher mutation rate in chimpanzees than in humans (Moorjani et al., 2016; Besenbacher et al., 2019). The estimate of total fixed mutations may be too high if you compare the reference genomes of humans and chimpanzees since some of the differences are polymorphic (i.e., not fixed). This overestimate is partly compensated by other considerations, so it doesn't significantly affect the total count. None of the other calculations make much of a difference in the estimated mutation rate, so they don't change the conclusion that most of our genome must be evolving at the neutral rate.
8 Nei (1987), p. 59.
9 Roach et al. (2010); The 1000 Genomes Project (2011); Kong et al. (2012); Jónsson et al. (2017).
10 Levy et al. (2007).
11 Telenti et al. (2016); Sudmant et al. (2015).
12 Actually, the DNA of identical twins will not be identical because of somatic cell mutations that have accumulated since the splitting of the early embryo. Twins can have different DNA fingerprints.
13 Graur (2017); Lynch (2010); Agrawal and Whitlock (2012).
14 Graur (2017); Galeota-Sprung et al. (2020).
15 See Fu et al. (2013). Some of these nonessential genes are redundant copies of important genes and others code for functions that are noticeable but not essential. See chapter 7 for some examples.
16 Lynch (2010); Lynch (2016).
17 Reviewed in Alkuraya (2015).
18 Plenge (2017); Saleheen et al. (2017); Narasimhan et al. (2016). Many of the dispensable genes are members of gene families where other similar genes can compensate for the missing gene (see chapter 7).
19 Lynch (2010).
20 Tomasetti and Vogelstein (2015); Gorski (2015); Nowak and Waclaw (2017).
21 Church et al. (2009).
22 A value of 8 percent is often quoted because it's in the title of a paper by Rands et al. (2014). However, the estimates from dozens of other papers range from about 5 to about 12 percent. Ponting and Hardison (2011) have published an excellent review of the difficulties in arriving at an accurate estimate of the fraction of the genome that's subject to purifying selection.
23 Graur (2016), p. 494.

24 See Brunet et al. (2021) for an example of such a paper. Philosophers prefer to restrict the definition of SE function to describing the origin of a particular trait. In that sense, it's only traits that have been positively selected over many generations that count as SE functions. For example, zebra stripes would be an SE function if, and only if, that trait was specifically selected over the allele for an all-white (or all-black) zebra. From the molecular viewpoint there are three problems with this narrow definition. First, we often don't know (or don't care) how a particular stretch of DNA came to be present in our genome. This means that unless you can provide good evidence for positive selection in the past, you can't say whether it is an SE function. Second, if you can't say for certain that it's a true SE function, does that mean that it's junk, or does it have some other kind of function? For example, if the all-white alleles of the zebra were eliminated by natural selection does that mean the genes were junk? Third, there are good reasons to believe that some functional parts of the genome arose by accident but have now become essential. A good example are the genes for spliceosome proteins, which can be plausibly explained by constructive neutral evolution. Many philosophers will argue that since these genes did not originate by positive natural selection they don't have an SE function. Nevertheless, they are not junk.
25 Linquist et al. (2020); Linquist (2022).
26 Hall et al. (2022).
27 Gould (2002).
28 See Doolittle (2022) for an excellent discussion of this issue.
29 Dawkins (1988).
30 Zuckerkandl (1986); Zuckerkandl (1997); Zuckerkandl (2002); Shapiro and (von) Sternberg (2005).
31 See Mattick and Dinger (2013) and Mattick (2018) for the most recent examples of this argument.
32 Ponting and Hardison (2011); Haerty and Ponting (2014); Ponting (2017).
33 Nóbrega et al. (2004).
34 Sudmant et al. (2015); Mills et al. (2011); Zarrei et al. (2015); Telenti et al. (2016); Bergström et al. (2020). As expected, a small number of these structural variants are loss of function deletions. The average person has three of these deleterious mutations (Abel et al., 2020).
35 Commoner (1964); Bennett (1972); Cavalier-Smith (1978).
36 Gregory (2005a); Kidwell (2005).
37 Orgel et al. (1980).
38 Eddy (2013).
39 Lynch (2007), pp. 39–42. One can imagine an argument based on the idea that the human genome is a "Goldilocks" genome that has just the right amount of bulk DNA – not too much and not too little. Doolittle and Brunet (2017) coined the term while pointing out that nobody actually makes such an argument for very good reasons that have something to do with fairy tales.
40 Kapusta et al. (2017); Wright et al. (2014).
41 Qiu (2015); Carey (2015).
42 Carey (2015), p. 35.
43 Bandea (1990); Bandea (2013).
44 See Dan Graur's blog post: "A Pre-Refuted Hypothesis on the Subject of 'Junk DNA.'" judgestarling.tumblr.com/post/67599627086/a-pre-refuted-hypothesis-on-the-subject-of-junk-dna.
45 Vihinen (2014); Huang (2016a); Huang (2016b).

46 "The mutational-hazard theory, which postulates that virtually all forms of excess DNA impose a weak mutational burden, provides a null hypothesis for interpreting patterns of genome evolution, yielding a potentially unifying explanation for the dramatic gradient in genome size and many aspects of gene-structural complexity across the tree of life" (Lynch et al., 2011).
47 McHughen (2020), p. 46.

5. The Big Picture

1 During the announcement of completion of the human genome sequence. Quoted in *The Genome War* by James Shreeve (2004), p. 356.
2 ncbi.nlm.nih.gov/gene/7167.
3 Zerbino et al. (2020).
4 GRC = Genome Reference Consortium; "h38" is the latest version of the reference sequence where "h" stands for human; p14 is patch 14. ncbi.nlm.nih.gov/projects/genome/assembly/grc/human/data/index.shtml.
5 Miga (2015); Miga (2019); Nurk et al. (2022).
6 Nurk et al. (2022).
7 Church (2022).
8 Venter et al. (2001).
9 See the next chapter for more detail on the size of introns.
10 Hatje et al. (2019) report a similar value of 41 to 42 percent.
11 Symonová (2019); Stults et al. (2008).
12 See Urban et al. (2015), Leonard and Méchali (2013), and Prioleau and MacAlpine (2016) for recent reviews.
13 Reviewed in Miga (2019). The best data on centromere sequences comes from the complete human genome sequence because the all the centromere regions were sequenced (Altemose et al., 2022).
14 Mirkovitch et al. (1984).
15 Mistrelli (2020).
16 Hoyt et al. (2022).

6. How Many Genes? How Many Proteins?

1 Kampourakis is using the word *encode* to refer to any information stored in a gene including information that specifies noncoding RNAs like ribosomal RNA. Many of us prefer the traditional definition of the word *encode* that restricts its use to sequences that specify proteins encoded in mRNA using the genetic code.
2 Ensembl website showing the latest gene count: ensemble.org/Homo_sapiens/Info/Annotation.
3 To be fair, there are textbooks that aren't precise about the meaning of *gene*, and there are others that are internally inconsistent. They often describe a gene as a stretch of DNA that encodes a protein but then go on to talk about genes for noncoding RNAs such as ribosomal RNA and tRNA. This ambiguity began in 1965 with James Watson's first textbook, *The Molecular Biology of the Gene*. In most of the book he refers to genes as synonymous with protein-coding regions, but then he also describes "genes" for ribosomal RNAs and tRNAs on page 329.
4 Orgogozo et al. (2016).
5 Dawkins (1976), p. 33.
6 https://royalsociety.org/news/2017/07/science-book-prize-poll-results.

7 The selfish gene perspective is extensively covered in the book *The Gene's-Eye View of Evolution* by J. Arvid Ågren (2021). Its goal is to cover all the controversies surrounding this perspective, but the only mention of the neutral theory occurs in two sentences on the second-last page of the book. Genetic drift is barely mentioned and never addressed as a viable alternative to natural selection. It's difficult to counter such adaptationist perspectives that are highly popular in books and articles for the general public.
8 Clamp et al. (2007).
9 Consensus Coding Sequence Database (CCDS): https://www.ncbi.nlm.nih.gov/CCDS/CcdsBrowse.cgi? REQUEST=SHOW_STATISTICS&DATA.
10 Abascal et al. (2018); Hatje et al. (2019).
11 Caballero et al. (2014).
12 Clamp et al. (2007).
13 Ezkurdia et al. (2014a); Wang et al. (2018); Wilhelm et al. (2014); Kim et al. (2014).
14 Omenn et al. (2020).
15 Paik et al. (2015).
16 Ramsköld et al. (2009); Wilhelm et al. (2014).
17 Müller et al. (2020) looked at 100 species representing all the main branches of the tree of life. They used fairly stringent criteria for identifying real proteins by mass spectrometry.
18 The work is mostly being done at the Craig Venter Institute run by Craig Venter – the man who was at the center of the Genome War from 1997 to 2000 (see Hutchison et al., 2016).
19 Ohno (1972a).
20 See International Human Genome Sequencing Consortium (2001) and Pertea and Salzberg (2010).
21 Schuler et al. (1996).
22 Aparicio (2000).
23 See Claverie (2001) for data on ESTs. The UniGene database is at www.ncbi.nlm.nih/UniGene/UGOrg.cgi?TAXID=9606.
24 Liang et al. (2000); Ewing and Green (2000); Roest Crollius et al. (2000).
25 There are very few accurate accounts of the history of gene counting, but one of the ones that get it right is Hatje et al. (2019).
26 *Nature* editorial, June 25, 2000.
27 Pennisi (2003).
28 Pennisi (2005), p. 80.
29 Harrow et al. (2012).
30 See Piovesan et al. (2019). Other values could be used (e.g., Hubé and Francastel, 2015), but it won't change the main conclusions.
31 Looking back at Figure 1.13, you can see that not all exon sequences are entirely made up of coding region, even in protein-coding genes. If you add up all the exons in protein-coding genes, you get a value closer to 2 percent of the genome, but much of it consists of untranslated regions (UTRs) at the beginning and end of the mature mRNA molecule. It is unlikely that all these UTRs are functional, but I've included some of them in my estimate of 1.0 percent coding region (Piovesan et al., 2019).
32 The latest direct estimates of coding regions are somewhat higher (see the earlier link to the CCDS and Omenn et al., 2020, and Omenn, 2021). You will see higher estimates of more than 40 percent for just the protein-coding genes (Hatje et al., 2019; Piovesan et al., 2019), but these higher estimates include many hypothetical sequences at the ends of genes that may or may not be part of real

genes. It's worth noting that there are several hundred genes that don't have introns, and it's not clear to me how they factor into the exon and intron averages that are reported. All these values, such as the average number of introns and exons, are subject to change, but they are good enough to give a rough estimate of the size of a gene. The true actual amount of the genome devoted to both types of genes is likely to end up somewhere between 35 and 45 percent.

33 Francis and Wörheide (2017).

34 UCSC Genome Browser: https://genome.ucsc.edu/cgi-bin/hgGateway.

35 The smallest known introns are about 18 to 20 bp, and there are a few others in the 20–30-bp class, but these tiny introns are only found in a restricted group of eukaryotes, where they are apparently under strong selection for the smallest size consistent with survival (Gilson and McFadden, 1996; Russell et al., 1994; Slamovits and Keeling, 2009). These exceptionally small introns lend further support to the idea that most of the DNA in large introns is nonessential junk.

36 Castillo-Davis et al. (2002).

37 Mewes at al. (1997).

38 Hooks et al. (2014).

39 Richardson et al. (2017); Parenteau et al. (2008); Parenteau et al. (2011); Morgan et al. (2018); Parenteau et al. (2019a); Parenteau et al. (2019b); Parenteau and Elela (2019).

40 The papers making these (false) claims are now 15 years old (Pan et al., 2008; Wang et al. 2008).

41 Hsu and Hertel (2009).

42 Stepankiw et al. (2015); Pickrell et al. (2010); Melamud and Moult (2009a); Melamud and Moult (2009b); Skandalis (2016).

43 Some of the junk RNAs are produced from mutant alleles where the mutant gene is nonfunctional. Cells containing one normal allele and one mutant allele will produce (at least) two different transcripts that could appear to be examples of alternative splicing (Mucaki et al. 2020).

44 There's plenty of evidence that for most protein-coding genes there's only one dominant transcript (e.g., Gonzàlez-Porta et al., 2013).

45 Abascal et al. (2018); Tress et al. (2017); Jiang et al. (2020).

46 Bhuïyan et al. (2018).

47 The authors of this study (Mudge et al. 2011) prefer a different interpretation. They believe that alternative splicing arises frequently in different lineages and that this leads to rapid gene evolution. They do not consider the view that splice variants may be due to splicing errors. As we will see throughout this book, the failure to detect conservation is often not taken as a falsification of the function hypothesis but, instead, is assumed to be evidence of rapid species-specific evolution.

48 Pickrell et al. (2010); Merkin et al. (2012); Reyes et al. (2013).

49 Harrow et al. (2012).

50 It's worth noting that cells contain a large number of complex structures formed by the association of multiple protein subunits. Examples include ribosomes, spliceosomes, RNA polymerase, and the large complexes in the mitochondrial membrane that are responsible for producing ATP. Each of the subunits is encoded by a separate gene. Proponents of alternative splicing propose that all or most of these genes produce a number of different proteins. It's difficult to imagine how highly organized multi-subunit structures could accommodate all those variants.

51 Despite the fact that species-specific alternative splicing is inconsistent with known rates of selection, the leading proponents of alternative splicing argue that each species has evolved its own set of alternatively spliced transcripts for most of their

genes. They claim that this is an important contribution to the differences between species (Pan et al., 2005; Barbosa-Morais, 2012).

52 Melamud and Moult (2009b).

53 This article has disappeared from the *Scitable* website, but a more recent article still promotes the idea that more than 90 percent of human genes are alternatively spliced; this presumably explains how humans can be so complex with only 20,000 to 30,000 genes.

54 For example, Brosius (2005).

55 Lynch (2010).

56 Li et al. (2016).

57 Rogaev et al. (2009).

58 Scotti and Swanson (2016).

59 Bush et al. (2017).

60 This is an example of a logical fallacy known as begging the question where the phrase "begging the question" is used in the traditional sense of avoiding the question. The logical fallacy occurs when someone assumes as a premise the point they are trying to prove. In this case, they assume that alternative splicing is real and then use their interpretation of genetic disease – based on this assumption – as evidence that alternative splicing is real.

61 Shirley et al. (2019).

62 Tourasse et al. (2017).

7. Gene Families and the Birth and Death of Genes

1 Your genome is 3.2 billion base pairs, but you are diploid, so you have two copies in every cell except sperm and eggs. There are also millions of copies of histone H1.

2 Histone genes are only transcribed during cell division because it would make no sense to produce histones if DNA was not being replicated. Histone genes don't have introns – presumably because there has been selection for deleting introns in order to make lots of mRNA in a very short amount of time. That helps, but it's not enough.

3 Marzluff et al. (2002). There are five distinct copies of the H1 gene (see Willcockson et al., 2021).

4 The "gene" is transcribed to produce a long 6000 nucleotide-bp precursor that's cut into three pieces corresponding to three of the ribosomal RNAs. The rate of transcription is about 60 nucleotides per second.

5 Meyer and Van de Peer (2003); Nei and Hughes (1992); Nei (2005); Nei and Rooney (2005); Nei (2013).

6 Dulai et al. (1999). One of the most common forms of color blindness occurs when one of the two X-chromosome genes is mutated. Since males have only one X chromosome, this form of color-blindness is much more common in males.

7 Kuzmin et al. (2021).

8 Gross (2007); Menashe et al. (2007).

9 Meiosis is a particular type of cell division that occurs during formation of germ cells where half the chromosomes are segregated into each daughter cell resulting in the formation of two haploid cells. The process requires aligning each pair of chromosomes to ensure proper segregation. Meiosis is also known as reduction division. Homologous recombination also occurs during normal cell division (mitosis). It's an essential part of DNA repair.

10 Dulai et al. (1999).

11 Demuth et al. (2006); Demuth and Hahn (2009).
12 Nei (2007); Nei (2013); Stoltzfus (2006); Stoltzfus and Cable (2014).
13 Gould (1989, 2002).
14 Lien et al. (2016); Xu et al. (2014); Session et al. (2016).
15 Ohno (1985).
16 Zarrei et al. (2015); Neumann et al. (2010).
17 Symonová (2019); Hall et al. (2022).
18 This statement is contentious because many scientists have looked for correlations between blood types and susceptibility to various diseases and infections. Such correlations have been found, but they are often not reproducible and/or the effects are very small.
19 Estrada-Meya et al. (2010).
20 I exaggerate somewhat. Variants of the *OR6A2* gene play only a small role in the preference for cilantro.
21 In theory, it should be easy to identify the "parent" gene that gave rise to the pseudogene, but in practice this is often difficult. About 10,000 pseudogenes have a known parent in the human genome, but many of the rest have ambiguous parents because there are multiple copies of the active gene, and it isn't clear which one gave rise to the pseudogene (Harrow et al., 2012).
22 Harrow et al. (2012).
23 Noncoding RNA genes will also give rise to pseudogenes, but they are more difficult to detect, so there's not a good estimate of the number of noncoding pseudogenes. It's safe to assume there are several thousand in the human genome.
24 Lynch and Conery (2000); Lynch (2007).
25 Lynch (2007).
26 Miller (2008).
27 For example, Tomkins (2013).
28 Moleirinho et al. (2013).
29 Milligan and Lipovich (2015).
30 There are only about 200 human pseudogenes that produce a protein (Kim et al., 2014).
31 Balakirev and Ayala (2003); Wen et al. (2012).
32 Xu and Zhang (2015).
33 Kaessman (2009); Parker et al. (2009).
34 Shreeve (2004), p. 49.
35 Telenti et al. (2016).

8. Noncoding Genes and Junk RNA

1 Cech and Steitz (2014); Nozawa and Kinjo (2016); Carvalho Barbosa et al. (2020).
2 There are multiple copies of the snRNA genes, but only a handful are thought to be functional. The rest are nonfunctional pseudogenes (Hoeppner et al., 2018).
3 All RISC complexes contain members of the Argonaute family of proteins and miRNA, siRNA, and piRNA bind to these proteins. Because of this common mechanism, some scientists prefer to group all three of these RNA families into a single category.
4 Haberle and Stark (2018); Bagchi and Iyer (2016).
5 Symonová (2019).
6 Riggs et al. (1972); Lin and Riggs (1972, 1975).
7 Yamamoto and Alberts (1975); Yamamoto and Alberts (1976). Specific high-affinity binding sites were eventually identified when it became possible to clone individual genes that were activated by estrogen.

8 The actual situation is even worse because there will be significant binding to sites like AGCTCA and AGGTTA that differ from the optimal site by only a single base pair. Some of these sites will be functional because the receptor will still bind strongly to sites that closely resemble the ideal biding site. Functional sites are known as "estrogen response elements."
9 Schmidt et al. (2010).
10 de Boer et al. (2014).
11 Yona et al. (2018).
12 Struhl (2007).
13 Transcription terminators in protein-coding genes are usually called polyadenylation sites because they are coupled to the addition of strings of As (poly A) to the 3′ end of the mRNA. The RNA precursor is cleaved at the polyadenylation site before addition of poly A. RNA polymerase II usually dissociates at the polyadenylation site but frequently it continues transcribing (Eaton and West, 2020). Genes transcribed by RNA polymerases I and III may also have inefficient terminators. More than half of all human protein-coding genes have multiple polyadenylation sites so that multiple transcripts with different lengths are produced. There are half a dozen mammalian genes in which the different transcripts have different functions due to the properties of the 3′ UTRs (Pereira-Castro and Moreira, 2021). However, it's unlikely that alternative 3′ UTRs are functional in most genes. The multiple polyadenylation sites are probably just there to increase the probability of termination.
14 van Bakel et al. (2010); Lai and Pugh (2017); Jin et al. (2017).
15 Haberle and Stark (2018).
16 Reviewed in Jensen et al. (2013); Kaikkonen and Adelman (2018); Field and Adelman (2020).
17 ENCODE Consortium (2007).
18 The ENCODE Project Consortium (2007), p. 812.
19 "Human Genome Further Unraveled," BBC News, June 14, 2007. For a brief summary of major headlines from around the world, see www.yumpu.com/en/document/view/19465624/media-clips-embl.
20 Mattick (2007); Taft et al. (2007). Mattick continues to promote the idea that thereare a huge number of genes for regulatory RNAs (Morris and Mattick, 2014; Mattick and Amaral, 2023).
21 Willingham and Gingeras (2006).
22 See the Lauerman article at www.yumpu.com/en/document/view/19465624/media-clips-embl.
23 Timmer (2007); Moran (2007); Moran (2007); Hunt (2007); Moran (2008).
24 For example, Paul and Gilmour (1968).
25 Hopkin (2009).
26 Wheeler et al. (2008).
27 Levy et al. (2007).
28 Heather and Chain (2015).
29 Xu et al. (2017) ; Hon et al. (2017) : Volders et al. (2019) ; Zerbino et al. (2020).
30 Zhou et al. (2021).
31 Goudarzi et al. (2019)
32 Moran (2016); Willingham and Gingeras (2006); Kapranov et al. (2007); Sverdlov (2017).
33 Kapranov et al. (2007).
34 van Bakel et al. (2010).
35 Clark et al. (2011).

36 van Bakel et al. (2011).
37 Ponting and Haerty (2022).
38 Parrington (2015), p. 91.
39 The Garvan Institute (2012); see also Phillips (2012).
40 See Note 39.
41 Morris and Mattick (2014).
42 Hurst (2013).
43 Ulitsky and Bartel (2013).
44 Pallazo and Lees (2015), my emphasis.
45 Ulitsky (2016).
46 Liu et al. (2017) ; Ramilowski et al. (2020).
47 Reviewed in Jensen et al. (2013).
48 Keep in mind that the problem isn't solved by referring to abundant transcripts as "dark matter," which means RNA of unknown function (Soares and Valcárcel, 2006; Johnson et al., 2005; Kapranov et al., 2010). That's just another way of saying that function, not junk, is the default assumption.
49 Wang et al. (2004).
50 Eddy (2013).
51 Eddy (2013).
52 Brosius (2005).
53 Brosius and Gould (1992).
54 Gould and Vrba (1982).
55 Guerzoni and McLysaght (2016); Vakirlis et al. (2022).
56 Ruse (2015).
57 For example, Neme and Tautz (2016).
58 Dawkins (1988); Gould (2002).
59 Stoltzfus (1999); Gray et al. (2010).
60 SeeLukeš et al. (2011).
61 Ruiz-Orera et al. (2020); Palazzo and Koonin (2020).
62 Carroll (2020).
63 Also called "The Lonely Fact Fallacy" (Ponting and Haerty, 2022).
64 Pennisi (2010).
65 Jacob (1977).
66 Dawkins (1982), p. 1.
67 Gould (1995), p. 93.
68 Moran (2011).

9. The ENCODE Publicity Campaign

1 This quotation is a personal communication cited in Graur et al. (2013), p. 578.
2 "The main set of papers were originally offered to *Science*, provided that the journal would accept a list of author demands. These included agreeing in advance that, while the authors would read all of the peer reviews, only they and not *Science* would decide on how (and whether) to respond to the reviews in the published manuscripts. *Science* replied that it would not agree to such a demand, and as a result the set of papers that were later published in *Nature* were never published there" (personal communication from Bruce Alberts, who at the time was editor in chief of *Science*).
3 Yong (2012).
4 Moran (2012b).
5 ENCODE Project Consortium, 2012), emphasis mine.
6 Pennisi (2012).

7 ENCODE Project Consortium (2012).
8 Maher (2012a).
9 Skipper et al. (2012).
10 https://youtube.com/watch?v=Y3V2thsJ1Wc.
11 Ecker et al. (2012).
12 https://youtube.com/watch?v=PsV_sEDSE2o.
13 youtube.com/watch?v=TwXXgEz9o4w. Adam Rutherford (2016) writes the following in his book *A Brief History of Everyone Who Ever Lived*: "I was working at Nature at that time, and with Ewan [Birney], and as such I suppose I was part of the hype engine. I wrote and produced an animation voiced by the actor/musician/comedian Tim Minchin, which was all part of the press surrounding the release of this data set. It's a good cartoon, and I stand by it, while accepting that some of the coverage from the time was a bit breathless and overreaching. So it goes." Readers are invited to watch the video and judge for yourself whether it accurately represents the ENCODE data.
14 https://www.youtube.com/watch?v=KiwXtHRfBC8.
15 Pennisi (2012).
16 Maher (2012b). The comments on this blog post make interesting reading, especially in hindsight.
17 Birney (2012).
18 Maher (2012b).
19 Maher (2012b).
20 Timmer (2012).
21 White (2012).
22 Gitschier (2015).
23 See Note 14.
24 The runners-up. Genomics beyond genes. (2012) *Science* 338, 1525–32. doi:10.1126/science.338.6114.1525.
25 Kellis et al. (2014)
26 Kellis et al. (2014).
27 Kellis et al. (2014).
28 Kellis et al. (2014).
29 Graur (2014).
30 ENCODE debate revived online. (2014) *Nature* 509, 137. doi:nature.com/articles/509137e
31 Casane et al. (2015).
32 The ENCODE Consortium et al. (2020). Many of the authors of the ENCODE III summary paper are the same ones who were on the 2012 summary paper claiming that 80 percent of the genome was functional.
33 Encode Project Consortium et al. (2020).
34 Gould (1987), p. 150.
35 Alberts et al. (2015).
36 Alberts et al. (2015).

10. Turning Genes On and Off

1 Monod and Jacob (1961). The original sentence was "Anything that is true of *Escherichia coli* must also be true of elephants, only more so." It dates from a lecture given by Jacques Monod in 1954 (see Friedman, 2004).
2 A surprising number of today's junk DNA supporters are alumni of the talk.origins newsgroup where these debates have been raging since the mid-1980s. Newsgroups

were the main social media platform before blogs, Twitter, Instagram, and Facebook were invented. The talk.origins server ("Darwin") is run by David Greig, and it used to be located in my university office before I retired.

3 Francis Collins speaking at the 3rd Annual J.P. Morgan Healthcare Conference in San Francisco, January 12–15, 2015.

4 Raj and van Oudenaarden (2008); Barroso et al. (2018).

5 It seems likely that the test tube measurements will be similar to what's happening inside the cell, but it's difficult to determine the in vivo rate constants. Recent work with RNA-binding proteins indicates that the on and off rates inside the cell are measured in seconds, thus confirming the in vitro data (Sharma et al., 2021).

6 Kothary et al. (1989).

7 I'm using the word *conserved* in a very loose sense here. Haerty and Ponting (2014) prefer *constrained* to indicate that a sequence is under negative selection. In order to assign function to a regulatory sequence you need to demonstrate that detrimental mutations are being eliminated by selection. They report that the truly constrained regions upstream of genes covers only 200 bp (see also Rands et al., 2014). I'm using 2000 bp of regulatory sequence as a generous estimate – it includes a lot of spacer DNA.

8 See Maleszka (2018).

9 See the excellent book *The Making of a Fly* (1992) by Peter A. Lawrence.

10 Gould (1977), p. 405, my emphasis.

11 Gould (1980) published in *The Panda's Thumb*.

12 Compe and Egly (2021).

13 Zuckerkandl (2005); Lynch and Hagner (2015).

14 This type of mutation is called a *homeotic* mutation after *homeosis*, or the transformation of one body part into another. The functions of homeotic genes were discovered by Edward B. Lewis who received the Nobel Prize in 1995.

15 The product of the *Antp* gene is a transcription factor that controls expression of a number of genes leading to development of a *Drosophila* leg. The inappropriate expression of this gene is a clear example of how small changes in the genome can lead to significant changes in the appearance of an organism. Genes such as *Antp* are known as homeotic genes, and most of these genes encode transcription factors that have a specific DNA binding region called a homeobox. The genes are collectively referred to as *Hox* genes (from homeobox). *Hox* genes have become very famous because they are found in all animals, including humans. There are about 40 *Hox* genes in our genome.

16 Reviewed in Henderson et al. (1999).

17 Susumu Tonegawa won the 1987 Nobel Prize in Physiology or Medicine for his discovery of V(D)J recombination and the generation of antibody diversity.

18 Part of the problem in dealing with chromatin rearrangements is that scientists aren't certain about the cause-and-effect relationships. Does the binding of a transcription factor cause chromatin modifications such as demethylation of DNA and histone modifications, or do these chromatin remodeling events occur first leading to an open chromatin domain that allows transcription factor binding? It seems likely that transcription factor binding occurs first.

19 About 50 percent of the best candidates for regulatory sites (enhancers) based on chromatin markers do not work as transcription activators when tested in a separate assay (see Field and Adelman, 2020).

20 Barr and Bertram (1949).

21 Tukiainen et al. (2017).

22 Quoted from *Herding Hemingway's Cats* (Arney 2016), p. 58.

23 See reviews by Henikoff and Shilatifard (2011) and Henikoff and Greally (2016).
24 Inactivation of an X chromosome in females is an exception to the rule of "breathing." Most of the heterochromatic X chromosome seems to be permanently unavailable for binding transcription factors. There are some other exceptions; they all seem to involve special mechanisms of chromosome shutdown.
25 Ptashne and Gann (1997); Ptashne (2013).
26 This is from the promotional video produced by the European Molecular Biology Laboratory (EMBL) when the ENCODE papers were published in September 2012. The video also states that there are 4 million switches in the human genome (youtube.com/watch?v=Y3V2thsJ1Wc).
27 ENCODE Project Consortium (2012).
28 "[E]ven using the most conservative estimates, the fraction of bases likely to be involved in direct regulation, even though incomplete, is significantly higher than that ascribed to protein codon exons (1.2%), raising the possibility that more information in the human genome may be important for gene regulation than for biochemical function" (ENCODE, 2012).
29 Maher (2012a).
30 Feynman (1985) in "Cargo Cult Science," p. 385.
31 Cusanovich et al. (2014).
32 White et al. (2013).
33 Doolittle (2013).
34 Cramer (2019).
35 Zuckerkandl et al. (1989); Shapiro and (von) Sternberg (2005).
36 Zuckerkandl (1986, 1987).
37 Waddington (1942); Holliday (1994); Deans and Maggert (2015); Ganesan (2018).
38 Berger et al. (2009).
39 X-chromosome inactivation is an exception to the rule. The heterochromatic state of the inactive X chromosome is readily passed on after cell division.
40 Mitchell (2018).
41 Ptashne (2013).
42 Ptashne (2007). In her book, Kat Arney (2016) quotes Adrian Bird, who is one of the authorities on methylation: "I'm not saying that it is impossible that there are stable changes in the way genes are used that are influenced by the environment. I'm just saying that the evidence seems to be mostly flaky. But there's a very strong desire to interpret it in a positive way, and the journals and the media love it. So the idea of transgenerational epigenetic effects gets a great press, despite the weakness of the data." She also has a chapter devoted to an interview with Mark Ptashne, and it's worth buying her book for that chapter alone.
43 Arney (2016), p. 101.

11. Zen and the Art of Coping with a Sloppy Genome

1 Pirsig (1974), p. 167.
2 Piovesan et al. (2019); Hatje et al. (2019).
3 One of my friends is John Wilkins, a philosopher and a veteran of the creationist wars. He has been studying the species concept for several decades. I highly recommend his latest book: *Species: The Evolution of the Idea* (2nd ed.).
4 Doolittle (2013); Graur et al. (2013); Germain et al. (2014); Doolittle et al. (2014); Doolittle and Brunet (2017); Linquist et al. (2020).
5 Christie et al. (2022).

6 Graur (2017); Brzović and Šustar (2020); Linquist et al. (2020); Linquist (2022).
7 These are the polymorphic pseudogenes. I discussed an example of such a gene in chapter 7: it's the gene responsible for ABO blood groups where the defective copy is responsible for blood type O. The gene is dispensable; therefore, it is junk DNA, but I'm uncomfortable saying that the A and B alleles of this gene don't have a function since they encode an enzyme that attaches sugars to a protein on the surface of a red blood cell and the gene is well conserved in mammals. Definitions are difficult.
8 Eddy (2013).
9 Gould and Lewontin (1979).
10 As a textbook author, I'm particularly sensitive to this issue of "rewriting the textbooks." Whenever I was working on the new edition of my biochemistry textbook, I was constantly on the lookout for emerging ideas and concepts that needed to be taught in undergraduate courses. There have been several "revolutions" over the past 50 years in the way we present some concepts in metabolism, bioenergetics, and protein structure, and the most interesting thing about these revolutions is how difficult it is to get them accepted by the teachers who adopt the textbook. Every new concept that gets into a textbook means fewer adoptions as teachers resist having to change their lectures. Authors learn pretty quickly that "rewriting the textbooks" is something you want to avoid at all costs – or at the very least you learn that you don't want to be the first one to make a change.
11 Wellcome Sanger Institute (2012).
12 Sagan (1995), p. 210, emphasis in the original.
13 Sagan (1979), p. 75.
14 Collins (2010), p. 293.
15 Francis Collins is the founder of BioLogos, an organization dedicated to showing that science is not in conflict with the Bible. He is the author of *The Language of God*.
16 Meyer (2013), p. 402.
17 That's an oblique reference to a famous statement by Richard Dawkins (1989b): "It is absolutely safe to say that if you meet someone who claims not to believe in evolution, that person is ignorant, stupid or insane (or wicked, but I'd rather not consider that)." It's also consistent with Hanlan's razor: "never attribute to malice that which can be adequately explained by stupidity."
18 Keep in mind that speculations such as alternative splicing and a plethora of regulatory RNAs only support human exceptionalism if they are restricted to humans and not found in "less complex" animals. But noisy splicing and spurious transcripts are seen in all species, so the argument makes no sense.

References

Abascal, F., Juan, D., Jungreis, I., Martinez, L., Rigau, M., Rodriguez, J.M., Vazquez, J., and Tress, M.L. (2018). Loose ends: Almost one in five human genes still have unresolved coding status. *Nucleic Acids Research, 46*(14), 7070–84. doi:10.1093/nar/gky587

Abel, H.J., Larson, D.E., Regier, A.A., Chiang, C., Das, I., Kanchi, K.L., Layer, R.M., Neale, B.M., Salerno, W.J., Reeves, C., Buyske, S., NHGRI Centers for Common Disease Genomics, Matise, T.C., Muzny, D.M., Zody, M.C., Lander, E.S., Dutcher, S.K., Stitziel, N.O., and Hall, I.M. (2020). Mapping and characterization of structural variation in 17,795 human genomes. *Nature, 583*, 83–9. doi:10.1038/s41586-020-2371-0

Adams, M.D., Celniker, S.E., Holt, R.A., Evans, C.A., Gocayne, J.D., Amanatides, P.G., Scherer, S.E., Li, P.W., Hoskins, R.A., Galle, R.F., George, R.A., Lewis, S.E., Richards, S., Ashburner, M., Henderson, S.N., Sutton, G.G., Wortman, J.R, Yandell, M.D., Zhang, Q., … Venter, J.C. (2000). The genome sequence of Drosophila melanogaster. *Science, 287*, 2185–95. doi:10.1126/science.287.5461.2185

Agrawal, A.F., and Whitlock, M.C. (2012). Mutation load: The fitness of individuals in populations where deleterious alleles are abundant. *Annual Review of Ecology, Evolution, and Systematics, 43*, 115–35. doi:10.1146/annurev-ecolsys-110411-160257

Alberts, B., Cicerone, R.J., Fienberg, S.E., Kamb, A., McNutt, M., Nerem, R.M., Schekman, R., Shiffrin, R., Stodden, V., and Suresh, S. (2015). Self-correction in science at work. *Science, 348*, 1420–2. doi:10.1126/science.aab3847

Alkuraya, F.S. (2015). Human knockout research: New horizons and opportunities. *Trends in Genetics, 31*, 108–15. doi:10.1016/j.tig.2014.11.003

Altemose, N., Logsdon, G.A., Bzikadze, A.V., Sidhwani, P., Langley, S.A., Caldas, G.V., Hoyt, S.J., Uralsky, L., Ryabov, F.D., Shew, C.J., Sauria, M.E.G., Borchers, M., Gershman, A., Mikheenko, A., Shepelev, V.A., Dvorkina, T., Kunyavskaya, O., Vollger, M.R., Rhie, A., … Miga, K.H. (2022). Complete genomic and epigenetic maps of human centromeres. *Science, 376*, 56. doi:10.1126/science.abl4178

Aparicio, S.A. (2000). How to count … human genes. *Nature Genetics, 25*, 129–30. doi:10.1038/75949

Arney, K. (2016). *Herding Hemingway's cats: Understanding how our genes work.* Bloomsbury Sigma.

Ayala, F. (2004). Design without designer: Darwin's greatest discovery. In W.A. Dembski and M. Ruse (Eds.), *Debating design: From Darwin to DNA* (pp. 55–80). Cambridge University Press.

Aziz, R.K., Breitbart, M., and Edwards, R.A. (2010). Transposases are the most abundant, most ubiquitous genes in nature. *Nucleic Acids Research, 38*, 4207–17. doi:10.1093/nar/gkq140

Bagchi, D.N., and Iyer, V.R. (2016). The determinants of directionality in transcriptional initiation. *Trends in Genetics, 32*, 322–33. doi:10.1016/j.tig.2016.03.005

Balakirev, E.S., and Ayala, F.J. (2003). DNA polymorphism in the β-esterase gene cluster of *Drosophila melanogaster*. *Genetics, 164*, 533–44. doi:10.1146/annurev.genet.37.040103.103949

Baldi, S., Korber, P. and Becker, P.B. (2020). Beads on a string—nucleosome array arrangements and folding of the chromatin fiber. *Nature Structural & Molecular Biology, 27*, 109–18. doi:10.1038/s41594-019-0368-x

Bandea, C.I. (1990). A protective function for noncoding, or secondary DNA. *Medical Hypotheses, 31*, 33–4. doi:10.1016/0306-9877(90)90050-O

Bandea, C.I. (2013). On the concept of biological function, junk DNA and the gospels of ENCODE and Graur et al. *bioRxiv*, 000588. doi:10.1101/000588

Barbosa-Morais, N.L., Irimia, M., Pan, Q., Xiong, H.Y., Gueroussov, S., Lee, L.J., Slobodeniuc, V., Kutter, C., Watt, S., Çolak, R., Kim, T., Misquitta-Ali, C.M., Wilson, M.D., Kim, P.M., Odom, D.T., Frey, B.J., and Blencowe, B.J. (2012). The evolutionary landscape of alternative splicing in vertebrate species. *Science, 338*, 1587–93. doi:10.1126/science.1230612

Bar-On, Y.M., Phillips, R., and Milo, R. (2018). The biomass distribution on Earth. *Proceedings of the National Academy of Sciences, 115*, 6506–11. doi:10.1073/pnas.1711842115

Barr, M.L., and Bertram, E.G. (1949). A morphological disanons between neurones of the male and female, and the behaviour of the nucleolar satellite during accelerated nucleoprotein synthesis. *Nature, 163*, 676–7.

Barroso, G.V., Puzovic, N., and Dutheil, J.Y. (2018). The evolution of gene-specific transcriptional noise is driven by selection at the pathway level. *Genetics, 208*, 173–89. doi:10.1534/genetics.117.300467

Bennett, M. (1972). Nuclear DNA content and minimum generation time in herbaceous plants. *Proceedings of the Royal Society B, 181*, 109–35. doi:10.1098/rspb.1972.0042

Berger, S. L:, Kouzarides, T., Shiekhattar, R., and Shilatifard, A. (2009). An operational definition of epigenetics. *Genes & development, 23*, 781–783. doi:10.1101/gad.1787609.

Bergström, A., McCarthy, S.A., Hui, R., Almarri, M.A., Ayub, Q., Danecek, P., Chen, Y., Felkel, S., Hallast, P., Kamm, J., Blanché, H., Deleuze, J-F., Cann, H., Mallick, S., Reich, D., Sandhu, M.S., Skoglund, P., Scally, A., Xue, Y., … Tyler-Smith, C. (2020). Insights into human genetic variation and population history from 929 diverse genomes. *Science, 367*: eaay5012. doi:10.1126/science.aay5012

Besenbacher, S., Hvilsom, C., Marques-Bonet, T., Mailund, T., and Schierup, M. (2019). Direct estimation of mutations in great apes reconciles phylogenetic dating. *Nature Ecology & Evolution, 3*, 286–92 doi:10.1038/s41559-018-0778-x

Bhattacharyya, M.K., Smith, A.M., Ellis, T.N., Hedley, C., and Martin, C. (1990). The wrinkled-seed character of pea described by Mendel is caused by a transposon-like insertion in a gene encoding starch-branching enzyme. *Cell, 60*, 115–22. doi:10.1016/0092-8674(90)90721-P

Bhuiyan, S.A., Ly, S., Phan, M., Huntington, B., Hogan, E., Liu, C.C., Liu, J., and Pavlidis, P. (2018). Systematic evaluation of isoform function in literature reports of alternative splicing. *BMC Genomics, 19*, 637. doi:10.1186/s12864-018-5013-2

Birney, E. (2012, September 5). ENCODE: My own thoughts. *Ewan's Blog: Bioinformatician at Large*. https://ewanbirney.com/2012/09/encode-my-own-thoughts.html

Blommaert, J. (2020). Genome size evolution: Towards new model systems for old questions. *Proceedings of the Royal Society B*, *287*, 20201441. doi:10.1098/rspb.2020.1441

Blommaert, J., Riss, S., Hecox-Lea, B., Welch, D.M., and Stelzer, C.-P. (2019). Small, but surprisingly repetitive genomes: Transposon expansion and not polyploidy has driven a doubling in genome size in a metazoan species complex. *BMC GenoMics*, *20*, 1–12. doi:10.1186/s12864-019-5859-y

Bloom, K.S. (2013) Centromeric heterochromatin: The primordial segregation machine. *Annual Review of Genetics*, *48*, 457–84. doi:10.1146/annurev-genet-120213-092033

The Brassica rapa Genome Sequencing Project, Wang, X., Wang, H., Wang, J., Sun, R., Wu, J., Liu, S., Bai, Y., Mun, J.-H., Bancroft, I., Cheng, F., Huang, S., Li, X., Hua, W., Wang, J., Wang, X., Freeling, M., Pires, J.C., Paterson, A.H., … Zhang, Z. (2011). The genome of the mesopolyploid crop species *Brassica rapa*. *Nature Genetics*, *43*, 1035–9. doi:10.1038/ng.919

Brenner, S. (1998). Refuge of spandrels. *Current Biology*, *8*, R669. doi:10.1016/s0960-9822(98)70427-0

Brenner, S. (2000) Biochemistry strikes back. *Trends in Biochemical Sciences*, *25*(12), 584. doi:10.1016/S0968-0004(00)01722-9

Britten, R.J. (1997). Mobile elements inserted in the distant past have taken on important functions. *Gene*, *205*, 177–82. doi:10.1016/S0378-1119(97)00399-5

Britten, R.J., and Davidson, E.H. (1969). Gene regulation for higher cells: A theory. *Science*, *165*, 349–57. doi:10.1126/science.165.3891.349

Britten, R.J., and Davidson, E.H. (1971). Repetitive and non-repetitive DNA sequences and a speculation on the origins of evolutionary novelty. *Quarterly Review of Biology*, *46*, 111–38. doi:10.1086/406830

Britten, R.J., and Kohne, D. (1968). Repeated sequences in DNA. *Science*, *161*, 529–40.

Brosius, J. (2005). Waste not, want not–transcript excess in multicellular eukaryotes. *Trends in Genetics*, *21*, 287–8. doi:10.1016/j.tig.2005.02.014

Brosius, J., and Gould, S.J. (1992). On "genomenclature": a comprehensive (and respectful) taxonomy for pseudogenes and other "junk DNA." *Proceedings of the National Academy of Sciences*, *89*, 10706–10. doi:10.1073/pnas.89.22.10706

Brunet, T.D., Doolittle, W.F., and Bielawski, J.P. (2021). The role of purifying selection in the origin and maintenance of complex function. *Studies in History and Philosophy of Science Part A*, *87*, 125–35. doi:10.1016/j.shpsa.2021.03.005

Brzović, Z., and Šustar, P. (2020). Postgenomics function monism. *Studies in History and Philosophy of Science Part C: Studies in History and Philosophy of Biological and Biomedical Sciences*, *80*, 101243. doi:10.1016/j.shpsc.2019.101243

Bush, S.J., Chen, L., Tovar-Corona, J.M., and Urrutia, A.O. (2017). Alternative splicing and the evolution of phenotypic novelty. *Philosophical Transactions of the Royal Society B*, *372*, 20150474. doi:10.1098/rstb.2015.0474

Caballero, J., Smit, A.F., Hood, L., and Glusman, G. (2014). Realistic artificial DNA sequences as negative controls for computational genomics. *Nucleic Acids Research*, *42*, e99. doi:10.1093/nar/gku356

Calabrese, F.M., Simone, D., and Attimonelli, M. (2012). Primates and mouse NumtS in the UCSC Genome Browser. *BMC Bioinformatics*, *13*(Suppl. 4), S15. doi:10.1186/1471-2105-13-S4-S15

Carroll, S.B. (2005). *Endless forms most beautiful*. W.W. Norton & Company.

Carroll, S.B. (2020). *A series of fortunate events: Chance and the making of the planet, life, and you*. Princeton University Press.

Carvalho Barbosa, C., Calhoun, S.H., and Wieden, H.-J. (2020). Non-coding RNAs: What are we missing? *Biochemistry and Cell Biology*, *98*, 23–30. doi:10.1139/bcb-2019-0037

Carey, N. (2015). *Junk DNA*. Columbia University Press.

Casane, D., Fumey, J., and Laurenti, P. (2015). L'apophénie d'ENCODE ou Pangloss examine le génome humain [ENCODE apophenia or a panglossian analysis of the human genome]. *Médecine/Sciences (Paris)*, *31*, 680–6. doi:10.1051/medsci/20153106023

Castillo-Davis, C.I., Mekhedov, S.L., Hartl, D.L., Koonin, E.V., and Kondrashov, F.A. (2002). Selection for short introns in highly expressed genes. *Nature Genetics*, *31*, 415–8. doi:10.1038/ng940

Cattoni, D. I., Valeri, A., Le Gall, A., and Nollmann, M. (2015). A matter of scale: how emerging technologies are redefining our view of chromosome architecture. *TRENDS in Genetics*, *31*, 454–464. doi:10.1016/j.tig.2015.05.011

Cavalier-Smith, T. (1978). Nuclear volume control by nucleoskeletal DNA, selection for cell volume and cell growth rate, and the solution of the DNA C-value paradox. *Journal of Cell Science*, *34*, 247–78. doi:10.1242/jcs.34.1.247

Cavalier-Smith, T. (1980). How selfish is DNA? *Nature*, *285*, 617–8. doi:10.1038/285617a0

Cech, T.R., and Steitz, J.A. (2014). The noncoding RNA revolution—trashing old rules to forge new ones. *Cell*, *157*, 77–94. doi:10.1016/j.cell.2014.03.008

Charlesworth, B., Sniegowski, P., and Stephan, W. (1994). The evolutionary dynamics of repetitive DNA in eukaryotes. *Nature*, *371*, 215–20. doi:10.1038/371215a0

Christie, J.R., Brusse, C., Bourrat, P., Takacs, P., and Griffiths, P.E. (2022). Are biological traits explained by their 'selected effect' functions? *Australasian Philosophical Review*. Advanced online publication. doi:philsci-archive.pitt.edu/19832/

Church, D.M. (2022). A next-generation human genome sequence. *Science*, *376*, 34–5. doi:10.1126/science.abo5367

Church, D.M., Goodstadt, L., Hillier, L.W., Zody, M.C., Goldstein, S., She, X., Bult, C.J., Agarwala, R., Cherry, J.L., DiCuccio, M., Hlavina, W., Kapustin, Y., Meric, P., Maglott, D., Birtle, Z., Marques, A.C., Graves, T., Zhou, S., Teague, B., … the Mouse Genome Sequencing Consortium. (2009). Lineage-specific biology revealed by a finished genome assembly of the mouse. *PLoS Biology*, *7*, e1000112. doi:10.1371/journal.pbio.1000112

Clamp, M., Fry, B., Kamal, M., Xie, X., Cuff, J., Lin, M.F., Kellis, M., Lindblad-Toh, K., and Lander, E.S. (2007). Distinguishing protein-coding and noncoding genes in the human genome. *Proceedings of the National Academy of Sciences*, *104*(49), 19428–33. doi:10.1073/pnas.0709013104

Clark, M.B., Amaral, P.P., Schlesinger, F.J., Dinger, M.E., Taft, R.J., Rinn, J.L., Ponting, C.P., Stadler, P.F., Morris, K.V., Morillon, A., Rozowsky, J.S., Gerstein, M.B., Wahlestedt, C., Hayashizaki, Y., Carninci, P., Gingeras, T.R., and Mattick, J.S. (2011). The reality of pervasive transcription. *PLoS Biology*, *9*(7), e1000625. doi:10.1371/journal.pbio.1000625

Claverie, J.-M. (2001). What if there are only 30,000 human genes? *Science*, *291*, 1255–7. doi:10.1126/science.1058969

Collins, F.S. (2006) *The language of God: A scientist presents evidence for belief*. Free Press

Collins, F.S. (2010). *The language of life: DNA and the revolution in personalized medicine*. HarperCollins.

Commoner, B. (1964). Roles of deoxyribonucleic acid in inheritance. *Nature*, *202*, 960–8. doi:10.1038/202960a0

Compe, E., and Egly, J.-M. (2021). The long road to understanding RNAPII transcription initiation and related syndromes. *Annual Review of Biochemistry*, *90*, 193–219. doi:10.1146/annurev-biochem-090220-112253

Cook, J. (2019). Got climate change misconceptions? John Cook can help. *Reports of the National Center for Science Education, 39*, 4–6.
Cordaux, R., Lee, J., Dinoso, L., and Batzer, M. A. (2006). Recently integrated Alu retrotransposons are essentially neutral residents of the human genome. *Gene, 373*, 138–44. doi:10.1016/j.gene.2006.01.020
Cosby, R.L., Judd, J., Zhang, R., Zhong, A., Garry, N., Pritham, E. J., and Feschotte, C. (2021). Recurrent evolution of vertebrate transcription factors by transposase capture. *Science, 371*. doi:10.1126/science.abc6405
Coursey, T.L., and McBride, A.A. (2019). Hitchhiking of viral genomes on cellular chromosomes. *Annual Review of Virology, 6*, 275–96. doi:10.1146/annurev-virology-092818-015716
Cramer, P. (2019). Organization and regulation of gene transcription. *Nature, 573*, 45–54. doi:10.1038/s41586-019-1517-4
Crow, J.F., and Kimura, M. (1970). *An introduction to population genetics theory*. Harper & Row.
Cusanovich, D.A., Pavlovic, B., Pritchard, J.K., and Gilad, Y. (2014). The functional consequences of variation in transcription factor binding. *PLoS Genetics, 10*, e1004226. doi:10.1371/journal.pgen.1004226
Dawkins, R. (1976). *The selfish gene*. Oxford University Press.
Dawkins, R. (1982). *The extended phenotype*. Oxford University Press.
Dawkins, R. (1988). The evolution of evolvability. In C.G. Langton (Ed.), *Artificial life: The proceedings of an interdisciplinary workshop on the synthesis and simulation of living systems held September 1987 in Los Alamos, New Mexico* (pp. 201–20). Addison-Wesley.
Dawkins, R. (1989a). *The selfish gene* (New ed.). Oxford University Press.
Dawkins, R. (1989b, April 9). Review of BLUEPRINTS Solving the Mystery of Evolution. *New York Times*.
Dawkins, R., and Wong, Y. (2016). *The Ancestor's Tale* (2nd ed). Weidenfeld & Nicolson.
Deans, C., and Maggert, K.A. (2015). What do you mean, "epigenetic"? *Genetics, 199*, 887–96. doi:10.1534/genetics.114.173492
de Boer, C.G., van Bakel, H., Tsui, K., Li, J., Morris, Q.D., Nislow, C., Greenblatt, J.F., and Hughes, T.R. (2014). A unified model for yeast transcript definition. *Genome Research, 24*, 154–66. doi:10.1101/gr.164327.113
De Koning, A., Gu, W., Castoe, T.A., Batzer, M.A., and Pollock, D.D. (2011). Repetitive elements may comprise over two-thirds of the human genome. *PLoS Genetics, 7*, e1002384. doi:10.1371/journal.pgen.1002384
Demuth, J.P., De Bie, T., Stajich, J.E., Cristianini, N., and Hahn, M.W. (2006). The evolution of mammalian gene families. *PLoS ONE, 1*, e85. doi:10.1371/journal.pone.0000085
Demuth, J.P., and Hahn, M.W. (2009). The life and death of gene families. *BioEssays, 31*, 29–39. doi:10.1002/bies.080085
De Parseval, N., and Heidmann, T. (2005). Human endogenous retroviruses: From infectious elements to human genes. *Cytogenetic and Genome Research, 110*, 318–32. doi:10.1159/000084964
DiCarlo, C. (2011). *How to become a really good pain in the ass*. Prometheus Books.
DiCarlo, C. (2017). *Six steps to better thinking*. FriesenPress.
Doolittle, W.F. (2013). Is junk DNA bunk? A critique of ENCODE. *Proceedings of the National Academy of Sciences, 110*, 5294–300. doi:10.1073/pnas.1221376110
Doolittle, W.F. (2022). All about levels: Transposable elements as selfish DNAs and drivers of evolution. *Biology & Philosophy, 37*(4), 1–20. doi:s10539-022-09852-3
Doolittle, W.F., and Brunet, T.D. (2017). On causal roles and selected effects: Our genome is mostly junk. *BMC Biology, 15*, 116. doi:10.1186/s12915-017-0460-9

Doolittle, W.F., Brunet, T.D., Linquist, S., and Gregory, T.R. (2014). Distinguishing between "function" and "effect" in genome biology. *Genome Biology and Evolution, 6,* 1234–7. doi:10.1093/gbe/evu098

Doolittle, W.F., and Sapienza, C. (1980). Selfish genes, the phenotype paradigm and genome evolution. *Nature, 284,* 601–3. doi:10.1038/284601a0

Dover, G. (1980). Ignorant DNA? *Nature, 285,* 618–9. doi:10.1038/285618a0

Dover, G., and Doolittle, W.F. (1980). Modes of genome evolution. *Nature, 288,* 646–7. doi:10.1038/288646a0

Dulai, K.S., von Dornum, M., Mollon, J.D., and Hunt, D.M. (1999). The evolution of trichromatic color vision by opsin gene duplication in New World and Old World primates. *Genome Research, 9,* 629–38. doi:10.1101/gr.9.7.629

Dunham, I., Hunt, A., Collins, J., Bruskiewich, R., Beare, D., Clamp, M., Smink, L., Ainscough, R., Almeida, J., and Babbage, A. (1999). The DNA sequence of human chromosome 22. *Nature, 402,* 489–95. doi:10.1038/990031

Eaton, J.D., and West, S. (2020). Termination of transcription by RNA polymerase II: BOOM! *TRENDS in Genetics, 36,* P664–75. doi:10.1016/j.tig.2020.05.008

Ecker, J.R., Bickmore, W.A., Barroso, I., Pritchard, J.K., Gilad, Y., & Segal, E. (2012). ENCODE explained. *Nature, 489,* 52–4. doi:10.1038/489052a

Eddy, S.R. (2012). The C-value paradox, junk DNA and ENCODE. *Current Biology, 22,* R898. doi:10.1016/j.cub.2012.10.002

Eddy, S.R. (2013). The ENCODE project: Missteps overshadowing a success. *Current Biology, 23,* R259–61. doi:10.1016/j.cub.2013.03.023

The ENCODE Project Consortium. (2007). Identification and analysis of functional elements in 1% of the human genome by the ENCODE pilot project. *Nature, 447,* 799–816. doi.org/10.1038/nature05874

The ENCODE Project Consortium. (2012). An integrated encyclopedia of DNA elements in the human genome. *Nature, 489,* 57–74. doi:10.1038/nature11247

The ENCODE Project Consortium, Snyder, M.P., Gingeras, T.R., Moore, J.E., Weng, Z., Gerstein, M.B., Ren, B., Hardison, R.C., Stamatoyannopoulos, J.A., Graveley, B.R., Feingold, E.A., Pazin, M.J., Pagan, M., Gilchrist, D.A., Hitz, B.C., Cherry, J.M., Bernstein, B.E., Mendenhall, E.M., Zerbino, D.R., … Myers, R.M. (2020). Perspectives on ENCODE. *Nature, 583,* 693–8. doi:10.1038/s41586-020-2449-8

Estrada-Mena, B., Estrada, F.J., Ulloa-Arvizu, R., Guido, M., Méndez, R., Coral, R., Canto, T., Granados, J., Rubí-Castellanos, R., Rangel-Villalobos, H., and García-Carrancá, A. (2010). Blood group O alleles in Native Americans: Implications in the peopling of the Americas. *American Journal of Physical Anthropology, 142,* 85–94. doi:10.1002/ajpa.21204

Ewing, B., and Green, P. (2000). Analysis of expressed sequence tags indicates 35,000 human genes. *Nature Genetics, 25,* 232–4. doi:10.1038/76115

Ezkurdia, I., Vázquez, J., Valencia, A., and Tress, M. (2014). Analyzing the first drafts of the human proteome. *Journal of Proteome Research, 13,* 3854–5. doi:10.1021/pr500572z

Feynman, R.P. (1985). *"Surely you're joking, Mr. Feynman": Adventures of a curious character.* W.W. Norton.

Field, A., and Adelman, K. (2020). Evaluating enhancer function and transcription. *Annual Review of Biochemistry, 89,* 213–34. doi:10.1146/annurev-biochem-011420-095916

Fleischmann, R., Adams, M., White, O., Clayton, R., Kirkness, E., Kerlavage, A., Bult, C., Tomb, J., Dougherty, B., Merrick, J., McKenney, K., Sutton, G., Fitzhugh, W., Fields, C., Gocayne, J.D., Scott, J., Shirley, R., Liu, L-L., Glodek, A., … Venter, J.C. (1995). Whole-genome random sequencing and assembly of *Haemophilus influenzae* Rd. *Science, 269,* 496–512. doi:10.1126/science.7542800

Francis, W.R., and Wörheide, G. (2017). Similar ratios of introns to intergenic sequence across animal genomes. *Genome Biology and Evolution, 9*, 1582–98. doi:10.1093/gbe/evx103

Friedman, H.C. (2004). From butyribacterium to *E. coli*: An essay on unity in biochemistry. *Perspectives in Biology and Medicine, 47*, 47–66. doi:10.1353/pbm.2004.0007

Fu, W., O'Connor, T.D., Jun, G., Kang, H.M., Abecasis, G., Leal, S.M., Gabriel, S., Rieder, M.J., Altshuler, D., Shendure, J., Nickerson, D.A., Bamshad, M.J., NHLBI Exome Sequencing Project, and Akey, J.M. (2013). Analysis of 6,515 exomes reveals the recent origin of most human protein-coding variants. *Nature, 493*, 216–20. doi:10.1038/nature11690

Galeota-Sprung, B., Sniegowski, P., and Ewens, W. (2020). Mutational load and the functional fraction of the human genome. *Genome Biology and Evolution, 12*, 273–81. doi:10.1093/gbe/evaa040

Ganesan, A. (2018). Epigenetics: The first 25 centuries. *Philosophical Transactions B, 373*, 20170067. doi:10.1098/rstb.2017.0067

The Garvin Institute. (2012, March). *Making something of junk earns geneticist top award* [Press release].

Germain, P.-L., Ratti, E., and Boem, F. (2014). Junk or functional DNA? ENCODE and the function controversy. *Biology & Philosophy*. Advanced online publication. doi:10.1007/s10539-014-9441-3

Gilson, P.R., and McFadden, G.I. (1996). The miniaturized nuclear genome of eukaryotic endosymbiont contains genes that overlap, genes that are cotranscribed, and the smallest known spliceosomal introns. *Proceedings of the National Academy of Sciences, 93*, 7737–42. doi:10.1073/pnas.93.15.7737

Gitschier, J. (2015). The philosophical approach: An interview with Ford Doolittle. *PLoS Genetics, 11*(5), e1005173. doi:10.1371/journal.pgen.1005173

Gonzàlez-Porta, M., Frankish, A., Rung, J., Harrow, J., and Brazma, A. (2013). Transcriptome analysis of human tissues and cell lines reveals one dominant transcript per gene. *Genome Biology, 14*, 1–11. doi:10.1186/gb-2013-14-7-r70

Goodier, J.L. (2016). Restricting retrotransposons: A review. *Mobile DNA, 7*, 16. doi:10.1186/s13100-016-0070-z

Gorski, D. (2015, January 5). *Is cancer due mostly to "bad luck"?* Science-Based Medicine. https://sciencebasedmedicine.org/is-cancer-due-mostly-to-bad-luck/

Goudarzi, M., Berg, K., Pieper, L.M., and Schier, A.F. (2019). Individual long non-coding RNAs have no overt functions in zebrafish embryogenesis, viability and fertility. *Elife, 8*, e40815. doi:10.7554/eLife.40815.001

Gould, S.J. (1977). *Ontogeny and phylogeny*. The Belknap Press of Harvard University Press.

Gould, S.J. (1980). *The panda's thumb*. W.W. Norton & Co.

Gould, S.J. (1983). *Hen's teeth and horse's toes*. W.W. Norton & Co.

Gould, S.J. (1987). "Nurturing Nature": A review of *Not in Our Genes: Ideology and Human Nature* by R.C. Lewontin, S. Rose, and L.J. Kamin [Reprint]. In *An urchin in the storm* (pp. 145–54). W.W. Norton.

Gould, S.J. (1989). *Wonderful life: The Burgess Shale and the nature of history*. W.W. Norton & Co.

Gould, S.J. (1995). *Dinosaur in a haystack*. Harmony Books.

Gould, S.J. (2002). *The structure of evolutionary theory*. The Belknap Press of Harvard University Press.

Gould, S.J., and Lewontin, R.C. (1979). The spandrels of San Marco and the Panglossian paradigm: a critique of the adaptationist programme. *Proceedings of the Royal Society of London. Series B. Biological Sciences, 205*, 581–98. doi:10.1098/rspb.1979.0086

Gould, S.J., and Vrba, E.S. (1982). Exaptation—a missing term in the science of form. *Paleobiology*, *8*, 4–15.

Graur, D. (2014, April 23). @ENCODE_NIH in PNAS 2014: In 2012, the dog ate our lab notebook and we had no laxative to retrieve it. *Judgestarling*. tumblr.com /judgestarling/83591181083/encode-nih-in-pnas-2014-in-2012-the-dog-ate-our-lab

Graur, D. (2016). *Molecular and genome evolution*. Sinauer Associates Inc.

Graur, D. (2017). An upper limit on the functional fraction of the human genome. *Genome Biology and Evolution*, *9*, 1880–5. doi:10.1093/gbe/evx121

Graur, D., Zheng, Y., Price, N., Azevedo, R.B., Zufall, R.A., and Elhaik, E. (2013). On the immortality of television sets: "Function" in the human genome according to the evolution-free gospel of ENCODE. *Genome Biology and Evolution*, *5*, 578–90. doi:10.1093/gbe/evt028

Gray, M.W., Lukeš, J., Archibald, J.M., Keeling, P.J., and Doolittle, W. (2010). Irremediable complexity? *Science*, *330*(6006), 920–1. doi:10.1126/science.1198594

Gregory, T.R. (2001). Coincidence, coevolution, or causation? DNA content, cell size, and the C-value enigma. *Biological Reviews*, *76*, 65–101. doi:10.1017/s146479 3100005595

Gregory, T.R. (2005a). Synergy between sequence and size in large-scale genomics. *Nature Reviews Genetics*, *6*, 699–708. doi:10.1038/nrg1674

Gregory, T.R. (2005b). Genome size evolution in animals. In T.R. Gregory (Ed.), *The evolution of the genome* (pp. 3–87). Elsevier Academic Press.

Gregory, T.R. (2007). The onion test. *Genomicron*. http://www.genomicron.evolverzone .com/2007/04/onion-test/

Gross, L. (2007). A genetic basis for hypersensitivity to "sweaty" odors in humans. *PLoS Biology*, *5*, e298. doi:10.1371/journal.pbio.0050298

Guerzoni, D., and McLysaght, A. (2016) De novo genes arise at a slow but steady rate along the primate lineage and have been subject to incomplete lineage sorting. *Genome Biology and Evolution*, *8*, 1222–32. doi:10.1093/gbe/evw074

Gymrek, M., Willems, T., Guilmatre, A., Zeng, H., Markus, B., Georgiev, S., Daly, M.J., Price, A.L., Pritchard, J.K., Sharp, A.J., and Erlich, Y. (2016) Abundant contribution of short tandem repeats to gene expression variation in humans. *Nature Genetics*, *48*, 22–9. doi:10.1038/ng.3461

Haberle, V., and Stark, A. (2018) Eukaryotic core promoters and the functional basis of transcription initiation. *Nature Reviews Molecular Cell Biology*, *19*, 621–37. doi:10.1038 /s41580-018-0028-8-8.

Haerty, W., and Ponting, C.P. (2014). No gene in the genome makes sense except in the light of evolution. *Annual Review of Genomics and Human Genetics*, *15*, 71–92. doi:10.1146/annurev-genom-090413-025621

Haldane, J.B.S. (1949). The rate of mutation of human genes. *Hereditas*, *35*(Suppl. 1), 267–73. doi:10.1111/j.1601-5223.1949.tb03339.x

Hall, A.N., Morton, E., and Queitsch, C. (2022). First discovered, long out of sight, finally visible: Ribosomal DNA. *Trends in Genetics*, *38*, P587–97 doi:10.1016 /j.tig.2022.02.005

Harrow, J., Frankish, A., Gonzalez, J.M., Tapanari, E., Diekhans, M., Kokocinski, F., Aken, B.L., Barrell, D., Zadissa, A., Searle, S., Barnes, I., Bignell, A., Boychenko, V., Hunt, T., Kay, M., Mukherjee, G., Rajan, J., Despacio-Reyes, G., Saunders, G., … Hubbard, T.J. (2012). GENCODE: the reference human genome annotation for the ENCODE Project. *Genome Research*, *22*, 1760–74. doi:10.1101/gr.135350.111

Hartley, G., and O'Neill, R.J. (2019). Centromere repeats: Hidden gems of the genome. *Genes*, *10*, 223. doi.org/10.3390/genes10030223

Hatje, K., Mühlhausen, S., Simm, D., and Kollmar, M. (2019). The protein-coding human genome: Annotating high-hanging fruits. *BioEssays, 41*, 1900066. doi:10.1002/bies.201900066

He, Z., Ji, R., Havlickova, L., Wang, L., Li, Y., Lee, H.T., Song, J., Koh, C., Yang, J., Zhang, M., Parkin, I.A.P., Wang, X., Edwards, D., King, G.J., Zou, J., Liu, K., Snowdon, R.J., Banga, S.S., Machackova, I. and Bancroft, I. (2021). Genome structural evolution in *Brassica* crops. *Nature Plants*. Advanced online publication. doi:10.1038/s41477-021-00928-8

Heather, J.M., and Chain, B. (2015). The sequence of sequencers: The history of sequencing DNA. *Genomics, 107*, 1–8. doi:10.1016/j.ygeno.2015.11.003

Henderson, I.R., Owen, P., and Nataro, J.P. (1999). Molecular switches—the ON and OFF of bacterial phase variation. *Molecular Microbiology, 33*, 919–32. doi:10.1046/j.1365-2958.1999.01555.x

Henikoff, S., and Greally, J.M. (2016) Epigenetics, cellular memory and gene regulation. *Current Biology, 26*, R644–8. doi:10.1016/j.cub.2016.06.011

Henikoff, S., and Shilatifard, A. (2011). Histone modification: cause or cog? *TRENDS in Genetics, 27*, 389–96. doi:10.1016/j.tig.2011.06.006

Hoeppner, M.P., Denisenko, E., Gardner, P.P., Schmeier, S., and Poole, A.M. (2018). An evaluation of function of multicopy noncoding RNAs in mammals using ENCODE/FANTOM data and comparative genomics. *Molecular Biology and Evolution, 35*, 1451–62. doi:10.1093/molbev/msy046

Holliday, R. (1994). Epigenetics: An overview. *Developmental Genetics, 15*, 453–7. doi:10.1002/dvg.1020150602

Hon, C.-C., Ramilowski, J.A., Harshbarger, J., Bertin, N., Rackham, O.J., Gough, J., Denisenko, E., Schmeier, S., Poulsen, T.M., Severin, J., Lizio, M., Kawaji, H., Kasukawa, T., Itoh, M., Burroughs, A.M., Noma, S., Djebali, S., Alam, T., Medvedeva, Y.A., … Forrest, A.R.R. (2017). An atlas of human long non-coding RNAs with accurate 5′ ends. *Nature, 543*, 199–204. doi:10.1038/nature21374

Hooks, K.B., Delneri, D., and Griffiths-Jones, S. (2014). Intron evolution in *Saccharomycetaceae*. *Genome Biology and Evolution, 6*, 2543–56. doi:10.1093/gbe/evu196

Hopkin, K. (2009). The evolving definition of a gene: With the discovery that nearly all of the genome is transcribed, the definition of a "gene" needs another revision. *BioScience, 59*, 928–31. 10.1525/bio.2009.59.11.3

Hoyt, S.J., Storer, J.M., Hartley, G.A., Grady, P.G., Gershman, A., de Lima, L.G., Limouse, C., Halabian, R., Wojenski, L., Rodriguez, M., Altemose, N., Rhie, A., Core, L.J., Gerton, J.L., Makalowski, W., Olson, D., Rosen, J. Smit, A.F.A., Straight, A.F., … O'Neill, R.J. (2022). From telomere to telomere: the transcriptional and epigenetic state of human repeat elements. *Science, 376*, 57. doi:10.1126/science.abk3112

Hsu, S.-N., and Hertel, K.J. (2009). Spliceosomes walk the line: Splicing errors and their impact on cellular function. *RNA Biology, 6*, 526–30. doi:10.4161/rna.6.5.9860

Huang, S. (2016a). Editorial: Special issue on the comprehensive functionality of genomic DNAs. *Genomics, 108*, 1–2. doi:10.1016.j_ygeno.2016.06.001

Huang, S. (2016b). New thoughts on an old riddle: What determines genetic diversity within and between species? *Genomics, 108*, 3–10. doi:10.1016/j.ygeno.2016.01.008

Hubby, J.L., and Lewontin, R.C. (1966). A molecular approach to the study of genic heterozygosity in natural populations. I. The number of alleles at different loci in *Drosophila pseudoobscura*. *Genetics, 54*, 577–94. doi:10.1093/genetics/54.2.577

Hubé, F., and Francastel, C. (2015). Mammalian introns: When the junk generates molecular diversity. *International Journal of Molecular Sciences, 16*, 4429–52. doi:10.3390/ijms16034429

Hunt, A. (2007, October 11). *Junk to the second power*. Panda's Thumb. pandasthumb.org/archives/2007/10/junk-to-the-sec.html

Hurst, L.D. (2013) Open questions: A logic (or lack thereof) of genome organization. *BMC Biology, 11*, 58. doi:10.1186/1741-7007-11-58

Hutchison, C.A., Chuang, R.-Y., Noskov, V.N., Assad-Garcia, N., Deerinck, T.J., Ellisman, M.H., Gill, J., Kannan, K., Karas, B.J., Ma, L., Pelletier, J.F., Qi, Z.-Q., Richter, R.A., Strychalski, E.A., Sun, L., Suzuki, Y., Tsvetanova, B., Wise, K.S., Smith, H.O., … Venter, J.C. (2016). Design and synthesis of a minimal bacterial genome. *Science, 351*(6280). doi:10.1126/science.aad6253

Ibarra-Laclette, E., Lyons, E., Hernández-Guzmán, G., Pérez-Torres, C.A., Carretero-Paulet, L., Chang, T.-H., Lan, T., Welch, A.J., Juárez, M.J.A., Simpson, J., Fernández-Cortés, A., Arteaga-Vázquez, M., Góngora-Castillo, E., Acevedo-Hernández, G., Schuster, S.C., Himmelbauer, H., Minoche, A.E., Xu, S., Lynch, M., … Herrera-Estrella, L. (2013). Architecture and evolution of a minute plant genome. *Nature, 498*, 94–8. doi:10.1038/nature12132

International Human Genome Consortium. (2004). Finishing the euchromatic sequence of the human genome. *Nature, 431*, 931–45. doi:10.1038/nature03001

International Human Genome Sequencing Consortium. (2001). Initial sequencing and analysis of the human genome. *Nature, 409*, 860–921. doi:10.1038/35057062

Jacob, F. (1977). Evolution and tinkering. *Science, 196*, 1161–7. doi:10.1126/science.860134

Jain, H.K. (1980). Incidental DNA. *Nature, 288*, 647–8. doi:10.1038/288647a0

Jensen, T.H., Jacquier, A., and Libri, D. (2013). Dealing with pervasive transcription. *Molecular Cell, 52*, 473–84. doi:10.1016/j.molcel.2013.10.032

Jiang, L., Wang, M., Lin, S., Jian, R., Li, X., Chan, J., Dong, G., Fang, H., Robinson, A.E., GTEx Consortium, and Snyder, M.P. (2020). A quantitative proteome map of the human body. *Cell, 183*, 269–83.e19. doi:10.1016/j.cell.2020.08.036

Jin, Y., Eser, U., Struhl, K., and Churchman, L.S. (2017). The ground state and evolution of promoter region directionality. *Cell, 170*, 889–98. e810. doi:10.1016/j.cell.2017.07.006

Johnson, J.M., Edwards, S., Shoemaker, D., and Schadt, E.E. (2005). Dark matter in the genome: evidence of widespread transcription detected by microarray tiling experiments. *Trends in Genetics, 21*, 93–102. doi:10.1016/j.tig.2004.12.009

Johnson, W.E. (2019). Origins and evolutionary consequences of ancient endogenous retroviruses. *Nature Reviews Microbiology, 17*, 355–70. doi:10.1038/s41579-019-0189-2

Jónsson, H., Sulem, P., Kehr, B., Kristmundsdottir, S., Zink, F., Hjartarson, E., Hardarson, M.T., Hjorleifsson, K.E., Eggertsson, H.P., Gudjonsson, S.A., Ward, J.D., Arnadottir, G.A., Helgason, E.A., Helgason, H., Gylfason, A., Jonasdottir, A., Jonasdottir, A., Rafnar, T., Frigge, M., … Stefansson, K. (2017). Parental influence on human germline de novo mutations in 1,548 trios from Iceland. *Nature, 549*, 519–22. doi:10.1038/nature24018

Judson, H.F. (1996). *The eight day of creation (expanded edition)*. Cold Spring Harbor Laboratory Press

Kaessmann, H. (2009). More than just a copy. *Science, 325*, 958–9. doi:10.1126/science.1178487

Kaikkonen, M.U., and Adelman, K. (2018). Emerging roles of non-coding RNA transcription. *Trends in Biochemical Sciences, 43*, 654–67. doi:10.1016/j.tibs.2018.06.002

Kampourakis, K. (2017). *Making sense of genes*. Cambridge University Press.

Kapranov, P., St Laurent, G., Raz, T., Ozsolak, F., Reynolds, C.P., Sorensen, P.H., Reaman, G., Milos, P., Arceci, R.J. Thompson, J.F., and Triche, T.J. (2010). The majority of total nuclear-encoded non-ribosomal RNA in a human cell is 'dark matter' un-annotated RNA. *BMC Biology, 8*, 149. doi:10.1186/1741-7007-8-149

Kapranov, P., Willingham, A.T., and Gingeras, T.R. (2007). Genome-wide transcription and the implications for genomic organization. *Nature Reviews Genetics, 8*, 413–23. doi:10.1038/nrg2083

Kapusta, A., Suh, A., and Feschotte, C. (2017). Dynamics of genome size evolution in birds and mammals. *Proceedings of the National Academy of Sciences, 114*, E1460–9. doi:10.1073/pnas.1616702114

Kellis, M., Wold, B., Snyder, M.P., Bernstein, B.E., Kundaje, A., Marinov, G.K., Ward, L.D., Birney, E., Crawford, G.E., Dekker, J., Dunham, I., Elnitski, L.L., Farnham, P.J., Feingold, E.A., Gerstein, M., Giddings, M.C., Gingeras, T.R., Green, E.D., Guigo, R., … Hardison, R.C. (2014). Defining functional DNA elements in the human genome. *Proceedings of the National Academy of Sciences, 111*, 6131–8. doi:10.1073/pnas.1318948111

Kidwell, M.G. (2005). Genome size evolution in plants. In T.R. Gregory (Ed.), *The evolution of the genome* (pp. 165–213). Elsevier.

Kim, J., Farré, M., Auvil, L., Capitanu, B., Larkin, D.M., Ma, J. and Lewin, H.A. (2017). Reconstruction and evolutionary history of eutherian chromosomes. *Proceedings of the National Academy of Sciences, 114*, E5379–88. doi:10.1073/pnas.1702012114

Kim, M.-S., Pinto, S.M., Getnet, D., Nirujogi, R.S., Manda, S.S., Chaerkady, R., Madugundu, A.K., Kelkar, D.S., Isserlin, R., Jain, S., Thomas, J.K., Muthusamy, B., Leal-Rojas, P., Kumar, P., Sahasrabuddhe, N.A., Balakrishnan, L., Advani, J., George, B., Renuse, S., … Pandey, A. (2014). A draft map of the human proteome. *Nature, 509*, 575–81. doi:10.1038/nature13302

Kimura, M. (1968) Evolutionary rate at the molecular level. *Nature, 217*, 624–6.

Kimura, M. (1989). The neutral theory of molecular evolution and the world view of the neutralists. *Genome, 31*, 24–31. doi:10.1139/g89-009

Kimura, M. (1991). Recent development of the neutral theory viewed from the Wrightian tradition of theoretical population genetics. *Proceedings of the National Academy of Sciences, 88*, 5969–73. doi:10.1073/pnas.88.14.5969

King, J.L., and Jukes, T.H. (1969). Non-Darwinian evolution. *Science, 164*, 788–98. doi:10.1126/science.164.3881.788

Kirchberger, P.C., Schmidt, M.L., and Ochman, H. (2020). The ingenuity of bacterial genomes. *Annual Review of Microbiology, 74*, 815–34. doi:10.1146/annurev-micro-020518-115822

Kong, A., Frigge, M.L., Masson, G., Besenbacher, S., Sulem, P., Magnusson, G., Gudjonsson, S.A., Sigurdsson, A., Jonasdottir, A., Jonasdottir, A., Wong, W.S.W., Sigurdsson, G., Walters, G.B., Steinberg, S., Helgason, H., Thorleifsson, G., Gudbjartsson, D.F., Helgason, A., Magnusson, O.T., … Stefansson, K. (2012). Rate of *de novo* mutations and the importance of father/'s age to disease risk. *Nature, 488*, 471–5. doi:10.1038/nature11396

Konkel, M.K., Walker, J.A., Hotard, A.B., Ranck, M.C., Fontenot, C.C., Storer, J., Stewart, C., Marth, G.T., the 1000 Genomes Consortium, and Batzer, M.A. (2015). Sequence analysis and characterization of active human *Alu* subfamilies based on the 1000 genomes pilot project. *Genome Biology and Evolution, 7*, 2608–22. doi:10.1093/gbe/evv167

Kothary, R., Clapoff, S., Darling, S., Perry, M.D., Moran, L.A., and Rossant, J. (1989). Inducible expression of an hsp68-lacZ hybrid gene in transgenic mice. *Development, 105*, 707–14. doi:10.1242/dev.105.4.707

Kronenberg, Z.N., Fiddes, I.T., Gordon, D., Murali, S., Cantsilieris, S., Meyerson, O.S., Underwood, J.G., Nelson, B.J., Chaisson, M.J., Dougherty, M.L., Munson, K.M., Hastie, A.R., Diekhans, M., Hormozdiari, F., Lorusso, N., Hoekzema, K., Qui, R., Clark, K., Raja, A., … Eichler, E.E. (2018). High-resolution comparative analysis of great ape genomes. *Science, 360*, 1085. doi:10.1126/science.aar6343

Kuzmin, E., Taylor, J.S., and Boone, C. (2021). Retention of duplicated genes in evolution. *Trends in Genetics*. Advanced online publication. doi:10.1016/j.tig.2021.06.016

Lai, W.K., and Pugh, B.F. (2017). Genome-wide uniformity of human 'open' pre-initiation complexes. *Genome Research, 27*, 15–26. doi:10.1101/gr.210955.116

Lang, M., Sazzini, M., Calabrese, F., Simone, D., Boattini, A., Romeo, G., Luiselli, D., Attimonelli, M., and Gasparre, G. (2012). Polymorphic NumtS trace human population relationships. *Human Genetics, 131*, 757–71. 10.1007/s00439-011-1125-3

Langergraber, K.E., Prüfer, K., Rowney, C., Boesch, C., Crockford, C., Fawcett, K., Inoue, E., Inoue-Muruyama, M., Mitani, J.C., Muller, M.N., Robbins, M.M., Schubert, G., Stoinski, T.S., Viola, B., Watts, D., Wittig, R.M., Wrangham, R.W., Zuberbühler, K., Pääbo, S., and Vigilant, L. (2012). Generation times in wild chimpanzees and gorillas suggest earlier divergence times in great ape and human evolution. *Proceedings of the National Academy of Sciences, 109*, 15716–21. doi:10.1073/pnas.1211740109

Lawrence, P.A. (1992). *The making of a fly: The genetics of animal design*. Blackwell Scientific Publications.

Leonard, A.C., and Méchali, M. (2013). DNA replication origins. *Cold Spring Harbor Perspectives in Biology*, 5, a010116. doi:10.1101/cshperspect.a010116

Levy, S., Sutton, G., Ng, P.C., Feuk, L., Halpern, A.L., Walenz, B.P., Axelrod, N., Huang, J., Kirkness, E.F., Denisov, G., Lin, Y., MacDonald, J.R., Pang, A.W.C., Shago, M., Stockwell, T.B., Tsiamouri, A., Bafna, V., Bansal, V., Kravitz, S.A., … Venter, J.C. (2007). The diploid genome sequence of an individual human. *PLoS Biology, 5*, e254. doi:10.1371/journal.pbio.0050254

Lewin, B. (1974a). Sequence organization of eukaryotic DNA: Defining the unit of gene expression. *Cell, 1*, 107–11 doi:10.1016/0092-8674(74)90125-1

Lewin, B. (1974b). *Gene expression-2: Eukarotic chromosomes*. John Wiley & Sons.

Lewin, B. (1975). Units of transcription and translation: The relationship between heterogeneous nuclear RNA and messenger RNA. *Cell, 4*, 11–20. doi:10.1016/0092-8674

Lewontin, R.C., and Hubby, J.L. (1966). A molecular approach to the study of genic heterozygosity in natural populations. II. Amount of variation and degree of heterozygosity in natural populations of *Drosophila pseudoobscura*. *Genetics, 54*, 595. doi:10.1093/genetics/54.2.595

Li, W.-H., and Graur, D. (1991). *Fundamentals of Molecular Evolution*. Sinauer Associates.

Li, Y.I., van de Geijn, B., Raj, A., Knowles, D.A., Petti, A.A., Golan, D., Gilad, Y., and Pritchard, J.K. (2016). RNA splicing is a primary link between genetic variation and disease. *Science, 352*, 600–4. doi:10.1126/science.aad9417

Liang, F., Holt, I., Pertea, G., Karamycheva, S., Salzberg, S.L., and Quackenbush, J. (2000). Gene Index analysis of the human genome estimates approximately 120,000 genes. *Nature Genetics, 25*, 239–40. doi:10.1038/76126

Lien , S., Koop, B.F., Sandve, S.R., Miller, J.R., Kent, M.P., Nome, T., Hvidsten, T.R., Leong, J.S., Minkley, D.R., Zimin, A., Grammes, F., Grove, H., Gjuvsland, A., Walenz, B., Hermansen, R.A., von Schalburg, K., Rondeau, E.B., Di Genova, A., Samy, J.K.A., … Davidson, W.S. (2016). The Atlantic salmon genome provides insights into rediploidization. *Nature, 533*, 200–5. doi:10.1038/nature17164

Lin, S.-y., and Riggs, A.D. (1972). *lac* represser binding to non-operator DNA: Detailed studies and a comparison of equilibrium and rate competition methods. *Journal of Molecular Biology, 72*, 671–90. doi:10.1016/0022-2836(72)90184-2

Lin, S.-y., and Riggs, A.D. (1975). The general affinity of *lac* repressor for *E. coli* DNA: Implications for gene regulation in procaryotes and eucaryotes. *Cell*, 4, 107–11. doi:10.1016/0092-8674(75)90116-6
Linquist, S. (2022). Causal-role myopia and the functional investigation of junk DNA. *Biology & Philosophy*, *37*, 1–23. doi:10.1007/s10539-022-09853-2
Linquist, S., Doolittle, W.F., and Palazzo, A.F. (2020). Getting clear about the F-word in genomics. *PLoS Genetics*, *16*, e1008702. doi:10.1371/journal.pgen.1008702
Liu, S.J., Horlbeck, M.A., Cho, S.W., Birk, H.S., Malatesta, M., He, D., Attenello, F.J., Villalta, J.E., Cho, M.Y., Chen, Y., Mandegar, M.A., Olvera, M.P., Gilbert, L.A., Conklin, B.R., Change, H.Y., Weissman, H.S., and Lim, D.A. (2017). CRISPRi-based genome-scale identification of functional long noncoding RNA loci in human cells. *Science*, *355*, 39. doi:10.1126/science.aah7111
Lukeš, J., Archibald, J.M., Keeling, P.J., Doolittle, W.F., and Gray, M.W. (2011). How a neutral evolutionary ratchet can build cellular complexity. *IUBMB Life*, *63*, 528–37. doi:10.1002/iub.489
Lynch, M. (2007). *The origins of genome architecture*, Sinauer Associates, Inc. Publishers.
Lynch, M. (2010). Rate, molecular spectrum, and consequences of human mutation. *Proceedings of the National Academy of Sciences*, *107*, 961–8. doi:10.1073/pnas.0912629107
Lynch, M. (2016). Mutation and human exceptionalism: Our future genetic load. *Genetics*, *202*, 869–75. doi:10.1534/genetics.115.180471
Lynch, M., Bobay, L.-M., Catania, F., Gout, J.-F., and Rho, M. (2011). The repatterning of eukaryotic genomes by random genetic drift. *Annual Review of Genomics and Human Genetics*, *12*, 347–66. doi:10.1146/annurev-genom-082410-101412
Lynch, M., and Conery, J.S. (2000). The evolutionary fate and consequences of duplicate genes. *Science*, *290*, 1151–5. doi:10.1126/science.290.5494.1151
Lynch, M., and Conery, J.S. (2003). The origins of genome complexity. *Science*, *302*, 1401–4. doi:10.1126/science.1089370
Lynch, M., and Hagner, K. (2015). Evolutionary meandering of intermolecular interactions along the drift barrier. *Proceedings of the National Academy of Sciences*, *112*, E30–8. doi:10.1073/pnas.1421641112
Maher, B. (2012a). The human encyclopedia. *Nature*, *489*, 46–8.
Maher, B. (2012b) Fighting about ENCODE and junk. Nature newsblog, Sept. 6, 2012. blogs.nature.com/news/2012/09/fighting-about-encode-and-junk.html
Maleszka, R. (2018). Beyond royalactin and a master inducer explanation of phenotypic plasticity in honey bees. *Communications Biology*, *1*, 1–7. doi:10.1038/s42003-017-0004-4
Marzluff, W.F., Gongidi, P., Woods, K.R., Jin, J., and Maltais, L.J. (2002). The human and mouse replication-dependent histone genes. *Genomics*, *80*, 487–98. doi:10.1006/geno.2002.6850
Mattick, J.S. (2007). A new paradigm for developmental biology. *Journal of Experimental Biology*, *210*, 1526–47. doi:10.1242/jeb.005017
Mattick, J.S. (2018). The state of long non-coding RNA biology. *Non-Coding RNA*, *4*, 17. doi:10.3390/ncrna4030017
Mattick, J.S., and Amaral, P. (2023). *RNA, the epicenter of genetic information*. CRC Press.
Mattick, J.S., and Dinger, M.E. (2013). The extent of functionality in the human genome. *The HUGO Journal*, *7*, 2. doi:10.1186/1877-6566-7-2
McHughen, A. (2020). *DNA demystified: Unravelling the double helix*. Oxford University Press.

Melamud, E., and Moult, J. (2009a). Structural implication of splicing stochastics. *Nucleic Acids Research, 37*, 4862–72. doi:10.1093/nar/gkp444

Melamud, E., and Moult, J. (2009b). Stochastic noise in splicing machinery. *Nucleic Acids Research, 37*, 4873–86. doi:10.1093/nar/gkp471

Menashe, I., Abaffy, T., Hasin, Y., Goshen, S., Yahalom, V., Luetje, C.W., and Lancet, D. (2007). Genetic elucidation of human hyperosmia to isovaleric acid. *PLoS Biology, 5*, e284. doi:10.1371/journal.pbio.0050284

Merkin, J., Russell, C., Chen, P., and Burge, C.B. (2012). Evolutionary dynamics of gene and isoform regulation in mammalian tissues. *Science, 338*, 1593–9. doi:10.1126/science.1228186

Mewes, H., Albermann, K., Bähr, M., Frishman, D., Gleissner, A., Hani, J., Heumann, K., Kleine, K., Maierl, A., Oliver, S., Pfeiffer, F., and Zollner, A. (1997). Overview of the yeast genome. *Nature, 387*, 7–8. doi:10.1038/42755

Meyer, A., and Van de Peer, Y. (2003). 'Natural selection merely modified while redundancy created'–Susumu Ohno's idea of the evolutionary importance of gene and genome duplications. *Journal of Structural and Functional Genomics, 3*, 7–9. doi:10.1023/a:1022684816803

Meyer, S.C. (2013). *Darwin's doubt: The explosive origin of animal life and the case for intelligent design*. HarperCollins.

Miga, K.H. (2015). Completing the human genome: The progress and challenge of satellite DNA assembly. *Chromosome Research, 23*, 421–6. doi:10.1007/s10577-015-9488-2

Miga, K.H. (2019). Centromeric satellite DNAs: Hidden sequence variation in the human population. *Genes, 10*, 352. doi:10.3390/genes10050352

Miller, K.R. (2008). *Only a theory: Evolution and the battle for America's soul*. Penguin Group.

Milligan, M.J., and Lipovich, L. (2015). Pseudogene-derived lncRNAs: Emerging regulators of gene expression. *Frontiers in Genetics, 5*, 476. doi:10.3389/fgene.2014.00476

Mills, R.E., Walter, K., Stewart, C., Handsaker, R.E., Chen, K., Alkan, C., Abyzov, A., Yoon, S.C., Ye, K., Cheetham, R.K., Chinwalla, A., Conrad, D.F., Fu, Y., Grubert, F., Hajirasouliha, I., Hormozdiari, F., Iakoucheva, L.M., Ibal, Z., Kang, S., … 1000 Genomes Project. (2011). Mapping copy number variation by population-scale genome sequencing. *Nature, 470*, 59–65. doi:10.1038/nature09708

Mirkovitch, J., Mirault, M.-E., and Laemmli, U.K. (1984). Organization of the higher-order chromatin loop: Specific DNA attachment sites on nuclear scaffold. *Cell, 39*, 223–32. doi:10.1016/0092-8674(84)90208-3

Misteli, T. (2020). The self-organizing genome: Principles of genome architecture and function. *Cell, 183*, 28–45. 10.1016/j.cell.2020.09.014

Mitchell, K. (2018, May 29). *Grandma's trauma—a critical appraisal of the evidence for transgenerational epigenetic inheritance in humans*. Wiring the Brain. wiringthebrain.com/2018/05/grandmas-trauma-critical-appraisal-of.html

Moleirinho, A., Seixas, S., Lopes, A.M., Bento, C., Prata, M.J., and Amorim, A. (2013). Evolutionary constraints in the β-globin cluster: The signature of purifying selection at the δ-globin (HBD) locus and its role in developmental gene regulation. *Genome Biology and Evolution, 5*, 559–71. doi:10.1093/gbe/evt029

Monod, J.L. (1975). On the molecular theory of evolution: Progress and obstacles to progress in the sciences. In R. Harré (Ed.), *Problems of scientific revolution. The Herbert Spencer lectures 1973* (pp. 11–24). Clarendon.

Monod, J.L., and Jacob, F. (1961). General conclusions: Teleonomic mechanisms in cellular metabolism, growth and differentiation. *Cold Spring Harbor Symposia on*

Quantitative Biology (Cellular Regulatory Mechanisms), 26, 389–401. doi:10.1101/sqb.1961.026.01.048
Moorjani, P., Amorim, C.E.G., Arndt, P.F., and Przeworski, M. (2016). Variation in the molecular clock of primates. *Proceedings of the National Academy of Sciences, 113*, 10607–12. doi:10.1073/pnas.1600374113
Moran, L.A. (2007, October 11). *Junk RNA*. Sandwalk. sandwalk.blogspot.com/2007/10/junk-rna.html
Moran, L.A. (2008, August 28) *Useful RNAs?* Sandwalk. sandwalk.blogspot.com/2008/08/useful-rnas.html
Moran, L.A. (2011, October 24) *The myth of junk DNA by Jonathan Wells*. Sandwalk. sandwalk.blogspot.com/2011/10/myth-of-junk-dna-by-jonathan-wells.html
Moran, L.A. (2012a). Review of "Evolution: A View from the 21st Century" by James Shaprio. *Reports of the National Center for Science Education, 32*, 3.
Moran, L.A. (2012b, September 5). *ENCODE leader says that 80% of our genome is functional*. Sandwalk. sandwalk.blogspot.com/2012/09/encode-leader-says-that-80-of-our.html
Moran, L.A. (2016, September 6). *How many lncRNAs are functional: Can sequence comparisons tell us the answer?* Sandwalk. sandwalk.blogspot.com/2016/09/how-many-lncrnas-are-functional-can.html
Morange, M. (2014). Genome as a multipurpose structure built by evolution. *Perspectives in Biology and Medicine, 57*, 162–71. doi:10.1353/pbm.2014.0008
Morgan, J.T., Fink, G.R., and Bartel, D.P. (2018). Excised linear introns regulate growth in yeast. *Nature, 565*, 606–11. doi:10.1038/s41586-018-0828-1
Morris, K.V., and Mattick, J.S. (2014). The rise of regulatory RNA. *Nature Reviews Genetics, 15*, 423–37. doi:10.1038/nrg3722
Morton, N.E. (1991). Parameters of the human genome. *Proceedings of the National Academy of Sciences, 88*(17), 7474–6. doi:10.1073/pnas.88.17.7474
Moschetti, R., Palazzo, A., Lorusso, P., Viggiano, L., and Massimiliano Marsano, R. (2020). "What you need, baby, I got it": Transposable elements as suppliers of cis-operating sequences in *Drosophila*. *Biology, 9*, 25. doi:10.3390/biology9020025
Mucaki, E.J., Shirley, B.C., and Rogan, P.K. (2020). Expression changes confirm genomic variants predicted to result in allele-specific, alternative mRNA splicing. *Frontiers in Genetics, 11*, 109. doi:10.3389/fgene.2020.00109
Mudd, A.B., Bredeson, J.V., Baum, R., Hockemeyer, D., and Rokhsar, D.S. (2019). Muntjac chromosome evolution and architecture. *Communications Biology, 3*, 480. doi:10.1038/s42003-020-1096-9
Mudge, J.M., Frankish, A., Fernandez-Banet, J., Alioto, T., Derrien, T., Howald, C., Reymond, A., Guigó, R., Hubbard, T., and Harrow, J. (2011). The origins, evolution, and functional potential of alternative splicing in vertebrates. *Molecular Biology and Evolution, 28*, 2949–59. doi:10.1093/molbev/msr127
Muller, H.J. (1950). Our load of mutations. *American Journal of Human Genetics, 2*, 111–75.
Muller, H.J. (1966). The gene material as the initiator and the organizing basis of life. *American Naturalist, 100*, 493–517. doi:10.1086/282445
Müller, J.B., Geyer, P.E., Colaço, A.R., Treit, P.V., Strauss, M.T., Oroshi, M., Doll, S., Winter, S.V., Bader, J.M., Köhler, N., Theis, F., Santos, and Mann, M. (2020). The proteome landscape of the kingdoms of life. *Nature, 582*, 592–6. doi:10.1038/s41586-020-2402-x
Nachman, M.W., and Crowell, S.L. (2000). Estimate of the mutation rate per nucleotide in humans. *Genetics, 156*, 297–304. doi:10.1093/genetics/156.1.297
Narasimhan, V.M., Hunt, K.A., Mason, D., Baker, C.L., Karczewski, K.J., Barnes, M.R., Barnett, A.H., Bates, C., Bellary, S., Bockett, N.A., Giorda, K., Griffiths, C.J., Hemingway, H., Jia, Z., Kelly, M.A., Khawaja, H.A., Lek, M., McCarthy, S., McEachan,

R., ... Van Heel, D.A. (2016). Health and population effects of rare gene knockouts in adult humans with related parents. *Science, 352*, 474–7. doi:10.1126/science.aac8624

Nei, M. (1987). *Molecular Evolutionary Genetics*. Columbia University Press.

Nei, M. (2005). Selectionism and neutralism in molecular evolution. *Molecular Biology and Evolution, 22*, 2318–42. doi:10.1093/molbev/msi242

Nei, M. (2007). The new mutation theory of phenotypic evolution. *Proceedings of the National Academy of Sciences, 104*, 12235–42. doi:10.1073/pnas.0703349104

Nei, M. (2013). *Mutation-Driven Evolution*. Oxford University Press.

Nei, M., and Hughes, A. (1992). Balanced polymorphism and evolution by the birth-and-death process in the MHC loci. In *11th histocompatibility workshop and conference* (pp. 27–38). Oxford University Press.

Nei, M., and Rooney, A.P. (2005). Concerted and birth-and-death evolution of multigene families. *Annual Review of Genetics, 39*, 121. doi:10.1146/annurev.genet.39.073003.112240

Neme, R., and Tautz, D. (2016). Fast turnover of genome transcription across evolutionary time exposes entire non-coding DNA to de novo gene emergence. *Elife*, 5, e09977. doi:10.7554/eLife.09977

Neumann, R., Lawson, V.E., and Jeffreys, A.J. (2010). Dynamics and processes of copy number instability in human γ-globin genes. *Proceedings of the National Academy of Sciences, 107*, 8304–9. doi:10.1073/pnas.1003634107

Niu, D.-K., and Jiang, L. (2013). Can ENCODE tell us how much junk DNA we carry in our genome? *Biochemical and Biophysical Research Communications, 430*, 1340–3. doi:10.1016/j.bbrc.2012.12.074

Nóbrega, M.A., Zhu, Y., Plajzer-Frick, I., Afzal, V., and Rubin, E.M. (2004). Megabase deletions of gene deserts result in viable mice. *Nature, 431*, 988–93. doi:10.1038/nature03022

Nowak, M.A., and Waclaw, B. (2017). Genes, environment, and "bad luck." *Science, 355*, 1266–7. doi:10.1126/science.aam9746

Nozawa, M., and Kinjo, S. (2016). Noncoding RNAs, origin and evolution of. In R.M. Kliman (Ed.), *Encyclopedia of evolutionary biology* (Vol. 3, pp. 130–5). Academic Press.

Nurk, S., Koren, S., Rhie, A., Rautiainen, M., Bzikadze, A.V., Mikheenko, A., Vollger, M.R., Altemose, N., Uralsky, L. Gershman, A., Aganezov, S., Hoyt, S.J., Diekhans, M., Logsdon, G.A., Alonge, M., Antonarakis, S.E., Borchers, M., Bouffard, G.G., Brooks, S., ... Phillippy, A.M. (2022). The complete sequence of a human genome. *Science, 376*, 44–57. doi:10.1126/science.abj6987

Ohno, S. (1972a). So much "junk" DNA in our genome. In H.H. Smith (Ed.), *Evolution of genetic systems* (Brookhaven Symposia in Biology, Vol. 23, pp. 366–70). Gordon and Breach.

Ohno, S. (1972b). An argument for the genetic simplicity of man and other mammals. *Journal of Human Evolution, 1*, 651–62. doi:10.1016/0047-2484(72)90011-5

Ohno, S. (1985). Dispensable genes. *Trends in Genetics, 1*, 160–4. doi:10.1016/0168-9525(85)90070-8

Ohta, T. (1973). Slightly deleterious mutant substitutions in evolution. *Nature, 246*, 96–8. doi:10.1038/246096a0

Ohta, T. (1983). Theoretical study on the accumulation of selfish DNA. *Genetical Research, 41*, 1–15. doi:10.1017/S0016672300021029

Ohta, T. (1996). The neutral theory is dead. The current significance and standing of neutral and nearly neutral theories. *BioEssays, 18*, 673–7. doi:10.1002/bies.950180811

Ohta, T., and Kimura, M. (1971). Functional organization of genetic material as a product of molecular evolution. *Nature, 233*, 118–19. doi:10.1038/233118a0

Omenn, G.S. (2021). Reflections on the HUPO Human Proteome Project, the flagship project of the Human Proteome Organization, at 10 Years. *Molecular & Cellular Proteomics, 20,* 100062. doi:10.1016/j.mcpro.2021.100062

Omenn, G.S., Lane, L., Overall, C.M., Cristea, I.M., Corrales, F.J., Lindskog, C., Paik, Y.-K., Van Eyk, J.E., Liu, S., Pennington, S.R., Snyder, M.P., Baker, M.S., Bandeira, N., Aebersold, R., Moritz, R.L., and Deutsch, E.W. (2020). Research on the human proteome reaches a major milestone: > 90% of predicted human proteins now credibly detected, according to the HUPO Human Proteome Project. *Journal of Proteome Research, 19,* 4735–46. doi:10.1021/acs.jproteome.0c00485

Orgel, L.E., and Crick, F.H. (1980). Selfish DNA: The ultimate parasite. *Nature, 284,* 604–7. doi:10.1038/284604a0

Orgel, L.E., Crick, F.H., and Sapienza, C. (1980). Selfish DNA. *Nature, 288,* 645–7. doi:10.1038/288645a0

Orgogozo, V., Peluffo, A.E., and Morizot, B. (2016). Chapter one—The "Mendelian gene" and the "molecular gene": Two relevant concepts of genetic units. *Current Topics in Developmental Biology, 119,* 1–26. doi:10.1016/bs.ctdb.2016.03.002

Pace, N.R. (2006). Time for a change. *Nature, 441,* 289. doi:10.1038/441289

Paik, Y.-K., Omenn, G.S., Overall, C.M., Deutsch, E.W., and Hancock, W.S. (2015). Recent advances in the chromosome-centric Human Proteome Project: Missing proteins in the spot light. *Journal of Proteome Research, 14,* 3409–14. 10.1021/acs.jproteome.5b00785

Palazzo, A.F., and Gregory, T.R. (2014). The case for junk DNA. *PLoS Genetics, 10,* e1004351. doi:10.1371/journal.pgen.1004351

Palazzo, A.F., and Koonin, E.V. (2020). Functional long non-coding RNAs evolve from junk transcripts. *Cell, 183,* 1151–61. doi:10.1016/j.cell.2020.09.047

Palazzo, A.F., and Lee, E.S. (2015). Non-coding RNA: What is functional and what is junk? *Frontiers in Genetics, 6,* Article 2. doi:10.3389/fgene.2015.00002

Pan, Q., Bakowski, M.A., Morris, Q., Zhang, W., Frey, B.J., Hughes, T.R., and Blencowe, B.J. (2005). Alternative splicing of conserved exons is frequently species-specific in human and mouse. *Trends in Genetics, 21,* 73–7. doi:10.1016/j.tig.2004.12.004

Pan, Q., Shai, O., Lee, L.J., Frey, B.J., and Blencowe, B.J. (2008). Deep surveying of alternative splicing complexity in the human transcriptome by high-throughput sequencing. *Nature Genetics, 40,* 1413–5. doi:10.1038/ng.259

Parenteau, J., Durand, M., Morin, G., Gagnon, J., Lucier, J.-F., Wellinger, R.J., Chabot, B., and Elela, S.A. (2011). Introns within ribosomal protein genes regulate the production and function of yeast ribosomes. *Cell, 147,* 320–31. doi:10.1016/j.cell.2011.08.044

Parenteau, J., Durand, M., Véronneau, S., Lacombe, A.-A., Morin, G., Guérin, V., Cecez, B., Gervais-Bird, J., Koh, C.-S., Brunelle, D., Wellinger, R.J., Chabot, B., and Elela, S.A. (2008). Deletion of many yeast introns reveals a minority of genes that require splicing for function. *Molecular Biology of the Cell, 19,* 1932–41. doi:10.1091/mbc.E07-12-1254

Parenteau, J., and Elela, S.A. (2019a). Introns: Good day junk is bad day treasure. *Trends in Genetics, 35,* 923–34. doi:10.1016/j.tig.2019.09.010

Parenteau, J., Maignon, L., Berthoumieux, M., Catala, M., Gagnon, V., and Elela, S.A. (2019b). Introns are mediators of cell response to starvation. *Nature, 565,* 612–7. doi:10.1038/s41586-018-0859-7

Parker, H.G., VonHoldt, B.M., Quignon, P., Margulies, E.H., Shao, S., Mosher, D.S., Spady, T.C., Elkahloun, A., Cargill, M., Jones, P.G., Maslen, C.L., Acland, G.M., Sutter, N.B., Kuroki, K., Bustamante, C.D., Wayne, R.K., and Ostrander, E.A. (2009). An expressed *fgf4* retrogene is associated with breed-defining chondrodysplasia in domestic dogs. *Science, 325,* 995–8. doi:10.1126/science.1173275

Parrington, J. (2015). *The deeper genome: Why there is more to the human genome than meets the eye.* Oxford University Press.

Paul, J., and Gilmour, R.S. (1968). Organ-specific restriction of transcription in mammalian chromatin. *Journal of Molecular Biology, 34*, 305–16. doi:10.1016/0022-2836(68)90255-6

Pennisi, E. (2003). A low number wins the GeneSweep pool. *Science, 300*, 1484. doi:10.1126/science.300.5625.1484b

Pennisi, E. (2005). Why do humans have so few genes? *Science, 309*, 80. doi:10.1126/science.309.5731.80

Pennisi, E. (2010). Shining a light on the genome's 'dark matter.' *Science, 330*, 1614. doi:10.1126/science.330.6011.1614

Pennisi, E. (2012). ENCODE Project writes eulogy for junk DNA. *Science, 337*, 1159–61. doi:10.1126/science.337.6099.1159

Pereira-Castro, I., and Moreira, A. (2021). On the function and relevance of alternative 3'-UTRs in gene expression regulation. *WIREs RNA*. Advanced online publication. doi:10.1002/wrna.1653

Pertea, M., and Salzberg, S. (2010). Between a chicken and a grape: Estimating the number of human genes. *Genome Biology, 11*, 206. doi:10.1186/gb-2010-11-5-206

Phillips, N. (2012, March 13). Making something of junk earns geneticist top award. *The Sydney Morning Herald*. https://www.smh.com.au/world/making-something-of-junk-earns-geneticist-top-award-20120312-1uwil.html

Pickrell, J.K., Pai, A.A., Gilad, Y., and Pritchard, J.K. (2010). Noisy splicing drives mRNA isoform diversity in human cells. *PLoS Genetics, 6*, e1001236. doi:10.1371/journal.pgen.1001236

Pigliucci, M. (2008). Is evolvability evolvable? *Nature Reviews Genetics, 9*, 75–82. doi:10.1038/nrg2278

Piovesan, A., Antonaros, F., Vitale, L., Strippoli, P., Pelleri, M.C., and Caracausi, M. (2019). Human protein-coding genes and gene feature statistics in 2019. *BMC Research Notes, 12*, 315. doi:10.1186/s13104-019-4343-8

Pirsig, R.M. (1974). Zen and the art of motorcycle maintenance: An inquiry into values. Twenty-fifth anniversary ed. (1999) William Morrow.

Plenge, R.M. (2017). Biomedicine: Human genes lost and their functions found. *Nature, 544*, 171–2. doi:10.1038/544171a

Ponting, C.P. (2017). Biological function in the twilight zone of sequence conservation. *BMC Biology, 15*, 1–9. doi:10.1186/s12915-017-0411-5

Ponting, C.P., and Haerty, W. (2022). Genome-wide analysis of human long noncoding RNAs: A provocative review. *Annual Review of Genomics and Human Genetics, 23*, 153–72. doi:10.1146/annurev-genom-112921-123710

Ponting, C.P., and Hardison, R.C. (2011). What fraction of the human genome is functional? *Genome Research, 21*, 1769–76. doi:10.1101/gr.116814.110

Prioleau, M.-N., and MacAlpine, D.M. (2016). DNA replication origins—where do we begin? *Genes & Development, 30*, 1683–97. doi:10.1101/gad.285114.116

Ptashne, M. (2007). On the use of the word 'epigenetic.' *Current Biology, 17*, R233–6. doi:10.1016/j.cub.2007.02.030

Ptashne, M. (2013). Epigenetics: Core misconcept. *Proceedings of the National Academy of Sciences, 110*, 7101–3. doi:10.1073/pnas.1305399110

Ptashne, M., and Gann, A. (1997). Transcriptional activation by recruitment. *Nature, 386*, 569–77. doi:10.1038/386569a0

Qiu, G.-H. (2015). Protection of the genome and central protein-coding sequences by non-coding DNA against DNA damage from radiation. *Mutation Research/Reviews in Mutation Research, 764*, 108–17. doi:10.1016/j.mrrev.2015.04.001

Raj, A., and van Oudenaarden, A. (2008). Nature, nurture, or chance: Stochastic gene expression and its consequences. *Cell, 135*, 216–26. doi:10.1016/j.cell.2008.09.050

Ramilowski, J.A., Yip, C.W., Agrawal, S., Chang, J.-C., Ciani, Y., Kulakovskiy, I.V., Mendez, M., Ooi, J.L.C., Ouyang, J.F., Parkinson, N., Petri, A., Roos, L., Severin, J., Yasuzawa, K., Abugessaisa, I., Akalin, A., Antonov, I.V., Arner, E., Bonetti, A., ... Carninci, P. (2020). Functional annotation of human long noncoding RNAs via molecular phenotyping. *Genome Research, 30,* 1060–72. doi:10.1101/gr.254219.119

Ramsköld, D., Wang, E.T., Burge, C.B., and Sandberg, R. (2009). An abundance of ubiquitously expressed genes revealed by tissue transcriptome sequence data. *PLoS Computational Biology, 5*(12), e1000598. doi:10.1371/journal.pcbi.1000598

Rands, C.M., Meader, S., Ponting, C.P., and Lunter, G. (2014). 8.2% of the human genome is constrained: Variation in rates of turnover across functional element classes in the human lineage. *PLoS Genetics, 10,* e1004525. doi:10.1371/journal.pgen.1004525

Rao, V., and Nanjundiah, V. (2011). JBS Haldane, Ernst Mayr and the beanbag genetics dispute. *Journal of the History of Biology, 44,* 233–81. doi:10.1007/s10739-010-9229-5

Reyes, A., Anders, S., Weatheritt, R.J., Gibson, T.J., Steinmetz, L.M., and Huber, W. (2013). Drift and conservation of differential exon usage across tissues in primate species. *Proceedings of the National Academy of Sciences, 110,* 15377–82. doi:10.1073/pnas.1307202110

Richardson, S.M., Mitchell, L.A., Stracquadanio, G., Yang, K., Dymond, J.S., DiCarlo, J.E., Lee, D., Huang, C.L.V., Chandrasegaran, S., Cai, Y., Boeke, J.D., and Bader, J.S. (2017). Design of a synthetic yeast genome. *Science, 355,* 1040–4. doi:10.1126/science.aaf4557

Ridley, M. (1997). *Evolution*. Oxford University Press.

Riggs, A., Lin, S., and Wells, R. (1972). *Lac* repressor binding to synthetic DNAs of defined nucleotide sequence. *Proceedings of the National Academy of Sciences, 69,* 761–4. doi:10.1016/0022-2836(72)90184-2

Roach, J.C., Glusman, G., Smit, A.F.A., Huff, C.D., Hubley, R., Shannon, P.T., Rowen, L., Pant, K.P., Goodman, N., Bamshad, M., Shendure, J., Drmanac, R., Jorde, L.B., Hood, L., and Galas, D.J. (2010). Analysis of genetic inheritance in a family quartet by whole-genome sequencing. *Science, 328,* 636–9. doi:10.1126/science.1186802

Roest Crollius, H., Jaillon, O., Bernot, A., Dasilva, C., Bouneau, L., Fischer, C., Fizames, C., Wincker, P., Brottier, P., Quetier, F., Saurin, W., and Weissenbach, J. (2000). Estimate of human gene number provided by genome-wide analysis using *Tetraodon nigroviridis* DNA sequence. *Nature Genetics, 25,* 235–8. doi:10.1038/76118

Rogaev, E.I., Grigorenko, A.P., Faskhutdinova, G., Kittler, E.L., and Moliaka, Y.K. (2009). Genotype analysis identifies the cause of the "royal disease." *Science, 326*(5954), 817. 10.1126/science.1180660

Ruiz-Orera, J., Villanueva-Cañas, J.L., and Albà, M.M. (2020). Evolution of new proteins from translated sORFs in long non-coding RNAs. *Experimental Cell Research, 391,* 111940. doi:10.1016/j.yexcr.2020.111940

Ruse, M. (2015). Evolutionary biology and the question of teleology. *Studies in History and Philosophy of Science Part C: Studies in History and Philosophy of Biological and Biomedical Sciences, 58,* 100–6. doi:10.1016/j.shpsc.2015.12.001

Russell, C.B., Fraga, D., and Hinrichsen, R.D. (1994). Extremely short 20–33 nucleotide introns are the standard length in *Paramecium tetraurelia*. *Nucleic Acids Research, 22,* 1221–5. doi:10.1093/nar/22.7.1221

Rutherford, A. (2016). *A brief history of everyone who ever lived: The stories in our genes.* Weidenfeld & Nicolson.

Sagan, C. (1979). *Broca's brain: Reflection on the romance of science*. Ballantine Books

Sagan, C. (1995). *The demon-haunted world: Science as candle in the dark*. Random House.

Saleheen, D., Natarajan, P., Armean, I.M., Zhao, W., Rasheed, A., Khetarpal, S.A., Won, H.-H., Karczewski, K.J., O'Donnell-Luria, A.H., Samocha, K.E., Weisburg, B., Gupta, N., Zaidi, M., Samuel, M., Imran, A., Abbas, S., Majeed, F., Ishaq, M., Akhtar, S., … Kathiresan, S. (2017). Human knockouts and phenotypic analysis in a cohort with a high rate of consanguinity. *Nature*, *544*, 235–9. doi:10.1038/nature22034

Schmidt, D., Wilson, M.D., Ballester, B., Schwalie, P.C., Brown, G.D., Marshall, A., Kutter, C., Watt, S., Martinez-Jimenez, C.P., Mackay, S., Talianidis, I., Flicek, P., and Odom, D.T. (2010). Five-vertebrate ChIP-seq reveals the evolutionary dynamics of transcription factor binding. *Science*, *328*, 1036–40. doi:10.1126/science.1186176

Schuler, G., Boguski, M., Stewart, E., Stein, L., Gyapay, G., Rice, K., White, R.E., Rodriguez-Tomé, P., Aggarwal, A., Bajorek, E., Bentlila, S., Birren, B.B., Butler, A., Castle, A.B., Chiannilkulchai, N., Chu, A., Clee, C., Cowles, S., Day, P.J.R., … Hudson, T.J. (1996). A gene map of the human genome. *Science*, *274*, 540–6.

Schueler, M.G., Higgins, A.W., Rudd, M.K., Gustashaw, K., and Willard, H.F. (2001). Genomic and genetic definition of a functional human centromere. *Science*, *294*, 109–15. doi:10.1126/science.1065042

Scotti, M.M., and Swanson, M.S. (2016). RNA mis-splicing in disease. *Nature Reviews Genetics*, *17*(1), 19–32. doi:10.1038/nrg.2015.3

Session, A.M., Uno, Y., Kwon, T., Chapman, J.A., Toyoda, A., Takahashi, S., Fukui, A., Hikosaka, A., Suzuki, A., Kondo, M., van Heeringen, S.J., Quigley, I., Heinz, S., Ogino, H., Ochi, H., Hellsten, U, Lyons, J.B., Simakov, O., Putnam, N., … Rokhsar, D.S. (2016). Genome evolution in the allotetraploid frog *Xenopus laevis*. *Nature*, *538*, 336–43. doi:10.1038/nature19840

Shapiro, J.A. (2011). *Evolution: A view from the 21st century*. FT Press.

Shapiro, J.A. (2022). What we have learned about evolutionary genome change in the past 7 decades. *Biosystems*, *115–116*, 104669. doi:10.1016/j.biosystems.2022.104669

Shapiro, J.A., and von Sternberg, R. (2005). Why repetitive DNA is essential to genome function. *Biological Reviews*, *80*, 227–50. doi:10.1017/S1464793104006657

Sharma, D., Zagore, L.L., Brister, M.M., Ye, X., Crespo-Hernández, C.E., Licatalosi, D.D., and Jankowsky, E. (2021). The kinetic landscape of an RNA-binding protein in cells. *Nature*, *591*, 152–6. doi:10.1038/s41586-021-03222-x

Shirley, B., Mucaki, E., and Rogan, P. (2019). Pan-cancer repository of validated natural and cryptic mRNA splicing mutations [version 3; peer review: 2 approved, 1 approved with reservations]. *F1000Research*, *7*, 1908. doi:10.12688/f1000research.17204.3

Shreeve, J. (2005). *The genome war*. Ballantine Books.

Simone, D., Calabrese, F.M., Lang, M., Gasparre, G., and Attimonelli, M. (2011). The reference human nuclear mitochondrial sequences compilation validated and implemented on the UCSC genome browser. *BMC Genomics*, *12*, 517–27. doi:10.1186/1471-2164-12-517

Skandalis, A. (2016). Estimation of the minimum mRNA splicing error rate in vertebrates. *Mutation Research/Fundamental and Molecular Mechanisms of Mutagenesis*, *784*, 34–8. doi:10.1098/rstb.2015.0474

Skipper, M., Dhand, R., and Campbell, P. (2012). Presenting ENCODE. *Nature*, *489*, 45. doi:10.1038/489045a

Slamovits, C.H., and Keeling, P.J. (2009). Evolution of ultrasmall spliceosomal introns in highly reduced nuclear genomes. *Molecular Biology and Evolution*, *26*, 1699–705. doi:10.1093/molbev/msp081

Smit, A.F. (1999). Interspersed repeats and other mementos of transposable elements in mammalian genomes. *Current Opinion in Genetics & Development*, *9*, 657–63. doi:10.1016/S0959-437X(99)00031-3

Soares, L.M.M., and Valcárcel, J. (2006). The expanding transcriptome: The genome as the 'Book of Sand.' *The EMBO Journal, 25*, 923–31. doi:10.1038/sj.emboj.7601023

Stepankiw, N., Raghavan, M., Fogarty, E.A., Grimson, A., and Pleiss, J.A. (2015). Widespread alternative and aberrant splicing revealed by lariat sequencing. *Nucleic Acids Research, 43*, 8488–501. doi:10.1093/nar/gkv763

Stoltzfus, A. (1999). On the possibility of constructive neutral evolution. *Journal of Molecular Evolution, 49*(2), 169–81. doi:10.1007/PL00006540

Stoltzfus, A. (2006). Mutationism and the dual causation of evolutionary change. *Evolution & Development, 8*, 304–17. doi:10.1111/j.1525-142X.2006.00101.x

Stoltzfus, A., and Cable, K. (2014). Mendelian-mutationism: The forgotten evolutionary synthesis. *Journal of the History of Biology, 47*, 501–46. doi:10.1007/s10739-014-9383-2-7

Struhl, K. (2007). Transcriptional noise and the fidelity of initiation by RNA polymerase II. *Nature Structural & Molecular Biology, 14*, 103–5. doi:10.1038/nsmb0207-103

Sudmant, P.H., Rausch, T., Gardner, E.J., Handsaker, R.E., Abyzov, A., Huddleston, J., Zhang, Y., Ye, K., Jun, G., Fritz, M.H.-Y., Konkel, M.K., Malhotra, A., Stütz, A.M., Shi, X., Casale, F.P., Chen, J., Hormozdiari, F., Dayama, G., Chen, K., Malig, M., ... Korbel, J.O. (2015). An integrated map of structural variation in 2,504 human genomes. *Nature, 526*, 75–81. doi:10.1038/nature15394

Stults, D.M., Killen, M.W., Pierce, H.H., and Pierce, A.J. (2008) Genomic architecture and inheritance of human ribosomal RNA gene clusters. *Genome Research, 18*, 13–18. doi:10.1101/gr.6858507

Suntsova, M.V., and Buzdin, A.A. (2020). Differences between human and chimpanzee genomes and their implications in gene expression, protein functions and biochemical properties of the two species. *BMC Genomics, 21*, 1–12. doi:10.1186/s12864-020-06962-8

Sverdlov, E. (2017). Transcribed junk remains junk if it does not acquire a selected function in evolution. *BioEssays, 39*, 1700164. doi:10.1002/bies.201700164

Symonová, R. (2019). Integrative rDNAomics—Importance of the oldest repetitive fraction of the eukaryote genome. *Genes, 10*, 345–60. 10.3390/genes10050345

Taft, R.J., Pheasant, M., and Mattick, J.S. (2007). The relationship between non-protein-coding DNA and eukaryotic complexity. *BioEssays, 29*, 288–99. doi:10.1002/bies.20544

Telenti, A., Pierce, L.C., Biggs, W.H., Di Iulio, J., Wong, E.H., Fabani, M.M., Kirkness, E.F., Moustafa, A., Shah, N., Xie, C., Brewerton, S.C., Bulsara, N., Garner, C., Metzker, G., Sandoval, E., Perkins, B.A., Och, F.J., Turpaz, Y., and Venter, J.C. (2016). Deep sequencing of 10,000 human genomes. *Proceedings of the National Academy of Sciences, 113*, 11901–6. doi:10.1073/pnas.1613365113

The 1000 Genomes Project. (2011). Variation in genome-wide mutation rates within and between human families. *Nature Genetics, 43*, 712–15. doi:10.1038/ng.862

Thomas, C.A., Jr. (1971). The genetic organization of chromosomes. *Annual Review of Genetics, 5*, 237–56. doi:10.1146/annurev.ge.05.120171.001321

Timmer, J. (2007, June 13). *ENCODE finds the human genome to be an active place*. Ars Technica. arstechnica.com/science/2007/06/encode-finds-the-human-genome-to-be-an-active-place/

Timmer, J. (2012, September 10). *Most of what you read was wrong: How press releases rewrote scientific history*. Ars Technica. arstechnica.com/staff/2012/09/most-of-what-you-read-was-wrong-how-press-releases-rewrote-scientific-history/

Tomasetti, C., and Vogelstein, B. (2015). Variation in cancer risk among tissues can be explained by the number of stem cell divisions. *Science, 347*, 78–81. doi:10.1126/science.1260825

Tomkins, J.P. (2013). The human beta-globin pseudogene is non-variable and functional. *Answers Research Journal, 6*, 293–301.

Tourasse, N.J., Millet, J.R., and Dupuy, D. (2017). Quantitative RNA-seq meta analysis of alternative exon usage in *C. elegans*. *Genome Research*, *27*, 2120–8. doi:10.1101/gr.224626.117

Tress, M.L., Abascal, F., and Valencia, A. (2017). Most alternative isoforms are not functionally important. *Trends in Biochemical Sciences*, *42*, 408–10. doi:10.1016/j.tibs.2017.04.002

Tukiainen, T., Villani, A.-C., Yen, A., Rivas, M.A., Marshall, J.L., Satija, R., Aguirre, M., Gauthier, L., Fleharty, M., Kirby, A., Cummings, B.B., Castel, S.E., Karczewski, K.J., Aguet, F., Byrnes, A., GTEx Consortuim, Lappalainen, T., Regev, A., Ardlie, K.G., … MacArthur, D.G. (2017). Landscape of X chromosome inactivation across human tissues. *Nature*, *550*, 244–8. doi:10.1038/nature24265

Ulitsky, I. (2016). Evolution to the rescue: Using comparative genomics to understand long non-coding RNAs. *Nature Reviews Genetics*, *17*, 601–14. doi:10.1038/nrg.2016.85

Ulitsky, I., and Bartel, D.P. (2013). lincRNAs: Genomics, evolution, and mechanisms. *Cell*, *154*, 26–46. doi:10.1016/j.cell.2013.06.020

Urban, J.M., Foulk, M.S., Casella, C., and Gerbi, S.A. (2015). The hunt for origins of DNA replication in multicellular eukaryotes. *F1000 Prime Reports*, *7*, 30. doi:10.12703/P7-30

Vakirlis, N., Vance, Z., Duggan, K. M., and McLysaght, A. (2022). De novo birth of functional microproteins in the human lineage. *Cell Reports*, *41*, 111808. doi:10.1016/j.celrep.2022.111808

van Bakel, H., Nislow, C., Blencowe, B.J., and Hughes, T.R. (2010). Most "dark matter" transcripts are associated with known genes. *PLoS Biology*, *8*(5), e1000371. doi:10.1371/journal.pbio.1000371

van Bakel, H., Nislow, C., Blencowe, B.J., and Hughes, T.R. (2011). Response to "the reality of pervasive transcription." *PLoS Biology*, *9*(7), e1001102. doi:10.1371/journal.pbio.1001102

Venter, J., Adams, M., Myers, E., Li, P., Mural, R., Sutton, G., Smith, H., Yandell, M., Evans, C., Holt, R., Gocayne, J., Amanatides, P., Ballew, R., Huson, D., Wortman, J., Zhang, Q., Kodira, C., Zheng, X., Chen, … Zhu, X. (2001). The sequence of the human genome. *Science*, *291*, 1304–51. doi:10.1126/science.1058040

Vihinen, M. (2014). Contribution of pseudogenes to sequence diversity. In L. Poliseno (Ed.), *Pseudogenes* (Methods in Molecular Biology, Vol. 1167, pp. 15–24). Humana Press. doi:10.1007/978-1-4939-0835-6_2

Volders, P.-J., Anckaert, J., Verheggen, K., Nuytens, J., Martens, L., Mestdagh, P., and Vandesompele, J. (2019). LNCipedia 5: towards a reference set of human long non-coding RNAs. *Nucleic Acids Research*, *47*, D135–9. doi:10.1093/nar/gky1031

Vollger, M.R., Guitart, X., Dishuck, P.C., Mercuri, L., Harvey, W.T., Gershman, A., Diekhans, M., Sulovari, A., Munson, K.M., Lewis, A.M., Hoekzema, K., Porubsky, D., Li, R., Nurk, S., Koren, S., Miga, K.H., Phillippy, A.M., Timp, W., Ventura, M., and Eichler, E.E. (2021). Segmental duplications and their variation in a complete human genome. *Science*, *276*, 55. doi:10.1126/science.abj6965

Waddington, C. (1942). The epigenotype. *Endeavour*, *1*, 18. doi:10.1093/ije/dyr184

Wade, N. (2012, June 12). A decade later, genetic map yields few new clues. *New York Times*.

Wang, E.T., Sandberg, R., Luo, S., Khrebtukova, I., Zhang, L., Mayr, C., Kingsmore, S.F., Schroth, G.P., and Burge, C.B. (2008). Alternative isoform regulation in human tissue transcriptomes. *Nature*, *456*, 470–6. doi:10.1038/nature07509

Wang, J., Zhang, J., Zheng, H., Li, J., Liu, D., Li, H., Samudrala, R., Yu, J., and Wong, G.K.-S. (2004). Mouse transcriptome: Neutral evolution of 'non-coding' complementary DNAs. *Nature*, *431*, 758–9. doi:10.1038/nature03016

Wang, M., Wang, J., Carver, J., Pullman, B.S., Cha, S.W., and Bandeira, N. (2018). Assembling the community-scale discoverable human proteome. *Cell Systems, 7,* 412–21, e415. doi:10.1016/j.cels.2018.08.004

Watson, J.D., and Crick, F.H. (1953). Molecular structure of nucleic acids. *Nature, 171,* 737–8. doi:10.1038/171737a0

Wellcome Sanger Institute. (2012, September 5). *An integrated encyclopedia of DNA elements in the human genome* [Press release]. www.sanger.ac.uk/news_item/2012-09-05-google-earth-of-biomedical-research/

Wen, Y.-Z., Zheng, L.-L., Qu, L.-H., Ayala, F.J. and Lun, Z.-R. (2012). Pseudogenes are not pseudo any more. *RNA Biology, 9,* 27–32. doi:10.4161/rna.9.1.18277

Wheeler, D.A., Srinivasan, M., Egholm, M., Shen, Y., Chen, L., McGuire, A., He, W., Chen, Y.-J., Makhijani, V., Roth, G.T., Gomes, X., Tartaro, K., Niazi, F., Turcotte, C.L., Irzyk, G.P., Lupski, J.R., Chinault, C., Xong, X.-z., Liu, Y., … Rothberg, J.M. (2008). The complete genome of an individual by massively parallel DNA sequencing. *Nature, 452,* 872. doi:10.1038/nature06884

White, M.A. (2012, September 13). A genome-sized media failure. *Huffington Post.* huffpost.com/entry/media-genome-science_b_1881788

White, M.A., Myers, C.A., Corbo, J.C., and Cohen, B.A. (2013). Massively parallel in vivo enhancer assay reveals that highly local features determine the cis-regulatory function of ChIP-seq peaks. *Proceedings of the National Academy of Sciences, 110,* 11952–7. doi:10.1073/pnas.1307449110

Wilhelm, M., Schlegl, J., Hahne, H., Gholami, A.M., Lieberenz, M., Savitski, M.M., Ziegler, E., Butzmann, L., Gessulat, S., Marx, H., Mathieson, T., Lemeer, S., Schnatbaum, K., Reimer, U., Wenschuh, H., Mollenhauer, M., Slotta-Huspenina, J., Boese, J.-H., Bantscheff, M., … Kuster, B. (2014). Mass-spectrometry-based draft of the human proteome. *Nature, 509,* 582–7. doi:10.1038/nature13319

Wilkins, J.S. (2013). The Salem region. In M. Pigliucci and M. Boudry (Eds.), *Philosophy of pseudoscience* (pp. 397–416). The University of Chicago Press.

Willcockson, M.A., Healton, S.E., Weiss, C.N., Bartholdy, B.A., Botbol, Y., Mishra, L.N., Sidhwani, D.S., Wilson, T.J., Pinto, H.B., Maron, M.I., Skalina, K.A., Norwood Toro, L., Zhao, J., Lee, C-H., Hou, H., Yusufova, N., Meydan, C., Osunsade, A., David, Y., … Skoultchi, A.I. (2021). H1 histones control the epigenetic landscape by local chromatin compaction. *Nature, 589,* 293–8. doi:10.1038/s41586-020-3032-z

Willingham, A.T., and Gingeras, T.R. (2006). TUF love for "junk" DNA. *Cell, 125,* 1215–20. doi:10.1016/j.cell.2006.06.009

Wright, N.A., Gregory, T.R., and Witt, C.C. (2014). Metabolic 'engines' of flight drive genome size reduction in birds. *Proceedings of the Royal Society B: Biological Sciences, 281,* 20132780. doi:10.1098/rspb.2013.2780

Xu, J., Bai, J., Zhang, X., Lv, Y., Gong, Y., Liu, L., Zhao, H., Yu, F., Ping, Y., Zhang, G., Lan, Y., Xiao, Y. and Li, X. (2017). A comprehensive overview of lncRNA annotation resources. *Briefings in Bioinformatics, 18,* 236–49. doi:10.1093/bib/bbw015

Xu, J., and Zhang, J. (2015). Are human translated pseudogenes functional? *Molecular Biology and Evolution, 33,* 755–60. doi:10.1093/molbev/msv268

Xu, P., Zhang, X., Wang, X., Li, J., Liu, G., Kuang, Y., Xu, J., Zheng, X., Ren, L., Wang, G., Zhang, Y., Huo, L., Zhao, Z., Cao, D., Lu, C., Li, C., Zhou, Y., Liu, Z., Fan, Z., … Yu, J. (2014). Genome sequence and genetic diversity of the common carp, *Cyprinus carpio. Nature Genetics, 46,* 1212–19. doi:10.1038/ng.3098

Yamamoto, K.R., and Alberts, B. (1975). The interaction of estradiol-receptor protein with the genome: an argument for the existence of undetected specific sites. *Cell, 4,* 301–10. doi:10.1016/0092-8674(75)90150-6

Yamamoto, K.R., and Alberts, B. (1976). Steroid receptors: Elements for modulation of eukaryotic transcription. *Annual Review of Biochemistry, 45,* 721–46. doi:10.1146/annurev.bi.45.070176.003445

Yona, A.H., Alm, E.J., and Gore, J. (2018). Random sequences rapidly evolve into de novo promoters. *Nature Communications, 9,* 1530. doi:10.1038/s41467-018-04026-w

Yong, M. (2012, September 5). ENCODE: The rough guide to the human genome. *Discovery.* discovermagazine.com/the-sciences/encode-the-rough-guide-to-the-human-genome

Zarrei, M., MacDonald, J.R., Merico, D., and Scherer, S.W. (2015). A copy number variation map of the human genome. *Nature Reviews Genetics, 16,* 172–83. doi:10.1038/nrg3871

Zerbino, D.R., Frankish, A., and Flicek, P. (2020). Progress, challenges, and surprises in annotating the human genome. *Annual Review of Genomics and Human Genetics, 21,* 55–79. doi:10.1146/annurev-genom-121119-083418

Zhou, B., Ji, B., Liu, K., Hu, G., Wang, F., Chen, Q., Yu, R., Huang, P., Ren, J., Guo, C., Zhao, H., Zhang, H., Zhao, D., Li, Z., Zeng, Q., Yu, J., Bian, Y., Cao, Z., Xu, S., … Wang, J. (2021). EVLncRNAs 2.0: An updated database of manually curated functional long non-coding RNAs validated by low-throughput experiments. *Nucleic Acids Research, 49,* D86–91. doi:10.1093/nar/gkaa1076

Zimmer, C. (2018). *She has her mother's laugh: The powers, perversions, and potentials of heredity.* Dutton/Penguin Random House.

Zuckerkandl, E. (1986). Polite DNA: Functional density and functional compatibility in genomes. *Journal of Molecular Evolution, 24,* 12–27. doi:10.1007/BF02099947

Zuckerkandl, E. (1987). On the molecular evolutionary clock. *Journal of Molecular Evolution, 26,* 34–46. doi:10.1007/BF02111280

Zuckerkandl, E. (1997). Junk DNA and sectorial gene repression. *Gene, 205,* 323–43. doi:10.1016/S0378-1119(97)00543-X

Zuckerkandl, E. (2002). Why so many noncoding nucleotides? The eukaryote genome as an epigenetic machine. *Genetica, 115,* 105–29. doi:1016080316076

Zuckerkandl, E. (2005). "Natural restoration" can generate biological complexity. *Complexity, 11,* 14–27. doi:10.1002/cplx.20104

Zuckerkandi, E., Latter, G., and Jurka, J. (1989). Maintenance of function without selection: *Alu* sequences as "cheap genes." *Journal of Molecular Evolution, 29,* 504–12. doi:10.1007/BF02602922

Index